Soil Health Series: Volume 2 Laboratory Methods for Soil Health Analysis

# Soil Health Series: Volume 2 Laboratory Methods for Soil Health Analysis

*Edited by Douglas L. Karlen, Diane E. Stott, and Maysoon M. Mikha*

Soil Science Society of America

WILEY

*Editorial Correspondence:*
Soil Science Society of America, Inc.
5585 Guilford Road, Madison, WI 53711-58011, USA
soils.org

*Registered Offices:*
John Wiley & Sons, Inc., 111 River Street, Hoboken, NJ 07030, USA

For details of our global editorial offices, customer services, and more information about Wiley products, visit us at www.wiley.com.

Wiley also publishes its books in a variety of electronic formats and by print-on-demand. Some content that appears in standard print versions of this book may not be available in other formats.

*Library of Congress Cataloging-in-Publication Data applied for*
Paperback: 9780891189824
doi: 10.2136/soilhealth.vol2

Cover Design: Wiley
Cover Image: © Negar Tafti, Yongqiang Zhang, Richard D. Bowden, Humberto Blanco, Martin C. Rabenhorst, Hailin Zhang, Brian Dougherty

Set in 9.5/12.5pt STIXTwoText by Straive, Pondicherry, India

## Dedication

These books are dedicated to Dr. John W. Doran, a retired USDA-ARS (Agricultural Research Service) Research Soil Scientist whose profound insight provided international inspiration to strive to understand the capacity of our fragile soil resources to function within ecosystem boundaries, sustain biological productivity, maintain environmental quality, and promote plant and animal health.

Understanding and quantifying soil health is a journey for everyone. Even for John, who early in his career believed soil quality was too abstract to be defined or measured. He initially thought soil quality was simply too dependent on numerous, uncontrollable factors, including land use decisions, ecosystem or environmental interactions, soil and plant management practices, and political or socioeconomic priorities. In the 1990s, John pivoted, stating he now recognized and encouraged the global soil science community to move forward, even though perceptions of what constitutes a *good* soil vary widely depending on individual priorities with respect to soil function. Continuing, he stated that to manage and maintain our soils in an acceptable state for future generations, *soil quality* (*soil health*) must be defined, and the definition must be broad enough to encompass the many facets of soil function.

John had profound impact on our careers and many others around the World. Through his patient, personal guidance he challenged everyone to examine soil biological, chemical, and physical properties, processes, and interactions to understand and quantify soil health. For Diane, this included crop residue and soil enzyme investigations, and for Maysoon, interactions between soil physical and biological processes mediated by water-filled pore space. Recognizing my knowledge of soil testing and plant analysis on Midwestern soils, as well as root-limiting, eluviated horizons and soil compaction in Southeastern U.S. soils, John encouraged me to develop a strategy to evaluate and combine the biological, chemical, and physical indicators that have become pillars for soil quality/health assessment. The Soil Management Assessment Framework (SMAF) was the first generation outcome of this challenge.

Throughout his life, John endeavored to involve all Earth's people, no matter their material wealth or status, in translating their lifestyles to practices that strengthen social equity and care for the earth we call home. Through development of the "soil quality test kit" John fostered transformation of soil quality into *soil health* by taking his science to farmers, ranchers, and other land managers. These two volumes have been prepared with that audience in mind to reflect the progress made during the past 25 years. Special thanks are also extended to John's life mate Janet, daughter Karin, son-in-law Michael, grandchildren Drew and Fayth, and all of his friends for their encouragement, patience and support as he continues his search for the "holy grail" of soil health. Without John's inspiration and dedication, who knows if science and concern for our fragile soil resources would have evolved as it has.

Thank you, John – you are an inspiration to all of us!

Douglas L. Karlen  Diane E. Stott  Maysoon M. Mikha

# Contents

# Foreword

Soil science receives increasing attention by the international policy arena and publication of this comprehensive "Soil Health" book by the Soil Science Society of America (SSSA) and Wiley International is therefore most welcome at this point in time. Striving for consensus on methods to assess soil health is important in positioning soil science in a societal and political discourse that, currently, only a few other scientific disciplines are deeply engaged in. Specifically, increasing the focus on sustainable development provides a suitable "point on the horizon" that provides a much needed focus for a wide range of activities. Sustainable development has long been a likeable, but still rather abstract concept. The United Nations General Assembly acceptance of seventeen Sustainable Development Goals (SDGs) by 193 Governments in 2015 changed the status of sustainable development by not only specifying the goals but also defining targets, indicators, and seeking commitments to reach those goals by 2030 (https://www.un.org/sustainabledevelopment-goals). In Europe, the Green Deal, accepted in 2019, has targets and indicators corresponding to those of the SDGs (https://ec.europa.eu/info/strategy/european-green-dealsoil).

So far, soil scientists have not been actively engaged in defining SDG targets, which is unfortunate considering soil functions contribute significantly to ecosystem services that, in turn, contribute to the SDGs. The connections are all too obvious for soil scientists, but not necessarily so for scientists in other disciplines, politicians, or the public at large. For example, adequate production of food (SDG2) is impossible without healthy soil. Ground- and surface-water quality (SDG6) are strongly influenced by the purifying and infiltrative capacities of soils. Carbon capture through increases in soil organic carbon (SOC) is a major mechanism contributing to the mitigation of an increasingly variable climate (SDG13) and living soils as an integral part of living landscapes are a dominant source of biodiversity (SDG15) (Bouma, 2014; Bouma et al., 2019). With complete certainty, we can show that healthy soils make better and more effective contributions to ecosystem services than unhealthy ones! This also applies when considering the recently introduced Soil Security concept, which articulates the 5 C's: soil capability, condition, capital, connectivity, and codification (Field et al., 2017). A given soil condition can be expressed in terms of soil health, whereas soil capability defines potential conditions, to be achieved by innovative soil management, thus increasing soil health to a characteristically attainable level for that particular soil. Healthy soils are a capital asset for land users; connectivity emphasizes interactions among land users, citizens, and politicians that are obviously important, especially when advocating measures to increase soil health that may initially lack societal support. Finally, codification is important because future land use rules and regulations could benefit by being based on quantitative soil health criteria, thus allowing a reproducible comparison between different soils.

These volumes provide an inspiring source of information to further evaluate the soil health concept, derive quantitative procedures that will allow more effective interaction among land users, and information needed to introduce soil science into laws and regulations. The introductory chapters of Volume 1 present a lucid and highly informative overview of the evolution of the soil health movement. Other chapters discuss data needs and show that modern monitoring and sensing techniques can result in a paradigm shift by removing the traditional data barriers. Specifically, these new methods can provide large amounts of data at relatively low cost. The valuable observation is made that systems focusing only on topsoils cannot adequately represent soil behavior in space and time. Subsoil properties, expressed in soil classification, have significant and very important effects on many soil functions. Numerous physical, chemical and biological methods are reviewed in Volume 2. Six chapters deal with soil biological methods, correctly reflecting the need to move beyond the traditional emphasis on physical and chemical assessment methods. After all, soils are very much alive!

The book *Soil Health* nicely illustrates the "roots" of the soil health concept within the soil science profession. It also indicates the way soil health can provide "wings" to the profession as a creative and innovative partner in future environmental research and innovation.

**Johan Bouma**
**Emmeritus Professor of Soil Science**
**Wageningen University**
**The Netherlands**

## References

Bouma, J. (2014). Soil science contributions towards Sustainable Development Goals and their implementation: Linking soil functions with ecosystem services. *J. Plant Nutr. Soil Sci.* 177(2), 111–120. doi:10.1002/jpln.201300646

Bouma, J., Montanarella, L., and Vanylo, G.E. (2019). The challenge for the soil science community to contribute to the implementation of the UN Sustainable Development Goals. *Soil Use Manage.* 35(4), 538–546. doi:10.1111/sum.12518

Field, D.J., Morgan, C.L.S., and Mc Bratney, A.C., editors. (2017). *Global soil security.* Progress in Soil Science. Springer Int. Publ., Switzerland. doi:10.1007/978-3-319-43394-3

# Preface

This two-volume series on Soil Health was written and edited during a very unique time in global history. Initiated in 2017, it was intended to simply be an update for the "Blue" and "Green" soil quality books entitled *Defining Soil Quality for a Sustainable Environment* and *Methods for Assessing Soil Quality* that were published by the Soil Science Society of America (SSSA) in the 1990s. In reality, the project was completed in 2020 as the United States and world were reeling from the Covid-19 coronavirus pandemic, wide-spread protest against discriminatory racial violence, and partisan differences between people concerned about economic recovery versus protecting public health.

Many factors have contributed to the global evolution of soil health as a focal point for protecting, improving, and sustaining the fragile soil resources that are so important for all of humanity. Building for decades on soil conservation principles and the guidance given by Hugh Hammond Bennett and many other leaders associated with those efforts, soil health gradually is becoming recognized by many different segments of global society. Aligned closely with soil security, improving soil health as a whole will greatly help the United Nations (UN) achieve their Sustainable Development Goals (SDGs). Consistent with soil health goals, the SDGs emphasize the significance of soil resources for food production, water availability, climate mitigation, and biodiversity (Bouma, 2019).

The paradox of completing this project during a period of social, economic, and anti-science conflicts associated with global differences in response to Covid-19, is that the pandemic's impact on economic security and life as many have known it throughout the 20th and early 21st centuries is not unique. Many of the same contentious arguments could easily be focused on humankind's decisions regarding how to use and care for our finite and fragile soil resources. Soil conservation leaders such as Hugh Hammond Bennett (1881–1960), "Founder of Soil Conservation," W. E. (Bill) Larson (1921–2013) who often stated that soil is "the thin layer covering the planet that stands between us and starvation," and many current conservationists can attest that conflict regarding how to best use soil resources is ancient. Several soil science textbooks, casual reading books, and other sustainability writings refer to the Biblical link between soil and humankind, specifically that the very name "Adam" is derived from a Hebrew noun of feminine gender (*adama*) meaning earth or soil (Hillel, 1991). Furthermore, Xenophon, a Greek historian (430–355 BCE) has been credited with recording the value of green-manure crops, while Cato (234–149 BCE) has been recognized for recommending the use of legumes, manure, and crop rotations, albeit with intensive cultivation to enhance productivity. At around 45 CE, Columella recommended using turnips (perhaps tillage radishes?) to improve soils (Donahue et al., 1971). He also suggested land drainage, application of ash (potash), marl (limestone), and planting of clover and alfalfa (N fixation) as ways to make soils more productive. But then, after Rome was conquered, scientific agriculture, the arts, and other forms of culture were stymied.

Advancing around 1500 yr, science was again introduced into agriculture through Joannes Baptista Van Helmont's (1577–1644 CE) experiment with a willow tree. Although the initial data were misinterpreted, Justice von Liebig (1803–1873 CE) eventually clarified that carbon (C) in the form of carbon dioxide ($CO_2$) came from the atmosphere, hydrogen and oxygen from air and water, and other essential minerals to support plant growth and development from the soil. Knowledge of soil development, mineralogy, chemistry, physics, biology, and biochemistry as well as the impact of soil management (tillage, fertilization, amendments, etc.) and cropping practices (rotations, genetics, varietal development, etc.) evolved steadily throughout the past 150 yr. **SO**, what does this history have to do with these 21st Century Soil Health books?

First, in contrast to the millennia throughout which humankind has been forewarned regarding the fragility of our soil resources, the concept of soil health (used interchangeably with soil quality) per se, was introduced only 50 yr ago (Alexander, 1971). This does not discount outstanding research and technological developments in soil science such as the physics of infiltration, drainage, and water retention; chemistry of nutrient cycling and availability of essential plant nutrients, or the biology of N fixation, weed and pest control. The current emphasis on soil health in no way implies a lack of respect or underestimation of the impact that historical soil science research and technology had and have for solving problems such as soil erosion, runoff, productivity, nutrient leaching, eutrophication, or sedimentation. Nor, does it discount contributions toward understanding and quantifying soil tilth, soil condition, soil security, or even sustainable development. All of those science-based accomplishments have been and are equally important strategies designed and pursued to protect and preserve our fragile and finite soil resources. Rather, soil health, defined as an integrative term reflecting the "capacity of a soil to function, within land use and ecosystem boundaries, to sustain biological productivity, maintain environmental quality, and promote plant animal, and human health" (Doran and Parkin, 1994), is another attempt to forewarn humanity that our soil resources must be protected and cared for to ensure our very survival. Still in its infancy, soil health research and our understanding of the intricacies of how soils function to perform numerous, and at times conflicting goals, will undoubtedly undergo further refinement and clarification for many decades.

Second, just like the Blue and Green books published just twenty years after the soil health concept was introduced, these volumes, written after two more decades of research, continue to reflect a "work in progress." Change within the soil science profession has never been simple as indicated by Hartemink and Anderson (2020) in their summary reflecting 100 yr of soil science in the United States. They stated that in 1908, the American Society of Agronomy (ASA) established a committee on soil classification and mapping, but it took 6 yr before the first report was issued, and on doing so, the committee disbanded because there was no consensus among members. From that perspective, progress toward understanding and using soil health principles to protect and preserve our fragile soil resources is indeed progressing. With utmost gratitude and respect we thank the authors, reviewers, and especially, the often-forgotten technical support personnel who are striving to continue the advancement of soil science. By developing practices to implement sometimes theoretical ideas or what may appear to be impossible actions, we thank and fully acknowledge all ongoing efforts. As the next generation of soil scientists, it will be through your rigorous, science-based work that even greater advances in soil health will be accomplished.

Third, my co-authors and I recognize and acknowledge soil health assessment is not an exact science, but there are a few principles that are non-negotiable. First, to qualify as a meaningful, comprehensive assessment, soil biological, chemical, and physical properties and processes must all be included. Failure to do so, does not invalidate the assessment, but rather limits it to an assessment of "soil biological health", "soil physical health", "soil chemical health", or some combination thereof. Furthermore, although some redundancy may occur, at least two different indicator measurements should be used for each indicator group (*i.e.*, biological, chemical, or physical). To aid indicator selection, many statistical tools are being developed and evaluated to help identify the best combination of potential measurements for assessing each critical soil function associated with the land use for which an evaluation is being made.

There is also no question that any soil health indicator must be fundamentally sound from all biological, chemical, physical and/or biochemical analytical perspectives. Indicators must have the potential to be calibrated and provide meaningful information across many different types of soil. This requires sensitivity to not only dynamic, management-induced forces, but also inherent soil properties and processes reflecting subtle differences in sand, silt, and clay size particles derived from rocks, sediments, volcanic ash, or any other source of parent material. Soil health assessments must accurately reflect interactions among the solid mineral particles, water, air, and organic matter contained within every soil. This includes detecting subtle changes affecting runoff, infiltration, and the soil's ability to hold water through *capillarity*– to act like a sponge; to facilitate gas exchange so that with the help of $CO_2$, soil water can slowly dissolve mineral particles and release essential plant nutrients– through *chemical weathering;* to provide water and dissolved nutrients through the soil *solution* to plants, and to support exchange between oxygen from air above the surface and excess $CO_2$ from respiring roots.

Some, perhaps many, will disagree with the choice of indicators that are included in these books. Right or wrong, our collective passion is to start somewhere and strive for improvement, readily accepting and admitting our errors, and always being willing to update and change. We firmly believe that starting with something good is much better than getting bogged down seeking the prefect. This does not mean we are discounting any fundamental chemical, physical, thermodynamic, or biological property or process that may be a critical driver influencing soil health. Rather through iterative and ongoing efforts, our sole desire is to keep learning until soil health and its implications are fully understood and our assessment methods are correct. Meanwhile, never hesitate to hold our feet to the refining fire, as long as collectively we are striving to protect and enhance the unique material we call soil that truly protects humanity from starvation and other, perhaps unknown calamities, sometimes self-induced through ignorance or failing to listen to what our predecessors have told us.

*Douglas L. Karlen (Co-Editor)*

## References


Alexander, M. (1971). Agriculture's responsibility in establishing soil quality criteria In: *Environmental improvement– Agriculture's challenge in the Seventies.* Washington, DC: National Academy of Sciences. p. 66–71.

Bouma, J. (2019). Soil security in sustainable development. *Soil Systems.* 3:5. doi:10.3390/soilsystems3010005

Donahue, R. L., J. C. Shickluna, and L. S. Robertson. 1971). *Soils: An introduction to soils and plant growth.* Englewood Cliffs, N.J.: Prentice Hall, Inc.

Doran, J.W., Coleman, D.C., Bezdicek, D.F., and Stewart, B.A., editors. (1994). *Defining soil quality for a sustainable environment. Soil Science Society of America (SSSA) Special Publication No. 35.* Madison, WI: SSSA Inc.

Doran, J.W., and Parkin, T.B. (1994). Defining and assessing soil quality. In: J.W. Doran, D.C. Coleman, D.F. Bezdicek, and B.A. Stewart, editors, *Defining soil quality for a sustainable environment. SSSA Special Publication No. 35*. Madison, WI: SSSA. p. 3–21. doi:10.2136/sssaspecpub35

Doran, J.W., and Jones, A.J. (eds.). (1996). *Methods for assessing soil quality. Soil Science Society of America (SSSA) Special Publication No. 49*. Madison, WI: SSSA Inc.

Hartemink, A. E. and Anderson, S.H. (2020). 100 years of soil science society in the U.S. *CSA News* 65(6), 26–27. doi:10.1002/csann.20144

Hillel, D. (1991). *Out of the earth: Civilization and the life of the soil*. Oakland, CA: University of California Press.

# 1

# Laboratory Methods for Soil Health Assessment: An Overview

*Steven R. Shafer, Douglas L. Karlen, Paul W. Tracy, Cristine L.S. Morgan, and C. Wayne Honeycutt*

The purpose for Volume II is to provide specific methods and guidelines available for individuals and laboratories to evaluate soil health indicators discussed in Volume I. This volume draws on and updates the 1996 Soil Science Society of America Special Publication Number 49 entitled Methods for Assessing Soil Quality that is commonly referred to as the "Green Book" for soil quality and soil health assessment. This volume, however, is not merely a revision of the 1996 book, but rather adds guidelines for several new soil health assessment tests and discusses advances in data interpretation made during the past two decades.

Soil health is defined widely as *the continued capacity of a soil to function as a vital living ecosystem that sustains plants, animals, and humans* (e.g., NRCS, 2020). In recent years, the concept of soil health has become better understood and more widely accepted in the United States and around the world. An important driver for increased interest and global acceptance of the concept is public recognition that to meet food, feed, fiber, and fuel demands associated with an increasing population, soil degradation through erosion and loss of soil organic carbon (SOC) must be stopped and reversed by enhancing desirable biological, chemical, and physical properties and processes within this living, dynamic resource. Thus, over the past 25 years, soil health has become a focal point for serious attention across a range of public- and private-sector agricultural, environmental, and conservation organizations. Collectively, these groups have identified numerous benefits to farmers; the agricultural industry as a whole; water, air and other natural resources; educators;

*Soil Health Series: Volume 2 Laboratory Methods for Soil Health Analysis*, First Edition.
Edited by Douglas L. Karlen, Diane E. Stott, and Maysoon M. Mikha.

and the general public. This includes identifying and implementing soil health-promoting practices (e.g., cover crops, reduced intensity and frequency of tillage, improvements in and expanded use of perennials, site-specific soil and crop management) that can increase SOC (Ismail et al., 1994; Karlen et al., 1994; Ussiri and Lal, 2009; Varvel and Wilhelm, 2010; Wander et al., 1998), thereby increasing available water holding capacity, enhancing drought resistance and resilience (Emerson, 1995; Hudson, 1994; Olness and Archer, 2005), reducing wind and water erosion, and reducing nutrient loss to surface waters (Langdale et al., 1985; Tonitto et al., 2006; Yoo et al., 1988; Zhu et al., 1989). Additional benefits associated with improvements in soil health include increased suppression of pests and pathogens, increased crop yield and quality, improved return on investment, and many broad, nonpoint environmental benefits. Agricultural productivity, economic return, and environmental goals all benefit from enhancing soil health.

The literature on soil health, including the implementation of practices and technologies that promote it, has exploded in recent decades. In a search of literature covered by Google Scholar, Brevik (2018) identified more than 20,000 references using the term "soil health" from January 2000 through February 2018. This represented more than 93% of the total number of references recovered. However, our understanding of soil health and its benefits did not develop over just the past 20 years; indeed, the idea of promoting good soil health is more than 100 years old and has a surprising history. Brevik writes, "The earliest clear reference to soil health found in this review was made by Wallace (1910), who wrote about the importance of humus, particularly as obtained from manure, in maintaining soil health". The author of this 1910 reference (a thesis submitted to Iowa State University) was a student who eventually would become President Franklin Roosevelt's Secretary of Agriculture, then Roosevelt's vice-president. The name of this earliest known user of the term "soil health" will be familiar to many, as it remains before us today on the largest research station operated by the U.S. Department of Agriculture: the Henry A. Wallace Agricultural Research Center in Beltsville, MD. Thus, the concept of soil health has a particularly illustrious pedigree in the history of agricultural science.

Research on soil properties– physical, chemical, and biological– has led to major advances in managing agricultural soils, contributing to significant crop yield increases throughout the 20th and 21st centuries. However, consensus on a holistic approach to understand, implement, and measure outcomes of soil management, with goals that include sustaining production and enhancing soil health, has escaped the scientific community. Reasons for this include: (i) ever-changing methods of measurement and how to interpret the data, especially for biological properties and processes; (ii) how to adapt analytical methods for soils having different properties, which may alter results and make data comparisons difficult;

(iii) the meaning of analytical results for different agricultural production systems and environments; (iv) unclear links among measurements, soil processes, and desired outcomes (ecosystem services such as agricultural yield, nutrient cycling, improved water quality, etc.); (v) complexity and costs for advanced measurement techniques; (vi) differences in sample handling and measurement protocols among analytical laboratories; (vii) producers' uncertainties about what the data mean and how to adjust management practices in response to the information, including potential risks and benefits; and (viii) inconsistent messaging about soil health and how to manage it to agricultural producers, natural resources managers, educators, policymakers, and other stakeholders.

Stakeholder diversity alone presents significant challenges to the community of scientists, practitioners, producers, and others who advocate making soil health the cornerstone of agricultural and environmental decision making. The needs of different segments of the community demand different kinds of data, information (interpretation of the data), and communication techniques. For example, the interests of a typical agricultural producer are unlikely to be met with a report on 20 to 30 laboratory measurements that quantify a range of physical, chemical, and biological properties of a soil. Such a report may be more than most producers would want to interpret. On the other hand, a small group of indicators, easily obtained and explained, might be helpful to a producer but insufficiently accurate, precise, and process-oriented for scientific research. The distinctly different needs of various stakeholders provide a critical starting point for any conversation about soil health.

## How Can a Farmer Assess Soil Health in the Field?

Many producers are keen to learn about soil health on their farms and how they can alter their current soil and crop management practices to sustain or improve it. This interest has greatly increased opportunities for agricultural experts who can successfully bridge researcher and producer communities and is a key factor driving development of public and private programs that strive to strive for clear communication about soil health. For example, pasture and range scientists affiliated with the Noble Research Institute in Ardmore, OK, often advise farmers to consider five indicators (Jeff Goodwin, personal communication, 2018), which we summarize here as "the Five C's of Soil Health". They are:

- **Color**– A healthy soil's dark brown color indicates the presence of a lot of carbon in the form of decomposed organic matter. In contrast, gray, yellow, or mottled colors indicate soil that has a low carbon content, is poorly drained and poorly aerated, and likely low in nutrients available to plants.

- **Crumbs**– A soil that is crumbly, like coffee grounds or cake crumbs, and holds that aggregate structure is likely in good physical condition supporting soil health. This is structure that allows water movement yet aeration, as well as root penetration. It holds up even when the soil is wet. If the dry soil can easily be ground to dust between the fingers, or it turns into a slick film when wet and rubbed between the forefinger and thumb, the aggregates are not stable and will not support a good crop.
- **Critters**– A healthy soil shows lots of evidence of life. Pulling the crop debris back from the surface should reveal earthworms, or their holes and castings. Turning over the soil with a shovel should uncover insects, pillbugs, and other arthropods essential in carbon and nutrient cycling. A low-power hand lens might allow observation of smaller arthropods such as mites that feed on debris and microbes, and perhaps even the filamentous hyphae of fungi or the near-microscopic worms that feed on them. A soil that lacks evidence of diverse life is not healthy.
- **Cooperation**, with roots, that is– A healthy soil does not constrain roots, its structure allows plant roots to grow vertically and laterally. When roots look stunted or turn at odd angles, it is likely that the soil is compacted or has a plow layer that obstructs root growth because it lacks good structure and aeration for a crop. Stubby, deformed, discolored, or rotten roots can also indicate the presence of parasitic nematodes, plant-feeding insect pests, or pathogenic microbes in the soil, none of which is desirable for a healthy soil.
- **Cologne**– A healthy soil has a fragrant, earthy aroma, indicative of the many aerated biological processes happening. A soil that has a sour or rotten-egg odor is poorly aerated, probably because of poor structure and poor drainage, and is not likely to be a hospitable environment conducive to plant root development or beneficial microbes.

## What Do Researchers Need, and Can They Reach Consensus?

The Five C's of Soil Health may be useful to a farmer, and they can use them to consider modifications to production practices that could push the soil toward more desirable characteristics. For research purposes, however, these indicators are insufficiently quantitative, repeatable, and explanatory for statistical analyses and hypothesis testing about soils at different locations, under different production systems, or subjected to different management practices. For those needs, measurements that are highly repeatable and based on standardized protocols and techniques within research laboratories are needed.

At this other extreme, a new set of challenges arises– how to get a representative sample, how to handle and store it before it can be analyzed, which properties to measure, which measurement method to use, how to report the data, how to develop recommendations from those data. Just as a physician cannot adequately describe the health of a human patient with a small number of measurements or distillation of many measurements into a single number, scientists must rely on multiple different indicator measurements to provide a scientifically meaningful assessment of a soil's health. Preferences regarding specific measurements to make and methods to use are no doubt numerically equal to the number of scientists wanting to assess soil health. Reaching consensus in the community has been and continues to be a difficult task.

Currently, there are two integrated and coordinated efforts to identify suitable soil health indicator measurement protocols and to assess their utility throughout the country. One, led by the U.S. Department of Agriculture– Natural Resources Conservation Service (USDA-NRCS) Soil Health Division (SHD), is Soil Health Technical Note No. 450–03 (Stott, 2019) entitled "Standard Indicators and Laboratory Procedures to Assess Soil Health." The other is a research project led by the Soil Health Institute (SHI), which is evaluating the utility of analytical methods to determine the usefulness of over 30 soil health indicators across much of North America. Methods addressed in this volume are applicable to both efforts and reinforce the concept that data on physical, chemical, and biological properties and processes all must be obtained for a full understanding of a soil's health.

Both the SHI and NRCS-SHD efforts obtained input from researchers, farmers, soil-testing laboratories, non-governmental organizations (NGOs), and representatives of state and federal agencies starting in 2013, when the two longest-serving agricultural foundations in the United States (Farm Foundation, established 1933; and The Samuel Roberts Noble Foundation, established 1945) partnered to design and initiate the Soil Renaissance effort. Several workshops were organized and facilitated to identify and strive for consensus regarding appropriate indicators of soil health. Each workshop was attended by a different mixture of university, government, and private industry scientists, field conservationists, and farmers. Technical discussion papers were written by teams of scientists from the U.S. Department of Agriculture and land-grant universities. Between 2014 and 2016, many measurement-related issues and challenges were assessed, including the status of existing soil health measurement frameworks; the benefits of a "tiered" approach for measurements at different stages of development and reliability; service lab adoption issues; data needs; communications plans; data interpretation, including issues related to different regions; sampling protocols; sample archives; quality assurance/quality control (QA/QC) protocols; and sampling frequency. To further the vision of the Soil Renaissance and implement its findings, the Soil Health Institute was created in 2015. In June 2017, the SHI used input from the Soil Renaissance effort to conduct a survey of 179 individuals who were active in the measurement-related workshops organized by the SHI and/or the Soil Renaissance over the three years. A consensus emerged among the 48 respondents that many of the measurements used to characterize soil conditions for many years are also valuable for assessing soil health. These measurements–physical and chemical, supplemented with a few key biological– are well-accepted in the scientific community and thus were designated by the Soil Renaissance participants as "Tier 1" indicators. They can be used directly or as ancillary factors needed to improve the interpretation of yet other measurements. They include:

- Physical:
  - Soil texture
  - Water-stable aggregation
  - Bulk density
  - Water penetration resistance
  - Visual rating of erosion
  - Infiltration
  - Available water holding capacity
- Chemical:
  - Routine inorganic chemical analysis (N, P, K, micronutrients, pH, cation exchange capacity, base saturation, electrical conductivity)
  - Soil organic carbon
- Biological:
  - Short-term carbon mineralization (respiration)
  - Nitrogen mineralization
  - Crop yield

The Soil Renaissance, SHI, and NRCS-SHD communities also identified a group of measurements that have been designated "Tier 2", mostly biological properties or processes in soil, for which there is scientific consensus that they are related to soil health but are less standardized with regard to measurement methods, interpretation, and known thresholds for management action. These indicators are identified in the SHI Action Plan (www.soilhealthinstitute.org, accessed February 20, 2020) as targets for research to develop sufficient response data to complete their development as reliable measurements. To achieve those goals, the Tier 2 indicators listed below need further development, testing, and evaluation on working farms so they can eventually be transferred and communicated to landowners, operators, and retailers as tools for improving soil and crop management practices. They include:

- Beta-glucosidase activity (organic matter decomposition)
- Macro-aggregate stability (water partitioning)

- Permanganate oxidizable carbon (carbon food source for microbes)
- Soil protein (bioavailable nitrogen)
- Ester-linked fatty acid methyl ester; phospholipid fatty acid (microbial community structure, diversity)
- Nematode population densities (trophic levels)
- Pathogenic fungi populations or bioassays (pathogen activities and host ranges)

The SHI, SHD and Soil Renaissance communities also identified a category of measurements designated "Tier 3", which are primarily measurements of soil biological properties or processes for which, again, there is scientific consensus that they are quite likely related to soil health, but they still require major research and development investments to determine whether they reveal information that can be used to improve soil and crop management decisions. Fundamental biological and agricultural principles suggest Tier 3 indicators may be very useful eventually for assessing soil health and making management decisions, provided significant research investments in their development are aggressively pursued. Therefore, Tier 3 measurements are worthy subjects of further research on long-term research sites and on-farm evaluations where there are detailed records of environmental conditions and management practices over enough years that Tier 3 measurements can be interpreted reliably. Prominent among such measurements are metagenomic analyses to reveal information about soil microbial populations, community structure, and diversity, as influenced by the status and trends of soil health and in relation to the history of environmental conditions and management practices on exceptionally well-characterized sites.

Consensus on *what* to measure is just part of the research associated with soil health measurements. There is also a need to reach consensus on *how* to measure each indicator, which can be very challenging and even contentious within the soil science and agronomic research communities. In the case of Tier 1 indicators, many analytical methods for measurements are widely accepted, for example, Soil Science Society of America Book Series 5, *Methods of Soil Analysis, second edition– Part 1, Physical and Mineralogical Methods* (1986); *Part 2, Chemical and Microbiological Properties* (1982); *Part 3, Chemical Methods* (1996). Variations in specific methods have been adapted in response to recommendations from research conducted in university, government, and private laboratories to obtain optimal, meaningful results for different soils collected from widely different locations and environments. These methods are in use for several different frameworks for soil health assessment (e.g., Karlen et al., 2014; Moebius-Clune et al., 2016).

Methods for Tier 2 indicators are under active development and evaluation, and the research community is not in full agreement on methods and interpretation. Tier 3 indicators, however, as might be expected, are still very much in development, and their interpretation and value as soil health indicators that can be used to guide soil and crop management practice decisions remain uncertain.

To develop consensus that would support research on Tier 1 and Tier 2 indicator evaluation, in early 2018, the SHI assembled a panel of experts in soil health measurement from USDA agencies (Agricultural Research Service, Natural Resources Conservation Service), universities, and a private laboratory to meet and recommend a specific protocol for each indicator listed below. The goal for this gathering was to assemble a definitive list of widely-applicable, effective indicators for evaluating soil health *and* the specific methods to use for each indicator in many production environments across a wide geographical scale. To accomplish this, SHI is partnering with numerous investigators at long-term agricultural research sites (with appropriate experimental designs, controls, documented management histories, production records, etc.) that are being sampled and analyzed for over 30 soil health indicators (www.soilhealthinstitute.org/north-american-project-to-evaluate-soil-health-measurements/) (accessed February 20, 2020). Together, the indicator methods described in USDA-SHD Technical Note 450–03 (Stott, 2019), information provided in this volume, and methods under evaluation in the wide-scale SHI project (Tables 1.1 and 1.2) offer researchers and others who need scientifically justifiable procedures a good selection for current use, comparison, testing in different locations and agricultural production conditions, and further refinement.

Measurements and methods in Tables 1.1 and 1.2 are the subjects of ongoing research being conducted by the SHI with university, government, and private-sector partners with funding (2017–2020) from the Foundation for Food and Agriculture Research, General Mills, The Samuel Roberts Noble Foundation, and matching-fund sources. The indicators under investigation by NRCS are a subset of those being evaluated by SHI, and both organizations coordinated to use the same methods for those specific indicators.

## What Do Commercial Analytical Laboratories Need?

The primary interest of researchers usually is a level of accuracy, precision, and explanatory linkage to processes occurring in soil, so that results can be used to explain and predict soil health in a way that leads to new ways of managing the soil resource. In most cases, the limits on accuracy and precision, and the QA/QC procedures to ensure desired data quality and curation, are specified by the individual researcher as needed for the goals of the research and as constrained by the research budget.

Analytical laboratories that measure soil properties for a fee are also concerned with accuracy and precision that reflect the reliability and reputation of their

**Table 1.1** Tier 1 Soil Health Indicators and Methods to be Assessed.

| Indicator | Method | Reference |
|---|---|---|
| **Soil pH** | 1:2 soil:water, standard pH electrode system | Thomas, 1996 |
| **Soil Electrical Conductivity (EC)** | 1:2 soil:water, standard electrical conductivity meter system | Rhoades, 1996 |
| **Cation Exchange Capacity (CEC)** | Sum of cations:<br>Soil pH ≥ 7.2: use ammonium acetate extractant;<br>Soil pH < 7.2: use Mehlich 3 extractant | Knudsen et al., 1982<br>Sikora and Moore, 2014 |
| **% Base Saturation (BS)** | Calculation: For soil pH ≥ 7.2: use ammonium acetate extractant; for soil pH < 7.2: use Mehlich 3 extractant | Knudsen et al., 1982<br>Sikora and Moore, 2014 |
| **Extractable Phosphorus** | Soil pH ≥ 7.2: use sodium bicarbonate extractant;<br>Soil pH < 7.2: use Mehlich 3 extractant | Olsen and Sommers, 1982<br>Sikora and Moore, 2014 |
| **Extractable Potassium, Calcium, Magnesium, Sodium** | pH ≥ 7.2: use ammonium acetate extractant;<br>Soil pH < 7.2: use Mehlich 3 extractant | Knudsen et al., 1982<br>Sikora and Moore, 2014 |
| **Extractable Iron, Zinc, Manganese, Copper** | DTPA extractant derivatives | Lindsay and Norvell, 1978 |
| **Total Nitrogen** | Dry combustion | Nelson and Sommers, 1996 |
| **Soil Organic Carbon (SOC)** | Dry combustion; corrected for inorganic C, if present, using pressure calcimeter | Nelson and Sommers, 1996<br>Sherrod et al., 2002 |
| **Soil Texture** | Pipette method with a minimum of 3 size classes. Weight/volume measurements | Gee and Bauder, 1986 |
| **Aggregate Stability** | Wet sieve procedure. Weight measurement<br>Water slaking image recognition | Kemper and Roseneau, 1986<br>Mikha and Rice, 2004<br>Fajardo et al., 2016 |

(Continued)

**Table 1.1** (Continued)

| Indicator | Method | Reference |
|---|---|---|
| **Available Water Holding Capacity** | Ceramic plate method measured at −33 kPa (−10 kPa for sandy soils) and −1500 kPa | Klute, 1986 |
| **Bulk Density (BD)** | Core method: diameter to be determined, (most likely 2-inch or 5.08 cm) | Blake and Hartge, 1986 |
| **Saturated Hydraulic Conductivity** | Two-ponding head method in field with Saturo | Reynolds and Elrick, 1990 |
| **Crop Yield** | Obtained from historical and current plot yield data provided by site manager | |
| **Short-Term Carbon Mineralization** | 4-d incubation followed by $CO_2$–C evolution and capture at 50% water-filled pore space. | Zibilske, 1994 |
| **Potentially Mineralizable Nitrogen** | Short-term anaerobic incubation with ammonium and nitrate measured colorimetrically pre- and post-incubation | Bundy and Meisinger, 1994 |

service. Relationships between measurements and soil processes elucidated in research laboratories underlie a service lab's analytical offerings, but in most cases, such relationships have been worked out by the research community. Although cost is certainly a consideration in a research budget, a service lab must offer analyses in a consistent, cost-effective, and competitive way to remain in business. Selection of specific methods often relies on recommendations from researchers at universities located within the general region from which a service lab draws customers; such methods are most likely to yield reliable results for the region in which they were developed.

Service labs must maintain consistent quality of data if they are to remain in business. A farmer must have confidence that analyses conducted in different years or on different parts of the farm reflect real properties of the soil, and if changes in a measurement are occurring, that these really do reflect changes in soil on the farm. Service labs may strive to achieve this reliability through associations with organizations that provide independent testing and verification of laboratory results.

One example of laboratory validation is offered through the North American Proficiency Testing (NAPT) Program delivered by the Soil Science Society of America. The NAPT program supports soil, plant, and water testing laboratories

**Table 1.2** Tier 2 Soil Health Indicators and Methods to be Assessed

| Indicator | Method | Reference |
|---|---|---|
| **Sodium Adsorption Ratio (SAR)** | Saturated paste extract followed by atomic absorption or inductively coupled plasma spectroscopy | Miller et al., 2013 |
| **Soil Stability Index** | Combination of wet and dry sieving at multiple sieve sizes | Franzluebbers et al., 2000 |
| **Active Carbon** | Permanganate oxidizable carbon (POXC). Digestion followed by colorimetric measurement | Weil et al., 2003 |
| **Soil Protein Index** | Autoclaved citrate extractable | Schindelbeck et al., 2016 |
| **B-Glucosidase** | Assay incubation followed by colorimetric measurement | Tabatabai, 1994 |
| **N-acetyl-B-D-glucosaminidase** | Assay incubation followed by colorimetric measurement | Deng and Popova, 2011 |
| **Phosphomonoesterase** | Assay incubation followed by colorimetric measurement | Acosta-Martínez and Tabatabai, 2011 |
| **Arylsulfatase** | Assay incubation followed by colorimetric measurement | Klose et al., 2011 |
| **Phospholipid Fatty Acid (PLFA)** | Bligh-Dyer extractant, solid phase extraction, transesterification; gas chromatography | Buyer and Sasser, 2012 |
| **Genomics** | 18S, 16S or ITS analysis or a combination of 16S and 18S/ITS and/or shotgun metagenomics | Earth Microbiome Project 500; Marotz et al., 2017<br>Thompson et al., 2017 |
| **Reflectance** | Diffuse reflectance spectroscopy | Veum et al., 2015 |

by providing interlaboratory sample exchanges and statistical analyses of data. American and Canadian experts from scientific organizations, state (U.S.) and provincial (Canada) departments of agriculture, regional working groups, and public and private analytical labs provide organization and oversight (SSSA, 2020).

Several layers of certifications or validations may be in place to ensure data quality for commercial analytical labs. For example, the Agricultural Laboratory Proficiency (ALP) Program provides soil samples to laboratories for analysis and then assesses that laboratory's results relative to the test soil's known properties. In turn, the ALP program is accredited by the American National Standards Institute/American Society of Quality Control National Accrediting Board (ANAB) through testing of ISO/IEC 17043, itself an international standard of the International Organization for Standardization and the International Electrotechnical Commission for laboratory proficiency testing that determines the performance of individual laboratories for specific tests or measurements and is used to monitor laboratories' continuing performance through interlaboratory comparisons. Thus, commercial laboratories conducting analyses for soil health indicators select measurements and indicators that are understood in the research community to relate to processes in soil important to soil health; and methods that are affordable to the farming community, as well as verifiable for accuracy and precision (= reliability to the producer) by independent tests.

## Summary and Conclusions

This introduction to Volume II provides those interested in soil health assessment methods a brief background regarding identification and selection of the indicators within this volume and to provide context regarding current and future soil health assessment efforts. It also demonstrates that multiple groups (*i.e.*, SHI and the NRCS-SHD) that engaged in workshops and conferences during 2013 to 2015 continue to coordinate soil health assessments to protect and enhance our fragile, yet life-sustaining soil resource for current and future generations.

## References

Acosta-Martínez, V., and Tabatabai, M.A. (2011). Phosphorus cycle enzymes. In R.P. Dick, (ed.), *Methods of soil enzymology*. Madison, WI: SSSA. doi:10.2136/sssabookser9.c8.

Blake, G.R., and Hartge, K.H. (1986). Bulk density. In: A. Klute (ed.), *Methods of soil analysis: Part 1. Physical and mineralogical methods* (p. 363–382). 2nd ed. Madison, WI: ASA and SSSA.

Brevik, E.C. (2018). A brief history of the soil health concept. The Profile. Madison, WI: Soil Science Society of America. Posted 18 Dec. https://profile.soils.org/files/soil-communication/documents/95_document1_d50a8524-dbf4-4856-8c62-4a03ede6c3d4.pdf (Accessed 20 Feb. 2020).

Bundy, L.G., and Meisinger, J.J. (1994). Nitrogen availability indices. In R.W. Weaver, S. Angle, P. Bottomley, D. Bezdicek, S. Smith, A. Tabatabai, and A. Wollum, (eds.), *Methods of soil analysis. Part 2* (p. 951–984). SSSA Book Ser. 5. SSSA, Madison, WI.

Buyer, J.S., and Sasser, M. (2012). High throughput phospholipid fatty acid analysis of soils. *Appl. Soil Ecol.* 61, 127–130. doi:10.1016/j.apsoil.2012.06.005

Deng, S., and Popova, I. (2011). Carbohydrate hydrolases. In R.P. Dick, (ed.), *Methods of soil enzymology* (p. 185–209). SSSA, Madison, WI.

Emerson, W.W. (1995). Water retention, organic-C, and soil texture. *Aust. J. Soil Res.* 33, 241–251. doi:10.1071/SR9950241

Fajardo, M., McBratney, A.B., Field, D.J., and Minasny, B. (2016). Soil slaking assessment using image recognition. *Soil Tillage Res.* 163, 119–129. doi:10.1016/j.still.2016.05.018

Franzluebbers, A.J., Wright, S.F., and Stuedemann, J.A. (2000). Soil aggregation and glomalin under pastures in the Southern Piedmont USA. *Soil Sci. Soc. Am. J.* 64, 1018–1026. doi:10.2136/sssaj2000.6431018x

Gee, G.W., and Bauder, J.W. (1986). Particle-size analysis. In A. Klute, (ed.), *Methods of soil analysis. Part 1– Physical and mineralogical methods* (p. 493–544). 2nd ed. Madison, WI: ASA and SSSA.

Hudson, B.D. (1994). Soil organic matter and available water capacity. *J. Soil Water Conserv.* 49, 189–194.

Ismail, I., Blevins, R.L., and Frye, W.W. (1994). Long-term notillage effects on soil properties and continuous corn yields. *Soil Sci. Soc. Am. J.* 58:193–198. doi:10.2136/sssaj1994.03615995005800010028x

Karlen, D.L., Wollenhaupt, N.C., Erbach, D.C., Berry, E.C., Swan, J.B., Eash, N.S., and Jordahl, J.L. (1994). Long-term tillage effects on soil quality. *Soil Tillage Res.* 32, 313–327. doi:10.1016/0167-1987(94)00427-G

Karlen, D.L., Stott, D.E., Cambardella, C.A., Kremer, R.J., King, K.W., and McCarty, G.W. (2014). Surface soil quality in five midwestern cropland Conservation Effects Assessment Project watersheds. *J. Soil Water Conserv.* 69, 393–401. doi:10.2489/jswc.69.5.393

Kemper, W.D., and Roseneau, R.C. (1986). Aggregate stability and size distribution. In: A. Klute, editor, *Methods of soil analysis: Part I. Physical and mineralogical methods* (p. 425–442). 2nd ed. Madison, WI: ASA and SSSA.

Klute, A. (1986). *Water retention: Laboratory methods. In: A. Klute, editor, Methods of soil analysis: Part 1. Physical and mineralogical methods (p. 635–662).* 2nd ed. Madison, WI: ASA and SSSA. doi:10.2136/sssabookser5.1.2ed

Klose, S., Bilen, S., Tabatabai, M.A., and Dick, W.A. (2011). Sulfur cycle enzymes. In R.P. Dick, (ed.), *Methods of soil enzymology* (p. 125–159). Madison, WI: SSSA.

Knudsen, D., Peterson, G.A., and Pratt, P.F. (1982). Lithium, sodium and potassium. In A.L. Page (ed.), *Methods of soil analysis. Part 2.* 2nd ed. Madison, WI: ASA and SSSA.

Langdale, G.W., Leonard, R.A., and Thomas, A.W. (1985). Conservation practice effects on phosphorus losses from Southern Piedmont watersheds. *J. Soil Water Conserv.* 40, 157–161.

Lindsay, W.L., and Norvell, W.A. (1978). Development of a DTPA soil test for zinc, iron, manganese, and copper. *Soil Sci. Soc. Am. J.* 42, 421–428. doi:10.2136/sssaj1978.03615995004200030009x

Marotz, C., Amir, A., Humphrey, G., Gaffney, J., Gogul, G., and Knight, R. (2017). DNA extraction for streamlined metagenomics of diverse environmental samples. *Biotech.* 62, 290–293. doi:10.2144/000114559

Miller, R.O., Gavlak, R., and Horneck, D. (2013). Saturated paste extract for calcium, magnesium, sodium and SAR. In *Soil, plant and water methods for the western region* (p. 21-22). 4th ed. WREP-125. Washington, D.C.: Wetlands Reserve Enhancement Program.

Mikha, M.M., and Rice, C.W. (2004). Tillage and manure effects on soil and aggregate-associated carbon and nitrogen. *Soil Sci. Soc. Am. J.* 68, 809–816. doi:10.2136/sssaj2004.8090

Moebius-Clune, B.N., Moebius-Clune, D.J., Gugino, B.K., Idowu, O.J., Schindelbeck, R.R., Ristow, A.J., van Es, H.M., Thies, J.E., Shayler, H.A., McBride, M.B., Kurtz, K.S.M., Wolfe, D.W., and Abawi, G.S. (2016). *Comprehensive Assessment of Soil Health– The Cornell Framework, edition 3.2,* Cornell University, Geneva, NY. http://soilhealth.cals.cornell.edu/training-manual/ (Accessed 7 Oct. 2018).

Natural Resources Conservation Service (2020). *Soil Health.* Washington, D.C.: Natural Resources Conservation Service. https://www.nrcs.usda.gov/wps/portal/nrcs/main/soils/health/ (Accessed 20 Feb. 2020).

Nelson, D.W., and Sommers, L.E. (1996). Total carbon, organic carbon, and organic matter. In D.L. Sparks, (ed.), *Methods of soil analysis: Part 3. Chemical methods* (p. 961–1010). Madison, WI: SSSA. doi:10.2136/sssabookser5.3.c34

Olness, A., and Archer, D.W. (2005). Effect of organic carbon on available water in soil. *Soil Sci.* 170, 90–101. doi:10.1097/00010694-200502000-00002

Olsen, S.R., and Sommers, L.E. (1982). Phosphorus. In A.L. Page, et al., editors, *Methods of soil analysis: Part 2. Chemical and microbiological properties* (p. 403–430). 2nd ed. Madison, WI: ASA and SSSA.

Reynolds, W.D., and Elrick, D.E. (1990). Ponded infiltration from a single ring: I. Analysis of steady flow. *Soil Sci. Soc. Am. J.* 54, 1233–1241. doi:10.2136/sssaj1990.03615995005400050006x

Rhoades, J.D. (1996). Salinity: Electrical conductivity and total dissolved solids. In: D.L. Sparks, editor, *Methods of soil analysis: Part 3. Chemical methods* (p. 417–435). Madison, WI: SSSA. doi:10.2136/sssabookser5.3.c14

Schindelbeck, R.R., Moebius-Clune, B.N., Moebius-Clune, D.J., Kurtz, K.S., and van Es, H.M. (2016). *Cornell university comprehensive assessment of soil health laboratory standard operating procedures.* Ithaca, NY: Cornell University. https://cpb-us-e1.wpmucdn.com/blogs.cornell.edu/dist/f/5772/files/2015/03/CASH-Standard-Operating-Procedures-030217final-u8hmwf.pdf

Sherrod, L.A., Dunn, G., Peterson, G.A., and Kolberg, R.L. (2002). Inorganic carbon analysis by modified pressure-calcimeter method. *Soil Sci. Soc. Am. J.* 66, 299–305. doi:10.2136/sssaj2002.2990

Sikora, F.S., and Moore, K.P. (2014). Soil test methods from the southeastern United States. Southern Cooperative Series Bulletin 419. Washington, D.C.: Wetlands Reserve Enhancement Partnership.

Soil Science Society of America. (2020). The North American Proficiency Test Program. Madison, WI: SSSA. https://www.naptprogram.org/ (*Accessed 20 Feb. 2020*).

Stott, D.E. (2019). *Recommended soil health indicators and associated laboratory procedures. Soil Health Technical Note No. 450-03.* Washington, D.C.: U.S. Department of Agriculture, Natural Resources Conservation Service.

Tabatabai, M.A. (1994). Soil enzymes. In: R.W. Weaver, S. Angle, P. Bottomley, D. Bezdicek, S. Smith, A. Tabatabai, A. Wollum, (eds.), *Methods of soil analysis: Part 2. Microbiological and biochemical properties* (p. 775–833). SSSA, Madison, WI.

Thomas, G.W. (1996). Soil pH and soil acidity. In D.L. Sparks, editor, *Methods of soil analysis: Part 3. Chemical methods* (p. 475–490). Madison, WI: SSSA.

Thompson, L.R., Sanders, J.G., McDonald, D., Amir, A., Jansson, J.K., Gilbert, J.A., and Knight, R., and The Earth Microbiome Project Consortium. (2017). A communal catalogue reveals earth's multiscale microbial diversity. *Nature* 551, 457–463. doi:10.1038/nature24621

Tonitto, C., David, M.B., and Drinkwater, L.E. (2006). Replacing bare fallows with cover crops in fertilizer-intensive cropping systems: A meta-analysis of crop yield and N dynamics. *Agric. Ecosyst. Environ.* 112, 58–72. doi:10.1016/j.agee.2005.07.003

Ussiri, D.A.N., and Lal, R. (2009). Long-term tillage effects on soil carbon storage and carbon dioxide emissions in continuous corn cropping system from an alfisol in Ohio. *Soil Tillage Res.* 104, 39–47. doi:10.1016/j.still.2008.11.008

Varvel, G.E., and Wilhelm, W.W. (2010). Long-term soil organic carbon as affected by tillage and cropping systems. *Soil Sci. Soc. Am. J.* 74, 915–921. doi:10.2136/sssaj2009.0362

Veum, K.S., Sudduth, K.E., Kremer, R.J., and Kitchen, N.R. (2015). Estimating a soil quality index with VNIR reflectance spectroscopy. *Soil Sci. Soc. Am. J.* 79, 637–649. doi:10.2136/sssaj2014.09.0390

Wallace, H.A. (1910). Relation between livestock farming and the fertility of the land. Thesis. Ames, IA: Iowa State University. doi:10.31274/rtd-180813-7404

Wander, M.M., Bidart-Bouzat, G., and Aref, S. (1998). Tillage impacts on depth distribution of total and particulate organic matter in three Illinois soils. *Soil Sci. Soc. Am. J.* 62, 1704–1711. doi:10.2136/sssaj1998.03615995006200060031x

Weil, R., Islam, K.R., Stine, M.A., Gruver, J.B., and Samson-Liebig, S.E. (2003). Estimating active carbon for soil quality assessment: A simple method for laboratory and field use. *Am. J. Altern. Agric.* 18, 3–17. doi:10.1079/AJAA2003003

Yoo, K.H., Touchton, J.T., and Walker, R.H. (1988). Runoff, sediment and nutrient losses from various tillage systems of cotton. *Soil Tillage Res.* 12, 13–24. doi:10.1016/0167-1987(88)90052-9

Zhu, J.C., Gantzer, C.J., Anderson, S.H., Alberts, E.E., and Buselinck, R.R. (1989). Runoff, soil and dissolved nutrient losses from no-tillage soybean and winter cover crops. *Soil Sci. Soc. Am. J.* 53, 1210–1214. doi:10.2136/sssaj1989.03615995005300040037x

Zibilske, L. (1994). Carbon mineralization. In: P.J. Bottomley, J.S. Angle, R.W. Weaver, (Eds.), *Methods of soil analysis: Part 2. Microbiological and biochemical properties* (p. 835–863). Madison, WI: SSSA.

# 2

# Sampling Considerations and Field Evaluations for Soil Health Assessment

*Mark A. Liebig, Dennis Chessman, Jonathan J. Halvorson, and Roberto Luciano*

Sampling approaches for soil health assessments will vary considerably depending upon the purpose for which an evaluation is undertaken. Decisions regarding sampling location, timing, and frequency will ultimately determine the quality and usefulness of collected data. Tradeoffs associated with balancing the quantity and quality of information obtained with the investment of time and resources must be addressed when selecting a sampling method. This chapter is intended to provide general guidelines for collecting samples and conducting select field evaluations for soil health assessments.

## Introduction

Soil health sampling approaches will vary considerably depending upon the evaluator's goals which may range from a qualitative understanding of near-surface soil conditions at a single point in time to detailed analytical characterizations of a suite of soil properties across broad landscapes to evaluate long-term change. Whether a land manager, consultant, conservationist, or scientist, each evaluator will need to balance the desire for useful information about a soil's status with the investment of time and resources to obtain that information (Dick et al., 1996).

Regardless of the chosen approach, a common requirement of all is to accurately assess the condition of a soil, which is an inherently complex medium varying across space and time. To address this complexity, evaluators must consider a series of questions dealing with location, timing, frequency, and sampling protocol that will affect the quality and usefulness of collected data.

*Soil Health Series: Volume 2 Laboratory Methods for Soil Health Analysis*, First Edition.
Edited by Douglas L. Karlen, Diane E. Stott, and Maysoon M. Mikha.

**Table 2.1** Synthesis of select resources addressing sampling considerations and field evaluations for soil health assessment.

| Citation | Title | Topics addressed |
|---|---|---|
| Pellant et al. (2020) | Interpreting indicators of rangeland health | Review of protocols for assessing ecosystem function on rangelands and woodlands. Provides assessment and interpretive guidelines for soil-associated measurements. |
| Ball et al. (2017) | Visual soil evaluation: A summary of some applications and potential developments for agriculture | Review of visual soil evaluation methods, with emphasis on the Visual Evaluation of Soil Structure (VESS). Addresses VESS applications to agricultural production and environmental quality. |
| USDA-NRCS (2001) | Soil quality test kit guide | Review of assessments made with the Soil Quality Test Kit. Background and interpretive guidelines provided for each assessment. |
| Boone et al. (1999) | Soil sampling, preparation, archiving, and quality control | Review of protocols for soil sampling and laboratory processing. Guidelines developed for the U.S. Long-Term Ecological Research Network. |
| Dick et al. (1996) | Standardized methods, sampling, and sample pretreatment | General guidelines for soil sampling, sample handling, and quality assurance/control. |
| Sarrantonio et al. (1996) | On-farm assessment of soil quality and health | Review and application of measurements made by the Soil Quality Test Kit. Results from case studies provided. |
| Petersen and Calvin (1986) | Sampling (Methods of soil analysis) | Addresses statistical considerations of different sampling approaches. |

This chapter provides general guidelines for collecting soil health samples and reviews select descriptive and analytical field evaluations. For more in-depth discussion, readers should see Pellant et al. (2020), Ball et al. (2017), USDA-NRCS (2001), Boone et al. (1999), Dick et al. (1996), Sarrantonio et al. (1996), or Petersen and Calvin (1986), synthesized in Table 2.1.

## Soil Variability

Spatial variation across landscapes (horizontal) and throughout the profile (vertical) is caused by differences in soil genesis and development, resulting in inherent differences in color, physical structure, texture, and chemical attributes (Soil

Science Division Staff, 2017). Inherent soil property differences provide the foundation for classifying soils using various taxonomic schemes (e.g., FAO, USDA). This "natural variation" can be gradual or abrupt across landscapes and depth increments, underscoring the necessity for preliminary site assessments before initiating full-scale sampling efforts (Boone et al., 1999).

In most agroecosystems, inherent spatial soil variation is coupled with management-induced variation, as reflected by horizontal and/or vertical zones having similar soil properties. Management-induced variation typically reflects long-term repeated use of tillage, chemical amendments, controlled traffic (vehicular and animal), irrigation practice, or crop residue removal (Boone et al., 1999). These induced characteristics can often mask inherent variation in soil properties (Wang et al., 2019). Consequently, sources of management-induced variation must be understood before conducting a soil health assessment. Furthermore, depending upon the evaluator's goals, it may be necessary to subdivide the sampling area into uniform zones to accurately assess management-induced variation (Dick et al., 1996).

All soil properties change over time in response to environmental- and management-related factors. Soil properties strongly influenced by temperature and moisture can fluctuate daily, while those reflecting inherent properties (e.g., texture, mineralogy) change slowly. Though land managers have negligible control over weather and soil forming factors, management decisions, including application of chemical amendments, tillage type and intensity, crop rotation, biomass harvest, and animal activity, can induce significant variation in soil properties (Wuest, 2015; Boone et al., 1999).

Among the portfolio of soil health indicators, those associated with soil biological activity are influenced by daily and seasonal weather changes and management practices that influence nutrient cycling, carbon balance, and physical conditions (Liebig et al., 2006; Dick et al., 1996). As soil health assessments have evolved to include more biological properties and processes (Bünemann et al., 2017), it is imperative evaluators account for temporal dynamics when collecting samples.

## Sampling Considerations

### Sources of Error

Obtaining an accurate depiction soil health underscores the importance of minimizing errors during each step of an evaluation. Errors are cumulative, beginning with decisions made during site selection and ending with the interpretation of collected data (Fig. 2.1). Understanding error types associated with an evaluation can guide decisions to reduce their influence on observed outcomes (Dick et al., 1996).

Selection error is associated with the over or under sampling of areas, depths, and/or times needed to accurately represent a site (Fig. 2.1). Sources associated

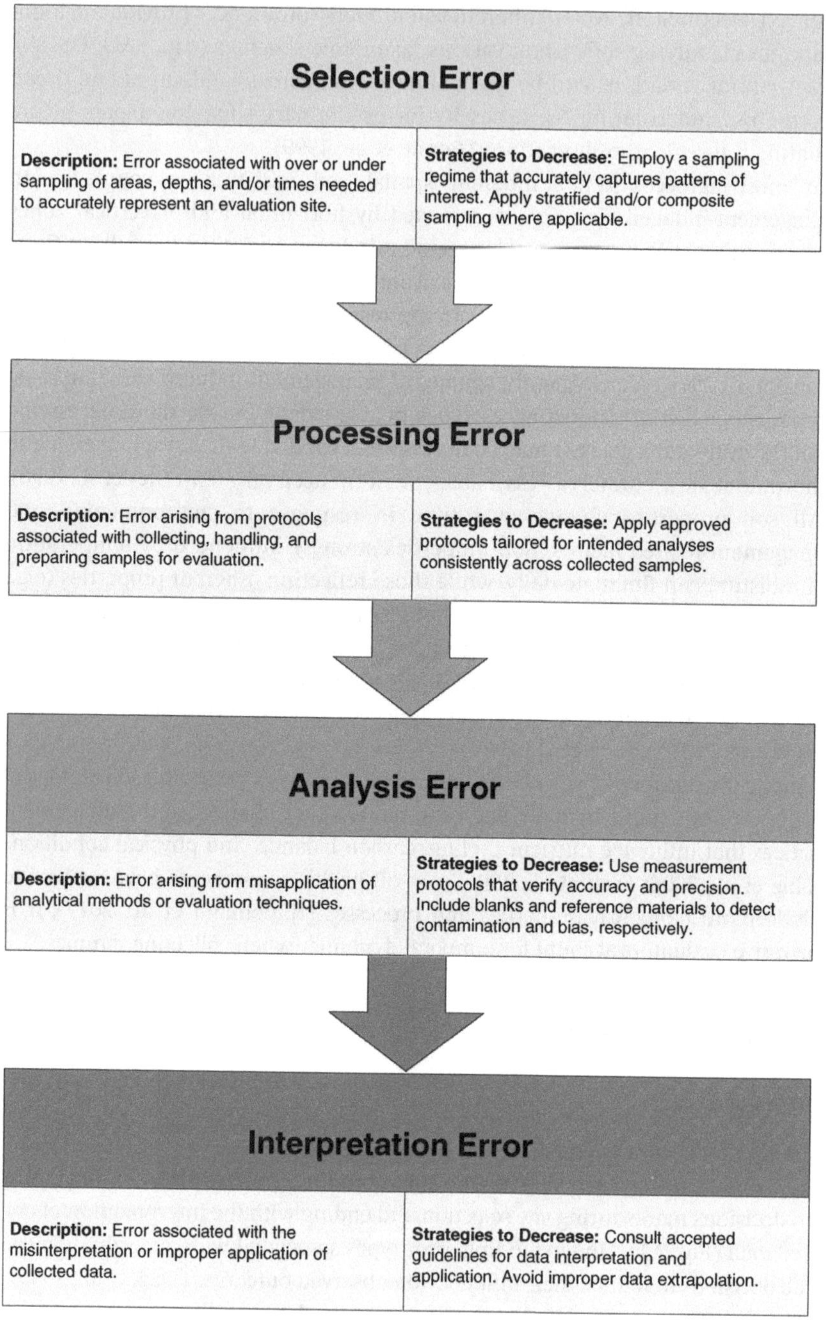


**Figure 2.1** Error components associated with soil property assessment.

with this type of error (also referred to as sampling error) are well known (Das, 1950). This error can be minimized by using a sampling plan that accurately captures patterns of interest. Selection error can also be reduced by increasing the number of samples collected within or among subareas if using a stratified sampling approach (see below).

Processing error are errors made while collecting, handling, and preparing samples for evaluation (Fig. 2.1). Reducing this error requires consistent application of approved protocols tailored for the specific type of analyses, inclusive of storage conditions. Closely aligned with processing error is measurement error, which arises from an improper application of analytical methods or evaluation techniques. Consistent use of consensus protocols will ensure accuracy and precision of each measurement. For laboratory analyses, use of blanks, internal standards, and reference samples is necessary to detect potential contamination and bias.

Once data have been collected, interpretation error can further confound errors from site selection, sample processing, and measurement (Fig. 2.1). Interpretation error results from accidental or systematic misinterpretation or improper application of data. Reduction of interpretation error relies on the evaluator's knowledge to accurately decipher data in context to the sampled site, while concurrently ensuring data outcomes are not extrapolated beyond inherent spatiotemporal constraints or methodological limitations.

## Site Characterization

Preliminary site characterization is important and encouraged, especially when spatial variation of inherent soil properties and/or previous land use is unknown. If the site is intended for long-term monitoring, preliminary site characterization is essential. Referencing maps and/or imagery of the site prior to in-field assessments may elicit attributes not visible from the ground. Preliminary field assessments can be made rapidly using hand probes, augers, or a shovel to identify variation in morphological attributes. More intensive profile assessments will provide additional information but will increase the investment of time and labor. Profile assessments are especially useful if framed by knowledge of pedogenic processes (Boone et al., 1999), recognizing such assessments are generally beyond the scope of most soil health assessments. Characterizing site geomorphology, however, may be helpful for interpreting landscape attributes and their relationship to ecosystem processes (Gringal et al., 1999).

Collection of metadata (Vol. 1, Chapter 4) is an important part of site characterization and will facilitate interpretation of measurements. At a minimum, geographical, landscape, soil, vegetation, climatic, and management attributes should be recorded and maintained for future reference (Table 2.2). When sampling as

**Table 2.2** Suggested minimum metadata for site characterization.

| Category | Metadata |
|---|---|
| Geography | • Site identifier/name<br>• State/Province, county, city, postal code<br>• Latitude, longitude, and elevation<br>• Regional characterization (e.g., Farm Resource Region, Hydrologic Unit Code, Major Land Resource Area) |
| Landscape, Soil, and Vegetation | • Landscape position, slope, and aspect<br>• Soil series and taxonomic description<br>• Native vegetation |
| Climate and Weather | • Mean annual temperature<br>• Mean annual precipitation<br>• Length of frost-free period |
| Management | • Current management<br>• Management history |

part of a long-term study, more metadata is required to accurately characterize each sampling site (see Table 1.2 in Boone et al., 1999).

## Sampling Designs

Following preliminary site characterization, the evaluator must decide how to sample the site. The design used for sampling should reflect the goals of the evaluator, recognizing and reconciling tradeoffs associated with different designs (Table 2.3). Typically, available time and resources dictate the selection of a sampling design.

### Judgement Sampling

Judgement sampling selects sites presumed to represent larger areas (Table 2.3). Since site selection is based on the evaluator's knowledge and skill, it is subject to considerable bias. As outcomes are inherent to the evaluator, application of this design for long-term use is limited as changes in evaluator judgement are possible over time. Data collected using judgement sampling has limited statistical validity especially when isolated from other sites sampled by the same evaluator.

Despite its drawbacks, judgement sampling can be useful when resources are scarce and demands for accuracy and statistical rigor are relaxed. This design may also be applicable when samples are difficult to collect, or an evaluation involves attributes with high analytical cost.

**Table 2.3** Definitions and attributes of different soil sampling designs.

| Description | Advantages | Disadvantages | Potential application |
|---|---|---|---|
| ***Judgement sampling*** <br> An approach to sampling that bases site selection on the knowledge and skill of the evaluator. Effective use of approach requires prior knowledge of site and attributes being assessed. Sites are generally selected to avoid nonrepresentative locations with the larger area. | • Uses expert knowledge of site.<br>• Sampling decisions informed by previous experience.<br>• Limited sampling can save time and resources. | • High degree of evaluator bias. Outcomes strongly associated with the individual doing site selection.<br>• Samplings over time are compromised by changes in evaluator judgement.<br>• Limited application to statistical analyses.<br>• Less legally defensible. | • Initial site screening when resources for sampling and analyses are limited and demands for accuracy and statistical rigor are relaxed.<br>• In some instances, this approach may be useful for the evaluation of very expensive attributes when only a restricted number of samples can be collected. |
| ***Simple random sampling*** <br> A sampling approach whereby each sample collected has an equal chance of being selected. If used, prior knowledge of site is not required. | • Approach minimizes bias associated with site selection.<br>• Relatively easy to apply.<br>• Approach is statistically valid. | • Approach treats all points as equally important regardless of inherent and/or management-induced stratification.<br>• Sampling locations may not adequately represent entire site (e.g., clustered sampling locations limit information on spatial distribution). | • Approach is useful when knowledge of the site is limited.<br>• Collected data may be amenable to modeling applications (e.g., sensitivity analysis). |

(*Continued*)

**Table 2.3** (Continued)

| Description | Advantages | Disadvantages | Potential application |
|---|---|---|---|
| ***Stratified random sampling*** | | | |
| An approach that uses random sampling within unique subareas to characterize a site. Prior knowledge of site stratification is helpful for identification of subareas. | • Acknowledges intra-site differences in inherent soil and/or management attributes. Accordingly, there is increased sensitivity to site nuances.<br>• Approach may increase sampling efficiency.<br>• Allows for understanding of presumed sources of site variability. | • Approach requires prior knowledge of site attributes.<br>• Requires a more complex data analysis.<br>• Delineation of site by stratified attribute can differ among evaluators.<br>• Approach may generate many samples, requiring more resources. | • Useful at sites with apparent soil, landscape, management, and/or contamination zones. |
| ***Systematic sampling*** | | | |
| An approach to sampling whereby an entire area is sampled at predetermined points, usually in a grid-like pattern. If used, prior knowledge of site is not required. | • Application of sampling approach is generally straightforward.<br>• Can return to the same location for future samplings with confidence.<br>• Approach allows for use of geostatistical data analyses. | • Selection of grid size can bias outcomes.<br>• Limited application of evaluator judgement.<br>• Approach may sacrifice sampling efficiency for the sake of following an established pattern.<br>• Depending on site, approach may be less cost effective than others. | • Useful for soil mapping purposes when knowledge of site is limited.<br>• Appropriate in situations where the variable of interest is not expensive to collect and analyze.<br>• Collected data may be amenable to modeling applications (e.g., sensitivity analysis). |

**Simple Random Sampling**

The simple random sampling design assumes each given area has an equal opportunity of being selected (Table 2.3), and this typically involves pre-selection of sampling points to avoid evaluator bias (Dick et al., 1996). Point selection may be done using a random number table and a coordinate grid overlain on an image of the sampling area, or with a random point generator and GIS (geographic information systems) software package (e.g., ArcGIS). Use of this sampling design minimizes bias associated with point selection.

Though easy to apply, simple random sampling has notable caveats since all points in an area are assumed to have equal importance, regardless of variation in inherent features and/or spatial stratification of management attributes. Spatial distribution of selected points therefore may not adequately represent an entire site, although this concern diminishes as the number of sample points increases. This design may be appropriate when knowledge of the site is limited.

**Stratified Random Sampling**

Stratified random sampling applies simple random sampling to preselected subareas of a site (Table 2.3). This design requires knowledge of site characteristics to accurately identify the location, extent, and boundary of subareas. This sampling design is most useful at sites with clearly defined soil, landscape, management, and/or contamination zones (Dick et al., 1996).

Assessing soil health indicators within defined subareas will provide a more complete understanding of inherent and/or management-induced variation. The drawback of this sampling design is a corollary to its strength, since subarea stratification will require more resources than simple random sampling.

**Systematic Sampling**

Systematic sampling involves selection of predetermined sampling points, typically in a grid- or transect-like pattern across the sampling area (Table 2.3). Use of fixed sampling points simplifies sampling, with the added advantage of being able to return to the same point in the future. The data are amenable to spatial analyses (e.g., geostatistics) which can be used to identify a range of soil properties within a site (Dick et al., 1996). Use of systematic sampling is best suited for soil mapping when knowledge of the site is limited but resources are abundant.

Caveats associated with systematic sampling include bias based on grid selection. If there are large distances between sampling points, nuances in soil condition that would be detected at finer scales can be missed. Furthermore, systematic sampling may also sacrifice sampling efficiency (e.g., sampling by stratified zone) because points are selected in an established pattern. Accordingly, systematic sampling may be less cost effective than other designs.

## Composite Sampling

Composite sampling can increase sampling accuracy and reduce analytical costs. Briefly, it involves collecting several homogenous samples in an area surrounding a sampling point and combining those samples into a single bulk (or composite) sample (Dick et al., 1996). For effective composite sampling, it is essential that each sub-sample contributes equally and that there are no interactions among sub-samples (Boone et al., 1999). Moreover, all sub-samples must be collected from the same soil type and depth to be homogenous.

Composite sampling can be used with any of the above sampling designs, although it is less suited to systematic sampling because of the fixed sampling point constraints. Also, since multiple sub-samples are combined and mixed, compositing samples in an area with a wide variation in physical disturbance should be avoided.

## Sampling Depth

Selection of appropriate sampling depths is directly related to the evaluator's goals. Evaluations of soil biological properties and processes as indictors of soil health, nutrient availability for plant growth, or profile carbon stocks will each require sampling at different but context-appropriate depth increments. The selection of sampling depth increments must also consider soil forming and/or management-induced changes, such as depth of tillage, to ensure important features are detected (Wienhold et al., 2006). Furthermore, since variability of some soil properties increases with depth, sampling intensity and the number of depth increments needed to detect differences among treatments or over time will vary by location (Kravchenko and Robertson, 2011). Selecting an appropriate number of depth increments requires a reasonable understanding of how variable a site is, how sensitive the indicators are, and the implications of over- or under-estimating the correct value.

Two common sampling strategies are to use genetic horizons or uniform depth increments. Sampling by horizon is well-suited for sites under native vegetation, where soil forming processes are readily evident throughout the profile due to distinct differences in morphology (i.e., texture, structure, color). Soils with distinct organic horizons benefit from this approach, because layers of surface litter are generally not amenable to fixed increment sampling (Soil Survey Division Staff, 2017). Sampling based on genetic horizons can significantly reduce observed variation throughout the profile (Boone et al., 1999), but requires knowledge of soil taxonomy.

Uniform depth sampling is distinctly different from horizon sampling because every increment starting at the soil surface is fixed (e.g., 0–5, 5–10, 10–20, 20–50, and 50–100 cm). This process facilitates consistency in sample collection across multiple sites, among different sampling crews, and over time (Boone et al., 1999). There is also less concern about biases associated with sampling depth provided the increments meet assessment objectives and accurately account for site characteristics and management attributes. Sampling by uniform depth increment also allows evaluators to know how many samples will be collected at the beginning of a project. Knowing this is beneficial for budgeting and for planning labor requirements (Boone et al., 1999).

One caveat associated with uniform sampling is the potential to miss important profile differences, especially if depth increments are large (e.g., ≥30 cm). Uniform sampling can also be problematic during near-surface assessments if soil properties are strongly stratified with depth (Bowman and Halvorson, 1998). Failure to adjust sampling depth for near-surface stratification can result in misleading management recommendations (Reeves and Liebig, 2016). Therefore, if near-surface stratification is suspected, using small depth increments for the top 30 cm of the soil profile is recommended.

Conversely, a distinct advantage of uniform sampling is the opportunity to quantify soil bulk density. In addition to being a useful measure of soil physical condition, soil bulk density enables conversion of concentration data to a volumetric basis, thereby permitting expression of results on an area basis for a given depth increment (Dick et al., 1996). Soil bulk density values are also essential for calculating nutrient stocks using the "equivalent soil mass" method (Ellert and Bettany, 1995). This method accounts for differences in genetic horizon thickness and/or soil bulk density differences among treatments by calculating a standard soil mass before computing nutrient stocks.

## Timing and Frequency of Sampling

Appropriate timing and frequency of sampling for soil health assessments will vary based on evaluator goals, indicators chosen, environmental conditions, management operations, and available resources. Evaluators must gauge tradeoffs associated with quantifying seasonal variability versus conducting an assessment at a single point in time. If done appropriately, the latter sampling option would select a time guided by knowledge of seasonal and/or annual variability and the timing of management operations. As shown by previous soil health evaluations, selecting an optimal sampling time is difficult (Wuest, 2015; Pikul et al., 2006; Mikha et al., 2006; Wienhold et al., 2006). Accordingly, sampling time decisions should be based by assessment objectives, recognizing that both management disturbances and environmental conditions can lead to misleading outcomes if not properly accounted for.

It is also important to recognize that due to temporal variability in every soil health indicator, the benefit to cost ratio will be low if sampling occurs too frequently. Annual sampling may be desirable for questions focusing on biological indicators, but for long-term soil health changes, sampling every three to five years or at the end of a recurring period inherent to the production system (e.g., following a crop rotation or grazing cycle) will generally be sufficient. It is also important to collect samples at approximately the same time of year or under similar weather patterns to minimize variation in soil water content and temperature (Dick et al., 1996).

## Sample Collection, Processing, and Archival

After deciding to implement a soil health assessment, it is important to carefully consider and document every aspect of the experimental design, sampling protocols, and how the samples will be handled. Each set of protocols is inherently project-specific, underscoring the importance of thorough documentation for future reference. As a guideline for this chapter, protocols adapted from the USDA-ARS GRACEnet (Greenhouse gas Reduction through Agricultural Carbon Enhancement network) project are listed below (Liebig et al, 2010). Additional guidelines can also be found in Boone et al. (1999) and Dick et al. (1996).

Sampling decisions will vary based on assessment objective(s), geographic location, investigator preferences, and/or agroecosystem attributes. For initial soil health samplings, extra care is warranted since those data will ultimately be referenced as baseline data against which long-term changes in soil properties are measured. After selecting an appropriate sampling design, the best approach for sample collection in order to meet project goals and soil conditions should be determined and documented in the metadata. For example, mechanical coring devices (handheld or machine-driven) will often be used because they permit rapid collection of soil samples with a uniform cross-sectional area. However, for soils with high near-surface sand content or excessive stones, a compliant cavity method may be a preferable approach (USDA-NRCS, 2004).

After carefully collecting soil samples, they should be placed in labeled plastic bags, sealed, stored in coolers with ice packs, and transported promptly to a laboratory where they can be held in cold storage (5 °C) until processed. Thick-gauge polyethylene or double bags may be required to limit moisture loss.

Processing protocols should minimize changes in soil properties. For biological attributes, storage time even at 5 °C should be minimized. If extended storage of biological samples is necessary, freezing at −20 °C is recommended over air-drying (Sun et al., 2015). For chemical soil health indicators, samples can be air-dried at 35 °C for 3 to 4 d before sieving to remove rocks, root fragments, and non-soil material. Some soil physical indicators (e.g. bulk density) should be determined using non-disturbed samples, while coarse (~8 mm) sieving can be used to prepare samples for aggregate stability analysis (Vol. 2, Chapters 4 and 5).

Though frequently overlooked, archiving of soil samples is critically important, especially for long-term studies. Archived soil samples provide 'time capsules' for assessing temporal changes in soil properties and are particularly valuable as new analytical capabilities are developed (Boone et al., 1999). The amount of soil archived will vary by evaluator goals, available storage space, and projected future needs (e.g. research vs production-scale monitoring), but in general several hundred grams of air-dried soil should be archived from 'time-zero' with additional amounts from key subsequent samplings. Archived soil samples should be kept in air-tight, non-reactive containers with secure lids and permanent labels. Samples should also be kept in a dry, secure location with moderate temperature conditions and a low probability of water or fire damage.

## Field Evaluations

Field evaluations of soil health can often provide timely insights into soil condition (Fig. 2.2). They can affirm the efficacy of current and previous soil and crop management decisions (e.g., tillage intensity, rotations, manure application) and thus help guide management changes to better align with goals of the land manager. Field evaluations can also help discern the value of more intensive, costly follow-up assessments, and may confirm (or refute) findings from previous laboratory analyses.

Soil-related information gathered during a field evaluation is strongly influenced by the approach taken, and much like the selection of sampling designs, the evaluator should be aware of tradeoffs associated with the selected approach. Therefore, attributes of field evaluations and their capacity to meet stated objectives should be carefully considered by the land manager prior to initiating assessments.

### General Field Observations

Field-scale soil health assessments should begin with general field observations such as aboveground biomass, plant growth characteristics and soil conditions. Since these observations are generally part of normal field management practices, they are a logical first step during soil health assessments to determine if more detailed, follow-up evaluations are warranted. Moreover, much of this information may be obtained through conversations with the land manager. Common field observations outlined by Magdoff and Van Es (2009) include:

- Are yields declining?
- Do crops perform less well than those on neighboring farms with similar soils?
- Do crops quickly show signs of stress or stunted growth during wet or dry periods?

**Figure 2.2** Generalized approaches to field evaluations of soil health. USDA-NRCS photo credits: Field observations – Roberto Luciano; Visual soil evaluations – Roberto Luciano; Test kits – Susan Samson-Liebig; Sensor-based measurements – Keith Anderson.

- Are there symptoms of nutrient deficiencies?
- Are there increased problems with diseases or weeds?
- Does the soil appear compacted?
- Does it take more power to run field equipment through the soil?
- Does the soil crust over easily?
- Are there signs of runoff and erosion?
- Are there changes in soil color?

On-the-ground observations of crop and soil conditions can be supplemented with aerial imagery to help identify potential production and environmental issues (Schepers et al., 2004). Doing so can efficiently guide follow-up evaluations and potential management interventions in affected areas if image locations are georeferenced.

Descriptive soil health field assessments can be translated into semi-quantitative formats using soil health scorecards. Developed in the early 1990s (Harris and Bezdicek, 1994), scorecards use stakeholder knowledge and field evaluations to identify relevant soil health indicators and assign an associated ranking as being healthy, impaired, or unhealthy (Romig et al., 1996). Scorecards rely on the evaluator's senses (e.g., sight, feel, smell) and manager input to discern the quality or character of an attribute, making assessments unique to the individual.

Soil health scorecards and other semi-quantitative evaluation approaches have been developed for many states in the USA following a collaborative process including farmers, conservationists, and scientists to identify relevant indicators and ranking criteria (USDA-NRCS, 1999). Property-specific pocket charts have also been developed based on associations between descriptive assessments and laboratory measurements (e.g, soil color categories with ranges of organic matter content) (Alexander, 1971). However, broad application of pocket charts for soil quality assessment has been limited because of the complexity and variability among soils. In the case of soil color, associations with soil organic matter content vary strongly with texture, soil depth, and land use (Wills et al., 2007), thereby constraining chart use to a limited geographical domain.

## Visual Soil Evaluations

Visual soil evaluations are important components of soil health assessment. If done thoroughly and in a quantitative manner, visual soil evaluations can be integrated with broader land evaluation frameworks (Mueller et al., 2012; Pellant et al., 2020). As there are multiple soil evaluation methods currently in use (Ball and Munkholm, 2015), evaluators need to mindful of their intended application and respective strengths and weaknesses. The Visual Evaluation of Soil Structure (VESS) (Ball et al., 2017) and rangeland health assessment protocols (Pellant et al., 2020) represent approaches for cropland and rangeland, respectively.

The VESS was designed for arable production systems (Ball et al., 2017) based on the Peerlkamp Spade Test (Peerlkamp, 1959). It classifies top- and sub-soil attributes into five scoring categories based on the size, shape, and visible porosity of soil aggregates, rooting characteristics, and presence/absence of macropores. VESS requires only a shovel and scoring guidelines (Scotland's Rural College, 2019), so evaluators can quickly assess soil structure and assign a score between 1 (best) and 5 (worst). The tactile "hands-on" nature of VESS enables it to work best when the soil is in a friable condition (neither too wet nor too dry). A limitation of VESS is that it does not work well in very sandy soils because of low structural cohesion (Franco et al., 2019).

The VESS has been used globally across a broad range of soil types and production systems to discern management impacts on soil health (Munkholm and

Holden, 2015). The scores have also been correlated to various soil physical properties, soil-atmosphere gas fluxes, and crop yield (Ball et al., 2017), thus confirming its value for soil health assessment.

Rangeland health assessments using three interrelated ecosystem attributes of soil/site stability, hydrological function, and biotic integrity (Pellant et al., 2020) have been shown to help characterize the status and trends of critical soil functions and effectively guide changes in management (Brown and Herrick, 2016). Using a combination of qualitative and quantitative indicators, Pellant et al. (2020) developed criteria for monitoring rangeland health using 17 indicators that include soil-related evaluations of bare ground, gullies, resistance to erosion/degradation, and compaction. Rangeland health assessments using these criteria have been adopted by public agencies and private landowners throughout the world and have served to improve the systematic understanding of soil quality in rangeland ecosystems (Brown and Herrick, 2016). Effective use of rangeland health assessments, however, requires the evaluator to recognize and correctly identify site characteristics including landscape and temporal variability since evaluations are made relative to an ecological site or its equivalent.

## Soil Health Test Kits

Soil test kits can provide land managers with expedient information on the status of soil physical, chemical, and biological properties. While many test kits focus on measurements of soil solution chemistry for horticultural applications, some test kits include a broad suite of measurements with linkages to soil functions relevant to maintaining productivity, regulating/partitioning water flow, and storing/cycling nutrients. The Soil Quality Test Kit (SQTK), developed by Dr. John Doran (USDA-ARS, retired), includes equipment and supplies for the measurement of select soil properties recognized as components of a minimum data set for monitoring soil health (NRCS, 2001). In the intervening 25 years since its development, the SQTK has been used by NRCS resource soil scientists throughout the USA as a screening tool to guide more in-depth soil health assessments. The SQTK also prompted the development of additional field-based assessments for use in rangeland (e.g., soil slake test; Seybold and Herrick, 2001). While test kits provide a means for receiving near immediate feedback about a soil's status, some tests can be time consuming. Moreover, measurements of soil solution chemistry by test kits require calibration against known standards to ensure accurate results.

## Sensor-based Measurements

Field-scale soil property mapping, generally used to improve nutrient use efficiencies, can also document the trajectory of some soil health indicators (Mulla and

Schepers, 1997; Smith et al., 1993). However, conventional soil sampling and laboratory analyses can be expensive, time consuming, and thus limit its value for making timely adjustments to management. In response, several novel sensor-based technologies have been developed (Vol. 1, Chapter 8), thus increasing the likelihood of making real-time soil health assessments.

Electromagnetic, optical, mechanical, electrochemical, airflow, and acoustic sensors for automated field measurements have been adapted to quantify changes in soil physical and chemical properties across agricultural landscapes (Adamchuk et al., 2004). Most sensors are property-specific, with electromagnetic, mechanical, electrochemical, and airflow types associated with measurements of electrical conductivity, soil resistance, nutrients or pH, and air permeability, respectively. Optical sensors are useful for predicting both chemical and physical properties (Thomasson et al., 2001). Sensor-based measurements can also be used to quantify field-scale distributions of inferred soil properties which can then be partitioned into distinct management zones. It is important to acknowledge, however, that sensor-based measurements provide either high-resolution spatial or temporal data, but usually not both. With further technological advancement, sensor-based measurements may ultimately provide real-time control of input applications.

## Summary

Soil health assessments should provide useful insights into the status of soil properties as they affect critical soil functions. Currently there is no single best method for soil health assessment because outcomes are intrinsically related to evaluator decisions related to method, time, location, and frequency of sampling. Multiple research endeavors are underway focused on improving assessments by tailoring collection, analysis, and data interpretation to inherent site attributes, project resources, intended data uses, and evaluator expertise.

## Acknowledgments

We thank Robyn Duttenhefner for her helpful edits to improve an earlier draft of the chapter.

The U.S. Department of Agriculture (USDA) prohibits discrimination in all its programs and activities on the basis of race, color, national origin, age, disability, and where applicable, sex, marital status, family status, parental status, religion, sexual orientation, genetic information, political beliefs, reprisal, or because all or part of an individual's income is derived from any public assistance program. (Not all prohibited bases apply to all programs.). USDA is an equal opportunity provider and employer. Mention of commercial products and organizations in this manuscript is solely to provide specific information. It does not constitute endorsement by USDA-ARS over other products and organizations not mentioned.

## References

Adamchuk, V.I., Hummel, J.W., Morgan, M.T., and Upadhyaya, S.K. (2004). On-the-go soil sensors for precision agriculture. *Comp. Elec. Agric.* 44, 71–91.

Alexander, J.D. (1971). Color chart for estimating organic carbon in mineral soils in Illinois. *Bull. AG-1941*. Champaign, IL: Univ. of Ill. Coop. Ext. Serv.

Ball, B.C., Guimarães, R.M.L., Cloy, J.M., Hargreaves, P.R., Shepherd, T.G., and McKenzie, B.M. (2017). Visual soil evaluation: A summary of some applications and potential developments for agriculture. *Soil Till. Res.* 173, 114–124.

Ball, B.C. and Munkholm, L.J. (Eds.) (2015). *Visual soil evaluation: Realizing potential crop production with minimum environmental impact*. Oxfordshire, UK: CABI.

Boone, R.D., Gringal, D.F., Sollins P., Ahrens, R.J., and Armstrong, D.E. (1999). Soil sampling, preparation, archiving, and quality control. In G.P. Robertson, D.C. Coleman, C.S. Bledsoe, and P. Sollins (Eds.) *Standard soil methods for long-term ecological research* (p. 3–28). New York: Oxford Univ. Press.

Bowman, R.A., and Halvorson, A.D. (1998). Soil chemical changes after nine years of differential N fertilization in a no-till dryland wheat-corn-fallow rotation. *Soil Sci.* 163, 241–247.

Brown, J.R., and Herrick, J.E. (2016). Making soil health a part of rangeland management. *J. Soil Water Conserv.* 71, 55A–60A.

Bünemann, E.K., Bongiorno, G., Bai, Z., Creamer, R.E., De Deyn, G., de Goede, R., Fleskens, L., Geissen, V., Kuyper, T.W., Mäder, P., Pulleman, M., Sukkel, W., van Groenigen, J.W., Brussaard, L. (2018). Soil quality – A critical review. *Soil Biol. Biochem.* 120, 105–125.

Das, A.C. (1950). Two-dimensional systematic sampling and associated stratified and random sampling. *Sankhya* 10, 95–108.

Dick, R.P., Thomas, D.R., and Halvorson, J.J. (1996). Standardized methods, sampling, and sample pretreatment. In J.W. Doran, and A.J. Jones (Eds.) *Methods for assessing soil quality* (p. 107–121). Soil Sci. Soc. Am. Spec. Publ. No. 49. Madison, WI: SSSA.

Ellert, B.H., and Bettany, J.R. (1995). Calculation of organic matter and nutrients stored in soils under contrasting management regimes. *Can. J. Soil Sci.* 75, 529–538.

Franco, H.H.S., Guimarães, R.M.L., Tormena, C.A., Cherubin, M.R., and Favilla, H.S. (2019). Global applications of the visual evaluation of soil structure method: A systematic review and meta-analysis. *Soil Till. Res.* 190, 61–69.

Gringal, D.F., Bell, J.C., Ahrens, R.J., Boone, R.D., Kelly, E.F., Monger, H.C., and Sollins, P. (1999). Site and landscape characterization for ecological studies. In G.P. Robertson, D.C. Coleman, C.S. Bledsoe, and P. Sollins (Eds.) *Standard soil methods for long-term ecological research* (p. 29–52). New York: Oxford Univ. Press.

Harris, R.F., and Bezdicek, D.F. (1994). Descriptive aspects of soil quality/health. In J.W. Doran, D.C. Coleman, D.F. Bezdicek, and B.A. Stewart (Eds.) *Defining soil quality for a sustainable environment* (p. 23–35). Soil Sci. Soc. Am. Spec. Publ. No. 35. Madison, WI: ASA and SSSA.

Kravchenko, A., and Robertson, G.P. (2014). Whole-profile soil carbon stocks: The danger of assuming too much from analyses of too little. *Soil Sci. Soc. Am. J.* 75, 235–240.

Liebig, M.A., Varvel, G.E., and Honeycutt, C.W. (2010). Sampling protocols. Chapter 1. Guidelines for site description and soil sampling, processing, analysis, and archiving. In R.F. Follett (Ed.), GRACEnet Sampling Protocols (p. 1-1 to 1-5). Available at https://www.ars.usda.gov/anrds/gracenet/gracenet-protocols (accessed 26 July 2019).

Liebig, M., Carpenter-Boggs, L., Johnson, J.M.F., Wright, S., and Barbour, N. (2006). Cropping system effects on soil biological characteristics in the Great Plains. *Renew. Agric. Food Systems* 21, 36–48.

Magdoff, F., and Van Es, H. (2009). How good are your soils? Field and laboratory evaluation of soil health. In *Building soils for better crops: Sustainable soil management* (p. 257–266). 3rd ed. Handbook Series Book No. 10. Washington, DC: Sustainable Agriculture Research & Education (SARE), Nat. Inst. Food Agric., USDA.

Mikha, M.M., Vigil, M.F., Liebig, M.A., Bowman, R.A., McConkey, B., Deibert, E.J., and Pikul, J.L., Jr. (2006). Cropping system influences on soil chemical properties and soil quality in the Great Plains. *Renew. Agric. Food Systems* 21, 26–35.

Mueller, L., Schindler, U., Shepherd, T.G., Ball, B.C., Smolentseva, E., Hu, C., Hennings, V., Schad, P., Rogasik, J., Zeitz, J., Schlindwein, S.L., Behrendt, A., Helming, K., and Eulenstein, F. (2012). A framework for assessing agricultural soil quality on a global scale. *Arch. Agron. Soil Sci.* 58, S76–S82.

Mulla, D.J., and Schepers, J.S. (1997). Key processes and properties for site-specific soil and crop management. In F.J. Pierce, and E.J. Sadler (Eds.) *The state of site-specific management for agriculture* (p. 1–18). Madison, WI: ASA, CSSA, and SSSA.

Munkholm, L.J., and Holden, N.M. (2015). Visual evaluation of grassland and aerable management impacts on soil quality. In B.C. Ball, and L.J. Munkholm (Eds.). *Visual soil evaluation: Realizing potential crop production with minimum environmental impact* (p. 49–65. Oxfordshire, UK: CABI.

Peerlkamp, P.K. (1959). A visual method of soil structure evaluation. *Meded. v.d. Landbouwhogeschool en Opzoekingsstations van de Staat te Gent.* 24, 216–221.

Pellant, M., Shaver, P.L., Pyke, D.A., Herrick, J.E., Lepak, N., Riegel, G., Kachergis, E., Newingham, B.A., Toledo, D., and Busby, F.E. (2020). *Interpreting indicators of rangeland health, Version 5.* Tech Ref 1734-6. Denver, CO: U.S. Department of the Interior, Bureau of Land Management, National Operations Center.

Peterson, R.G., and Calvin, L.D. (1986). Sampling. In A. Klute (Ed.) *Methods of soil analysis, Part 1* (p. 33–51). 2nd ed. Agrono. Monogr. 9. Madison, WI: ASA and SSSA.

Pikul, J.L. Jr., Schwartz, R.C., Benjamin, J.G., Baumhardt, R.L., and Merrill, S. (2006). Cropping system influences on soil physical properties in the Great Plains. *Renewable Agric. Food Systems* 21, 15–25.

Reeves, J.L., and Liebig, M.A. (2016). Depth matters: Soil pH and dilution effects in the northern Great Plains. *Soil Sci. Soc. Am. J.* 80, 1424–1427.

Romig, D.E., Garlynd, M.J., and Harris, R.F. (1996). Farmer-based assessment of soil quality: A soil health scorecard. In J.W. Doran and A.J. Jones (Eds.) *Methods for assessing soil quality* (p. 39–60). Soil Sci. Soc. Am. Spec. Publ. No. 35. Madison, WI: SSSA.

Sarrantonio, M., Doran, J.W., Liebig, M.A., and Halvorson, J.J. (1996). On-farm assessment of soil quality and health. In J.W. Doran, and A.J. Jones (Eds.) *Methods for assessing soil quality* (p. 83–105). Soil Sci. Soc. Am. Spec. Publ. No. 49. Madison, WI: SSSA.

Schepers, A., Shanahan, J.F., Liebig, M.A., Schepers, J.S., Johnson, S.H., and Luchiari, A., Jr. (2004). Appropriateness of management zones for characterizing spatial variability of soil properties and irrigated corn yields across years. *Agron. J.* 96, 195–203.

Scotland's Rural College. (2019). Visual evaluation of soil structure. Available at: https://www.sruc.ac.uk/info/120625/visual_evaluation_of_soil_structure (accessed 24 July 2019).

Seybold, C.A., and Herrick, J.E. (2001). Aggregate stability kit for soil quality assessments. *Catena* 44, 37–45.

Smith, J.L., Halvorson, J.J., and Papendick, R.I. (1993). Using multiple-variable indicator kriging for evaluating soil quality. *Soil Sci. Soc. Am. J.* 57, 743–749.

Soil Science Division Staff. (2017). *Soil survey manual.* C. Ditzler, K. Scheffe, and H.C. Monger (Eds.). USDA Handbook 18. Washington, DC: U.S. Government Printing Office.

Sun, S.-Q., Cai, H.-Y., Chang, S.X., and Bhatti, J.S. (2015). Sample storage-induced changes in the quantity and quality of soil labile organic carbon. *Sci. Rep.* 5, 17496. doi:10.1038/srep17496.

Thomasson, J.A., Sui, R., Cox, M.S., and Al–Rajehy, A. (2001). Soil reflectance sensing for determining soil properties in precision agriculture. *Trans. Am. Soc. Agric. Eng.* 44, 1445–1453.

USDA-NRCS. (2004). Compliant cavity (3B3). In R. Burt (Ed.) *Soil survey laboratory methods manual* (p. 98–100). Soil survey investigations report no. 42, version 4.0. Lincoln, NE: USDA-NRCS National Soil Survey Laboratory.

USDA-NRCS. (2001). Soil quality test kit guide. Available at: https://www.nrcs.usda.gov/Internet/FSE_DOCUMENTS/nrcs142p2_050956.pdf (accessed 24 July 2019).

USDA-NRCS. (19990. Soil quality card design guide: A guide to develop locally adapted conservation tools. Available at: https://www.nrcs.usda.gov/wps/portal/nrcs/main/soils/health/assessment/ (accessed 22 July 2019).

Wang, X., Jelinski, N.A., Toner, B., and Yoo, K. (2019). Long-term agricultural management and erosion change soil organic matter chemistry and association with minerals. *Sci. Total Environ.* 648, 1500–1510.

Wienhold, B.J., Pikul, J.L., Liebig, M.A., Mikha, M.M., Varvel, G.E., Doran, J.W., and Andrews, S.S. (2006). Cropping system effects on soil quality in the Great Plains: Synthesis from a regional project. *Renew. Agric. Food Systems* 21, 49–59.

Wills, S.A., Burras, C.L., and Sandor, J.A. (2007). Prediction of soil organic carbon content using field and laboratory measurements of soil color. *Soil Sci. Soc. Am. J.* 71, 380–388.

Wuest, S.B. (2015). Seasonal variation in soil bulk density, organic nitrogen, available phosphorus, and pH. *Soil Sci. Soc. Am. J.* 79, 1188–1197.

# 3

# Soil Organic Carbon Assessment Methods

*Charles W. Rice, Carlos B. Pires, James Lin, and Marcos V. M. Sarto*

## Summary

Globally, the amount of soil organic carbon (SOC) is more than twice that in the atmosphere or living vegetation. Soil organic carbon is an extremely important soil health indicator because it influences almost all soil biological, chemical, and physical properties and processes. Loss of SOC accelerates soil health problems such as soil erosion and decreases soil aggregation. This chapter explores those issues and discusses various SOC measurement methods.

## Introduction

Soils contain several important C pools and play an essential role in the global C cycle. Total soil C consists of organic C and inorganic C, with organic C being part of soil organic matter (SOM). Estimated amounts of organic C stored in world soils range from 1100 to 1600 Pg, more than twice that in living vegetation (560 Pg) or the atmosphere (750 Pg) (Sundquist, 1993; Lehmann and Kleber, 2015). Most of the world's productive soils are in cultivated agriculture, which through plow-based management, has increased SOM loss, accelerated soil erosion, and decreased soil aggregation (Lal, 2015; Karlen and Rice 2015). Intensive tillage, low crop yield, lack of crop diversification, and excessive removal of crop residues have reduced C replenishment to the soil and thus negatively affected soil health (Rice, 2005; Bonini Pires et al., 2020). Those practices have decreased SOC by as

*Soil Health Series: Volume 2 Laboratory Methods for Soil Health Analysis*, First Edition.
Edited by Douglas L. Karlen, Diane E. Stott, and Maysoon M. Mikha.

much as 26% in the upper 30 cm and 16% in the top 100 cm of many soil profiles (Sanderman et al., 2017).

The Soil Science Society of America (SSSA) defines SOM as the organic fraction of soil exclusive of undecayed plant and animal residues (SSSA, 1997). Measurements of SOM include decayed plant residues, soil microorganisms, soil fauna, and byproducts of decomposition that lead to the production of humic substances in a process called humification (Horwath, 2007). SOM can be classified in two distinct pools, active and passive, based on their chemical composition, stage of decomposition, and turnover time (Cambardella and Elliot, 1994; Gougoulias et al., 2014). Active pools exhibit a turnover in months to years, while turnover in passive pools occurs in decades to millennium (Magdoff, 1996). Soil organic matter can also be divided into three major categories: particulate organic matter (POM), humus, and resistant organic matter (ROM) (Bell and Lawrence, 2009). Each SOM classification has a method of measurement and converts the data to SOM using conversion factors that vary. Therefore for consistency, we recommend reporting the values as organic C.

## Origin and Factors Affecting SOC

Soil organic carbon is derived primarily from plant residues with transformations and storage of SOC being a function of biotic, chemical, and physical properties and processes interacting with plant residue quality or biochemistry as well as its accessibility to organisms. Plant residues are decomposed by soil microorganisms and most of the plant C is released to the atmosphere as $CO_2$. Approximately 10 to 20% of the C in plant residue becomes SOM, sometimes referred to as "humus." A portion of this C can persist in soils for hundreds to thousands of years. The theoretical potential for soil C storage is a function of climate and basic soil characteristics, while the amount of C residing in the soil is a function of plant and soil management.

Tillage and organic residue input are two primary drivers influencing organic carbon levels in soils. Soil disturbance (*i.e.*, tillage) disrupts soil aggregates and decreases physical protection by exposing C within soil aggregates to microbial decomposition, which results in a conversion of organic carbon to carbon dioxide.

Development of conservation agriculture practices (e.g., no-till, permanent soil cover, and crop rotation) have increased SOC. These and other advancements in crop and soil management practices (e.g., drought and insect resistance) have the potential to restore and improve soil health. A reduction in tillage intensity, with no-tillage being the least disruptive, allows macroaggregate formation and reduces degradation of SOC. Minimum- or conservation-tillage also conserves soil C through decreased oxidation and less soil erosion than with more intensive tillage practices. Cropping intensity and diversification can also impact SOC levels by varying the amount and quality (e.g., C/N ratio) of plant residue added to the soil. Likewise, the presence of cover crop roots during the intercrop period can release exudates into the soil and contribute to greater macroaggregate formation by increasing microbial biomass and fungal networks.

In agriculture, high residue producing crops, such as corn, wheat, and sorghum tend to sustain or increase soil C. Eliminating summer fallow, a practice traditionally used in the U.S. Great Plains, can increase soil C by providing more plant material on an annual basis. In those areas where sufficient soil water is available, double cropping to produce two crops or more crops per year may be a viable practice. In grassland systems, perennial crops also tend to increase SOC because there is no-tillage, and organic C is added through root turnover.

## Functions of SOC

Soil carbon supports essential soil and associated ecosystem functions. Concentrations of SOM range from 0.2% in mineral soils to over 80% in peat soils (Smith et al., 1993). Albrecht (1938) stated "soil organic matter is one of our most important national resources. . . and it must be given its proper rank in any conservation policy". Larson and Pierce (1994) listed SOC as one of five key measures of soil that included soil cation exchange capacity (CEC), bulk density (BD), water retention, and aeration.

Furthermore, although SOC comprises a small portion of agricultural soils, it significantly affects soil health (Rice et al., 1996; Doran and Zeiss, 2000; Kibblewhite et al., 2008, Bünemann et al., 2018). Currently, the USDA Natural Resource Conservation Service (NRCS) defines soil health as the capacity of a specific kind of soil to function, within natural or managed ecosystem boundaries, to sustain plant and animal productivity, maintain or enhance water and air quality, and support human health and habitation. Soil organic matter imparts many beneficial biological, chemical, and physical properties to soil, specifically improving its structure (Six et al., 2000; Dexter et al., 2008); supporting water infiltration and retention (Boyle et al., 1989; Yang et al., 2014); reducing erosion through increased infiltration, decreased runoff, and more large aggregates (King et al., 2019; Barthès and Roose, 2002); increasing crop yield through water and nutrient supply (Cambardella and Elliot (1992; Pan et al., 2009; Oldfield et al., 2018); and storing C for climate change mitigation (Paustian et al., 2016; Lal and Follett, 2009).

### Biological Effects

The biological benefits of SOC primarily relate to nutrient cycling by soil microorganisms for carbon and energy. Soil microorganisms convert complex plant and animal materials into simpler compounds. The primary SOM decomposers (i.e., consumers) include bacteria, fungi, earthworms, insects, protozoa, and nematodes. All are influenced by soil microbial diversity, which is affected by SOC and the associated land management decisions. For example, no-till results in SOC being concentrated in the topsoil surface (Doran, 1980; Lynch and Panting, 1980; Carter and Rennie, 1982; Balesdent et al., 2000). The combination of higher SOC and greater biological activity results in greater organism diversity, and potentially to greater biological control of plant diseases and pests.

**Chemical Effects**

Chemical benefits of SOC relate to the transformation and flow of soil nutrients, such as mineralization and immobilization of N, P, and S. The nine essential macronutrients for plant growth are C, O, H, P, K, S, Ca, and Mg. The first three are in greatest abundance within the SOM structure (Schnitzer and Khan, 1975). However, plants need N for enzymes, proteins, and chlorophyll, which are mainly derived from SOM (Schnitzer and Khan, 1975). Nearly all N in SOM is inaccessible to plants until it is converted by microbes into ammonium ($NH_4$) and nitrate ($NO_3$). The SOM is also a source of P, generally making up 15 to 80% of total soil P (Mortensen and Himes, 1964). Soil S is estimated to be 50 to 70% in or adsorbed onto SOM (Schnitzer and Khan, 1975). Conversely, K, Ca, and Mg are primarily derived from insoluble inorganic compounds and have not been demonstrated to be obtained from SOM (Schnitzer and Khan, 1975). However, K, Ca, and Mg can form complexes with SOM (Broadbent and Ott, 1957; Schnitzer and Skinner, 1963) and the capacity of soils to store mineral nutrients (*i.e.*, CEC), increases with SOM.

**Physical Effects**

The physical benefits of SOC relate to the formation and stabilization of soil aggregates. Several studies have reported high correlation between soil aggregation and SOC (Wilson et al., 2009; McVay et al., 2006; Six et al., 2002). Degradation of organic materials by soil organisms leads to the formation of humified materials associated primarily with mucilage that surrounds and binds to clay particles, thus developing and binding the particles into microaggregates (Balesdent et al., 2000). The decomposition of protected SOC can become slow due to the clay barrier, thus promoting soil carbon sequestration. Soil health benefits of greater soil aggregation include less crusting, compaction, and bulk density (Diaz-Zorita and Grosso, 2000); enhanced soil structure for greater water infiltration and water holding capacity (Hudson, 1994; Emerson, 1995; Gupta and Larson, 1979, Yang et al., 2014); decreased soil erosion (Schertz et al., 1994; Benito and Diaz-Fierros, 1992); and improved aeration for root growth and microbial activity. As tillage intensity increases, soil microbial activity increases right after tillage and microaggregates are dispersed, thus releasing SOM from protection (Puget et al., 1995, 2000).

### Crop Productivity

In addition to supporting biological control of crop pests and disease, SOM contributes to agricultural crop yields in various ways by overcoming negative soil conditions. Adequate amounts of SOM can enhance: (i) release of nutrients from decaying organic materials and thus reduce commercial fertilizer requirements; (ii) soil porosity which increases plant available water retention and aeration for root development; (iii) soil structure which reduces soil erosion potential and increases aggregation; and (iv) storage of other nutrients, including an increase in cation exchange sites for plant nutrient retention.

## Measurement of SOC

Recognizing there are multiple methods for measuring SOC, dry combustion, loss-on-ignition (LOI) (Schulte and Hopkins, 1996), and the Walkley–Black method (Nelson and Sommers, 1996) were evaluated for SOM are evaluated to examine the advantages and disadvantages of each method (Table 3.1). The traditional method for SOM analysis was wet oxidation in potassium dichromate ($K_2Cr_2O_7$), better known as the Walkley–Black method. The potassium dichromate solution oxidizes soil organic material through a chemical reaction that generates heat when two volumes of sulfuric acid are mixed with one volume of dichromate. The remaining dichromate is titrated with ferrous sulfate, with the titer being inversely related to the amount of C present in the soil sample (Nelson and Sommers, 1996; Meersmans et al., 2009).

Although Walkley-Black was the standard method for SOM analysis for decades, its use has diminished due to the high potential for environmental pollution during disposal and exposure of personnel to hazardous chemicals, such as potassium dichromate and sulfuric acid. In contrast, the LOI method is a simple, inexpensive method for SOM estimation that involves the combustion of samples at high temperatures and measuring weight loss after ignition. The ability of LOI to quantify SOM content has been considered reliable, but optimal heating temperatures and duration to maximize SOM combustion, while minimizing inorganic carbon combustion, are challenging to determine. Both of those variables can substantially affect LOI results (Salehi et al., 2011). Dry combustion using elemental carbon analyzers is now considered the gold standard for SOM measurement. This method is based on thermal oxidation of a soil sample at ~ 1000 °C and determining the quantity of $CO_2$ produced by gas chromatography or infrared analysis. This method is considered the most accurate but is also the most expensive method for determining total carbon. Furthermore, to accurately measure SOC, inorganic carbon must be removed by pre-treating samples with hydrochloric, phosphoric, or sulfuric acid before combustion. It is also important to point out that dry combustion measures SOC instead of SOM. Thus, a conversion factor is

**Table 3.1** Advantages and disadvantages of the most common methods for measuring soil organic matter.

| Method | Advantages | Disadvantages |
|---|---|---|
| Walkley-Black | Relatively simple, accurate, and quick | High environmental pollution potential, uses hazardous chemicals |
| Loss-On-Ignition | Simple, inexpensive, convenient | Heating temperature and time significantly affect accuracy. It has to be calibrated. |
| Dry combustion by elemental analyzer | Most accurate, quick | Elemental analyzers are expensive to purchase and maintain. |

For this chapter, we chose to present both the LOI and dry combustion methods since they are environmentally safe, simple, and the most widely used methods in commercial laboratories.

used to convert SOC into estimates of SOM. Historically, the most used conversion factor has been 1.724, which assumes organic matter is composed of 58% of carbon. However, recent studies have shown that this factor is too low for most soils. In a review of previously published data, the median conversion factor value was found to be 1.9 from empirical studies and 2 from more theoretical considerations. We concur that using a factor of 2, based on the assumption that organic matter is 50% carbon, will, in almost all cases, be more accurate than 1.724 (Pribyl, 2010).

## Methods for SOC Analysis

### Soil Sampling and Preparation

The same general principles that apply to soil sampling for nutrient evaluation apply to soil sampling for SOM determination. Soil samples should be collected to consistent soil depth(s), a consistent number of soil cores collected per composite samples should be maintained, thatch or mulch from the soil surface must be removed prior to sampling, composite samples should be inspected after collection and any obvious pieces of crop residue should be removed. Georeferenced points should also be collected within the field when possible. For all methods described below, soil samples should be dried at 35 to 40 °C in an oven and passed through a 2000-μm sieve.

### Dry Combustion

***Apparatus***

1) Forceps
2) 2000 μm sieve

3) 250 μm sieve
4) Aluminum tin capsule
5) Analytical balance with 0.001 precision
6) C/N elemental analyzer

***Reagents***

1) Concentrated $H_2SO_4$ or HCl
2) 4N phosphoric acid solution

***Procedures***

1) Air-dry or oven-dry soil at 35 to 40 °C.
2) Carefully remove all plant and animal materials from the soil using forceps.
3) Pass the soils through a 2000-μm sieve and grind the soil to a powder to pass through a 250-μm sieve.
4) Proceed to step 11 for soil with pH < 7.00.

***Pre-treatment for soils with inorganic carbon content***

1) For soils that have a pH > 7.50
2) Test for the presence of inorganic carbon or carbonate ($CaCO_3$) using concentrated $H_2SO_4$ or HCl.
3) Weigh 0.50 mg of the soil into a silver vessel and add 5 drops or (5 mL) of concentrated $H_2SO_4$ or HCl.
4) If effervescence is observed, inorganic carbon is present in the soil sample. If inorganic C is not present, go to Step 11.
5) Treat soil containing inorganic C with 0.1 mL of 4N phosphoric acid solution.
6) The treated sample should be allowed to dry for one or two days before analysis. Repeat steps for removal of inorganic C if necessary. Be cautious when handling the tins and do not use Al vessels as they will disintegrate when acid is added.
7) Weigh 50 mg of soil into an aluminum tin capsule using an analytical scale with 0.001 precision.
8) Fold the aluminum tin capsule containing the soil sample and load it on to the analyzer.
9) Weigh a known standard to serve as a check, using one standard every 12 to 20 samples. Standards can either be EDTA, aspartic acid, or soil with a known concentration of C.

   * The operation of the analyzer should be operated according to the guidelines of the machine.

***Loss-On-Ignition (LOI)*** **(Salehi et al., 2011; Schulte and Hopkins, 1996)**

***Apparatus***

1) 20 mL porcelain crucibles
2) Forceps

3) 2000 μm sieve
4) 50 mL beaker
5) Desiccator
6) Muffle
7) Analytical balance

***Procedures***

1) Air dry or oven-dry soil at 35 to 40 °C.
2) Carefully remove all plant and animal materials from the soil using forceps.
3) Pass the soil through a 2000 μm sieve.
4) Weigh 10 ( ± 0.05) g of the soil sample in a tared 50 mL beaker and oven-dry at 105 °C overnight. This process will ensure the removal of all water from the gypsum in gypsiferous soils.
5) Using a desiccator, cool the samples and then record the weight.
6) Transfer the oven-dry soil (~10 ( ± 0.05) g) into a porcelain crucible, heat the samples in a muffle furnace for 2 h at 360 °C (after the temperature reaches 360 °C).
7) After the combustion, cool the samples to 150 °C in a desiccator before reweighing and recording the value

***Calculation***

8) $SOM_{LOI}$ = [(soil weight after combustion – oven – dry soil weight) / (oven – dry soil weight) * 100]

**Soil Inorganic C**

***Dry combustion Procedure***

1) Follow step 11 to 13 of the method for SOC to determine total soil carbon.
2) Follow step 1 to 13 of the method for SOC.
3) Soil inorganic carbon is calculated as the difference between total soil carbon and SOC.

## Other Analytical Methods and Issues Affecting Soil Carbon

***Gravimetric method for loss of carbon dioxide*** **(Loeppert and Suarez, 1996)**

***Apparatus***

1) 50 mL Erlenmeyer flask
2) Analytical balance
3) 10 mL and 250 mL pipettes or dispensers

***Reagents***

1) Hydrochloric acid (HCl), 3 *M*. Transfer 250 mL of concentrated HCl to 500 mL of deionized water and dilute to a total volume of 1 L.

### *Procedure*

1) Weigh a stoppered, 50 mL Erlenmeyer flask containing 10 mL of 3 *M* HCl.
2) Transfer a 1 to 10 g air-dried soil sample (containing 0.1–0.3 g of $CaCO_3$ equivalent) to the container, a little at a time, to prevent excessive frothing.
3) After effervescence has subsided, replace the stopper loosely on the flask and swirl the flask occasionally for about 15 min.
4) At intervals of about 15 min, remove the stopper and swirl the flask for 10 to 20 s to displace any accumulated $CO_2$ with air.
5) Replace the stopper and then weigh the flask and its contents to the nearest 0.1 mg.
6) Repeat the agitation and weighing procedure until the weight of the container does not change by more than 1 to 2 mg. The reaction is usually complete within 1 h.

### *Calculation*

1) Weight of $CO_2$ = Difference between initial and final weights (flask + stopper + acid + soil)

$$CO_3 - C\% = (g\ CO_2 lost)/(g\ soil) * (0.272) * (100)$$
$$CaCO_3\% = (g\ CO_2 lost)/(g\ soil) * (2.273) * (100)$$

## Biochar

Biochar is a bioproduct from a thermochemical conversion (pyrolysis) of biomass (Koide et al., 2011). Biochar may be used as a soil amendment due to its chemical and physical characteristics, and because of its potential to help mitigate climate change. Sequestering biochar carbon in soil contributes greatly to reducing greenhouse gas emissions (Leng et al., 2019). Biochar has a high concentration of recalcitrant carbon that is resistant to decomposition (Glaser et al., 2002), but since a portion of most fresh biochar is easily decomposable (Lehmann and Joseph, 2015), and repeated analyses and applications of biochar may be necessary.

Due to its chemical complexity and large variability, a simple routine way to quantify biochar in soils is still scarce and under investigation, mainly because of the difficulty in distinguishing biochar from other forms of soil organic matter. Methods to evaluate biochar in soils are extremely labor intensive or need specialized instrumentation. According to Leng et al. (2019), biochar in soils has been quantified by scanning calorimetry, nuclear magnetic resonance spectroscopy or infrared spectroscopy, analysis of molecular markers, or by preferential removal of inorganic and non-biochar organic C by selective oxidation or acid treatment followed by the analysis of residual organic material by NMR, optical, or mass spectroscopy, or thermal conductivity.

### SOC Stocks: Considerations on Sampling Depth and Mass Corrections

Soil carbon stocks are commonly quantified at fixed depths as the product of soil bulk density, depth, and organic carbon (OC) concentration. Soil C stocks are estimated using the following equation:

$$C = OC \times Ds \times E$$

C is the organic C stock in the soil layer (Mg $ha^{-1}$); OC is the total organic C content in the soil layer (%); Ds is the soil density in the soil layer (Mg $m^{-3}$), and E is the thickness of the sampled layer (cm). However, this method systematically overestimates SOC stocks in samples from areas with a higher bulk density, such as minimum tillage, thus exaggerating their benefits. This is not to discourage the use of such calculations, but rather to build awareness that estimates of real SOC change can be greatly compromised if bulk densities differ among treatments or over time.

## References


Albrecht, W. 1938. Soils and men: Yearbook of agriculture. USDA, Washington, D.C.

Balesdent, J., C. Chenu, and M. Balabane. 2000. Relationship of soil organic dynamics to physical protection and tillage. *Soil Tillage Res.* 53:215–230. doi:10.1016/S0167-1987(99)00107-5

Barthès, B., and E. Roose. 2002. Aggregate stability as an indicator of soil susceptibility to runoff and erosion; validation at several levels. *Catena* 47:133–149. doi:10.1016/S0341-8162(01)00180-1

Bell, M., and D. Lawrence. 2009. Soil carbon sequestration—myths and mysteries. *T.G. Trop. Grassl.* 43:227–231.

Benito, E. and F. Diaz-Fierros. 1992. Effects of cropping on the structural stability of soils rich in organic matter. *Soil and Tillage Research*, 23:153–161. doi.org/10.1016%2F0167-1987%2892%2990011-y

Bonini Pires, C., T. Amado, G. Reimche, R. Schwalbert, M. Sarto, R. Nicoloso, J.E. Fiorin, and C.W. Rice. 2020. Diversified crop rotation with no-till changes microbial distribution with depth and enhances activity in a subtropical Oxisol. *Eur. J. Soil Sci.* 71: (In Press). doi:10.1111/ejss.12981

Boyle, M., W.T. Frankenberger, and L.H. Stolzy. 1989. The influence of organic matter on soil aggregation and water infiltration. *J. Prod. Agric.* 2:290–299. doi:10.2134/jpa1989.0290

Broadbent, F.E., and J.B. Ott. 1957. Soil organic matter-metal complexes: 1. Factors affecting retention of various cations. *Soil Sci.* 83:419–428. doi:10.1097/00010694-195706000-00001

Bünemann, E., G. Bongiorno, Z. Bai, R. Creamer, G. De Deyn, R. de Goede, L. Fleskens, V. Geissen, T.W. Kuyper, P. Mäder, M. Pulleman, W. Sukkel, J.W. van Groenigen, and L. Brussaard. 2018. Soil quality– A critical review. *Soil Biol. Biochem.* 120:105–125. doi:10.1016/j.soilbio.2018.01.030

Cambardella, C., and E. Elliott. 1992. Particulate soil organic matter changes across a grassland cultivation sequence. *Soil Sci. Soc. Am. J.* 56:777–783. doi:10.2136/sssaj1992.03615995005600030017x

Cambardella, C., and E. Elliott. 1994. Carbon and nitrogen dynamics of soil organic matter fractions from cultivated grassland soils. *Soil Sci. Soc. Am. J.* 58:123–130. doi:10.2136/sssaj1994.03615995005800010017x

Carter, M.R. and Rennie, D.A. 1982. Changes in soil quality under zero tillage farming systems: distribution of microbial biomass and mineralizable C and N potentials. Canadian Journal of Soil Science, 62: 587–597. doi.org/10.4141%2Fcjss82-066

Dexter, A.R., G. Richard, D. Arrouays, E.A. Czyz, C. Jolivet, and O. Duval. 2008. Complexed organic matter controls soil physical properties. *Geoderma* 144:620–627. doi:10.1016/j.geoderma.2008.01.022

Dıaz-Zorita, M. and Grosso, G.A. 2000. Effect of soil texture, organic carbon and water retention on the compactability of soils from the Argentinean pampas . *Soil and Tillage Research*, 54: 121–126. doi.org/10.1016%2Fs0167-1987%2800%2900089-1

Doran, J., and M. Zeiss. 2000. Soil health and sustainability: Managing the biotic component of soil quality. *Appl. Soil Ecol.* 15:3–11. doi:10.1016/S0929-1393(00)00067-6

Doran, J.W. 1980. Soil microbial and biochemical changes associated with reduced tillage. Soil Science Society of America Journal 44: 765–771. doi.org/10.2136%2Fsssaj1980.03615995004400040022x

Emerson, W.W., 1995. Water-retention, organic-C and soil texture. *Soil Research*, 33: 241–251. doi.org/10.1071%2Fsr9950241

Glaser, B., J. Lehmann, and W. Zech. 2002. Ameliorating physical and chemical properties of highly weathered soils in the tropics with charcoal- A review. *Biol. Fertil. Soils* 35:219–230. doi:10.1007/s00374-002-0466-4

Gougoulias, C., J.M. Clark, and L.J. Shaw. 2014. The role of soil microbes in the global carbon cycle: Tracking the below-ground microbial processing of plant-derived carbon for manipulating carbon dynamics in agricultural systems. *J. Sci. Food Agric.* 94:2362–2371. doi:10.1002/jsfa.6577

Gupta, S.C. and Larson, W.E., 1979. A model for predicting packing density of soils using particle-size distribution. *Soil Science Society of America Journal*, 43:758–764. doi.org/10.2136%2Fsssaj1979.03615995004300040028x

Horwath, W. 2007. Carbon cycling and formation of soil organic matter. In: E.A. Paul, editor, *Soil microbiology, ecology, and biochemistry*. 3rd ed. Academic Press, Cambridge, MA. p. 303–339. doi:10.1016/B978-0-08-047514-1.50016-0

Hudson, B.D. 1994. Soil organic matter and available water capacity. Journal of Soil and Water Conservation, 49:189–194. doi.org/10.1201/9780429445552-36

Karlen, D.L., and C.W. Rice. 2015. Soil degradation: Will humankind ever learn? *Sustainability* 7:12490–12501. doi:10.3390/su70912490

Kibblewhite, M.G., Ritz, K. and Swift, M.J., 2008. Soil health in agricultural systems. Philosophical Transactions of the Royal Society B: Biological Sciences, 363: 685–701. doi.org/10.1098%2Frstb.2007.2178

King, A.E., K.A. Congreves, B. Deen, K.E. Dunfield, R.P. Voroney, and C. Wagner-Riddle. 2019. Quantifying the relationships between soil fraction mass, fraction carbon, and total soil carbon to assess mechanisms of physical protection. *Soil Biol. Biochem.* 135:95–107. doi:10.1016/j.soilbio.2019.04.019

Koide, R.T., K. Petprakob, and M. Peoples. 2011. Quantitative analysis of biochar in field soil. *Soil Biol. Biochem.* 43:1563–1568. doi:10.1016/j.soilbio.2011.04.006

Lal, R. & Follett, R. F. (2009). Priorities in soil carbon research in response to climate change. In R. Lal and R.F. Follett, editors, *Soil carbon sequestration and the greenhouse effect*, 57, 401-410. doi:10.2136/sssaspecpub57.2ed.c23.

Lal, R. 2015. Sequestering carbon and increasing productivity by conservation agriculture. *J. Soil Water Conserv.* 70:55A–62A. doi:10.2489/jswc.70.3.55A

Larson, W.E., Pierce, F.J. 1994. The dynamics of soil quality as a measure of sustainable management. In: J.W. Doran, D.C. Coleman, D.F. Bezdicek, B.A. Stewart, (Eds.), Defining soil quality for a sustainable environment. SSSA, Madison, WI, pp. 37-51.

Lehmann, J., and M. Kleber. (2015) The contentious nature of soil organic matter. *Nature*, 528, S60–S67. doi:10.1038/nature16069.

Leng, L., Huang, H., Li, H., Li, J., & Zhou, W. (2019). Biochar stability assessment methods: A review. *The Science of the total environment*, 647, 210–222. doi:10.1016/j.scitotenv.2018.07.402.

Loeppert, R.H., and D.L. Suarez. (1996) Carbonate and Gypsum. In: D.L. Sparks, A.L. Page, P.A. Helmke, & R.H. Loeppert, editors, *Methods of soil analysis, part 3: Chemical methods.* Soil Science Society of America, American Society of Agronomy, Madison, WI. p. 437–474. doi:10.2136/sssabookser5.3.c15

Lynch, J.M. and Panting, L.M. 1980. Cultivation and the soil biomass. Soil Biology and Biochemistry, 12:29–33. doi.org/10.1016%2F0038-0717%2880%2990099-1

Magdoff, F.R. 1996. Soil organic matter fractions and implications for interpreting organic matter tests. In: F.R. Magdoff, M.A. Tabatabai, and E.A. Hanlon, Jr., editors, *Soil organic matter: Analysis and interpretation. Spec. Publ. 46.* Soil Science Society of America, Madison, WI., doi:10.2136/sssaspecpub46.

McVay, K.A., Budde, J.A., Fabrizzi, K., Mikha, M.M., Rice, C.W., Schlegel, A.J., Peterson, D.E., Sweeney, D.W. and Thompson, C. 2006. Management effects on soil physical properties in long-term tillage studies in Kansas. Soil Science Society of America Journal, 70: 434–438. doi.org/10.2136%2Fsssaj2005.0249

Meersmans, J., B. Van Wesemael, and M. Van Molle. 2009. Determining soil organic carbon for agricultural soils: A comparison between the Walkley & Black and the dry combustion methods (north Belgium). *Soil Use Manage.* 25:346–353. doi:10.1111/j.1475-2743.2009.00242.x

Mortensen, J.L., and F.L. Himes. 1964. Soil organic matter. *Chemistry of the Soil* 2:206–241.

Nelson, D.W., and L.E. Sommers. 1996. Total carbon, organic carbon, and organic matter. In: D.L. Sparks, Page, A. L., Helmke, P. A., & Loeppert, R. H. (Eds.). *Methods of soil analysis, part 3: Chemical methods (5, 961-1010).* SSSA, ASA, Madison, WI.

Oldfield, E.E., M.A. Bradford, and S.A. Wood. 2019. Global meta-analysis of the relationship between soil organic matter and crop yields. *Soil* 5:15–32. doi:10.5194/soil-5-15-2019

Pan, G., P. Smith, and W. Pan. 2009. The role of soil organic matter in maintaining the productivity and yield stability of cereals in China. *Agric. Ecosyst. Environ.* 129:344–348. doi:10.1016/j.agee.2008.10.008

Paustian, K., J. Lehmann, S. Ogle, D. Reay, G.P. Robertson, and P. Smith. 2016. Climate-smart soils. *Nature* 532:49–57. doi:10.1038/nature17174

Pribyl, D. 2010. A critical review of the conventional SOC to SOM conversion factor. *Geoderma* 156:75–83. doi:10.1016/j.geoderma.2010.02.003

Puget, P., C. Chenu, and J. Balesdent. 1995. Total and young organic matter distributions in aggregates of silty cultivated soils. *Eur. J. Soil Sci.* 46:449–459. doi:10.1111/j.1365-2389.1995.tb01341.x

Puget, P., C. Chenu, and J. Balesdent. 2000. Dynamics of soil organic matter associated with particle-size fractions of water-stable aggregates. *Eur. J. Soil Sci.* 51:595–605. doi:10.1111/j.1365-2389.2000.00353.x

Rice, C.W., T.B. Moorman, and M. Beare. 1996. Role of microbial biomass carbon and nitrogen in soil quality. In: J.W. Doran and A.J. Jones, editors, *Methods for assessing soil quality.* Soil Science Society of America, Madison, WI. doi:10.2136/sssaspecpub49.c12.

Rice, C.W. 2005. Carbon cycling in soils: Dynamics and management. In: D. Hillel, editor, *Encyclopedia of soils in the environment.* Elsevier Ltd., Oxford, U.K. p. 164–170. doi:10.1016/B0-12-348530-4/00183-1

Salehi, M.H., O.H. Beni, H.B. Harchegani, I.E. Borujeni, and H.R. Motaghian. 2011. Refining soil organic matter determination by loss-on-ignition. *Pedosphere* 21:473–482. doi:10.1016/S1002-0160(11)60149-5

Sanderman, J., T. Hengl, and G. Fiske. 2017. Soil carbon debt of 12,000 years of human land use. Proc. Natl. Acad. Sci. USA 114:9575–9580. doi:10.1073/pnas.1706103114 [erratum: 115(7):E1700]

Schertz, D.L., Moldenhauer, W.C., Livingston, S.J., Weesies, G.A. and Hintz, E.A. 1989. Effect of past soil erosion on crop productivity in Indiana. *Journal of Soil and Water Conservation*, 44: 604–608.

Schnitzer, M., and S.I.M. Skinner. 1963. Organo-metallic interactions in soils: 1. Reactions between a number of metal ions and the organic matter of a podzol Bh horizon. *Soil Sci.* 96:86–93. doi:10.1097/00010694-196308000-00003

Schnitzer, M., and S.U. Khan, editors. 1975. *Soil organic matter*. Elsevier, Amsterdam, The Netherlands.

Schulte, E.E., and B.G. Hopkins. 1996. Estimation of soil organic matter by weight loss-on-ignition In: *Soil organic matter: Analysis and interpretation. SSSA Special Publication 46*. SSSA, Madison, WI. p. 21–31. doi:10.2136/sssaspecpub46.c3.

Six, J., E.T. Elliott, and K. Paustian. 2000. Soil structure and soil organic matter. II. A normalized stability index and the effect of mineralogy. *Soil Sci. Soc. Am. J.* 64:1042–1049. doi:10.2136/sssaj2000.6431042x

Six, J., R.T. Conant, E.A. Paul, and K. Paustian. 2002. Stabilization mechanisms of soil organic matter: Implications for C-saturation of soils. *Plant Soil* 241:155–176. doi:10.1023/A:1016125726789

Smith, J.L., Papendick, R.I., Bezdicek, D.F., Lynch J.M. 1993. Soil organic matter dynamics and crop residue management. In: *Metting*, Blane F., Jr. (ed), Soil microbial ecology – Applications in agricultural and environmental management. Marcel Dekker, Inc., New York. p. 65–94.

Soil Science Society of America. 1997. Glossary of Soil Science Terms. SSSA, Madison, WI.

Sundquist, E.T. 1993. The global carbon dioxide budget. Science 259: 934–941. doi.org/10.1126/science.259.5097.934

Wilson, G.W.T., C.W. Rice, M.C. Rillig, A.C. Springer, and D.C. Hartnett. 2009. Arbuscular mycorrhizal fungal abundance controls soil aggregation and carbon sequestration. *Ecol. Lett.* 12:452–461. doi:10.1111/j.1461-0248.2009.01303.x

Yang, F., Zhang, G.L.,Yang, J.L., Li, D.C., Zhao, Y.G., Liu, F., Yang, R.M., and Yang, F. (2014). Organic matter controls of soil water retention in an Alpine grassland and Its significance for hydrological processes. *J. Hydrol.* 519, 3086–3093. doi:10.1016/j.jhydrol.2014.10.054

# 4

# Water-Stable Soil Aggregate Assessment

*Maysoon M. Mikha and Skye Wills*

Soil aggregates are naturally forming clusters of sand-, silt-, and clay-sized soil particles that are held together by forces much stronger than those holding adjacent aggregates together. Collectively they create and influence many soil structural properties that mediate numerous soil physical, chemical, and biological processes including water entry, retention and release to plants; air exchange; and resistance to wind and water erosion. This chapter examines soil sampling and preparation, methods for assessing water stable aggregation, and various ways to interpret the data for soil health assessments.

## Introduction

A soil aggregate is "a naturally occurring cluster or group of soil particles in which the forces holding the particles together are much stronger than the forces between adjacent aggregates" (Martin et al., 1955). Meanwhile, soil structure describes how different size aggregates or classes of soil particles (i.e., sand, silt, and clay fractions) are arranged and bonded together by organic and inorganic materials (Edwards and Bremner, 1967; Tisdall and Oades, 1982; Tisdall 1996). Stable soil structure requires aggregates to stay intact when subjected to anthropogenic (tillage, crop sequence, crop residue management) or environmental (dry-wet or freeze-thaw cycles) stressors (Lynch and Bragg, 1985; Tisdall 1996, Mikha et al, 2005). Aggregation is an important soil health indicator because of its impact on soil structure and the many physical, chemical, and biological processes it mediates (Elliott and Coleman, 1988; Oades, 1988; Van Veen and Kuikman, 1990; Cambardella and Elliott, 1993; Mikha and Rice, 2004; Six et al., 2000a, 2000b).

The size, shape, and stability of soil aggregates also determine pore size distribution (Lynch and Bragg, 1985), which influences crop production through

*Soil Health Series: Volume 2 Laboratory Methods for Soil Health Analysis*, First Edition.
Edited by Douglas L. Karlen, Diane E. Stott, and Maysoon M. Mikha.

indirect effects on flow and diffusion of soil water and air. Soil aggregate size distribution is important because as macroaggregate size increases, susceptibility to wind erosion decreases (Kemper and Chepil, 1965). Previous research has documented that irrigation or rainfall can disrupt unstable, dry surface soil macroaggregates causing them to slake and allowing the detached microaggregates and clay particles to fill soil pores, thus making them narrower or discontinuous. (Kemper and Chepil, 1965; Lynch and Bragg, 1985). When air-dried macroaggregates are not stable enough to withstand the pressure resulting from entrapped air when soil is rapidly wetted, slaking occurs (Gäth and Frede, 1995). Unstable surface aggregates can also result in the formation of crusts, an undesirable soil structure that inhibits water infiltration and air movement (Rathore et al., 1982; Tisdall and Oades, 1982; Lynch and Bragg, 1985; LeBissonnais, 1996) and can affect seed germination and crop growth. As macroaggregates are fractured into microaggregates and primary soil particles by trapped air pressure, mean geometric size decreases and the fraction of microaggregates increases (Six et al., 2000b). Quantifying soil aggregate size (macroaggregates vs. microaggregates), especially with regard to potential wind or water erosion (Kemper and Chepil, 1965) has resulted in this indicator becoming an important factor for evaluating soil health.

Physical binding of soil particles by fungal hyphae and plant roots are the primary drivers for soil aggregate formation, while the quantity and type of clay minerals, soil organic matter (SOM) content, and type of inorganic minerals associated with the parent material are responsible for aggregate stabilization (Lynch and Bragg, 1985; Miller and Jastrow, 1990; Degens, 1997; Angers, 1998; Miller and Jastrow, 2000). Recognizing the importance of SOM as a bonding agent for soil aggregate formation and stabilization (Tisdall and Oades, 1982; Dormaar, 1983; Chaney and Swift, 1984; Haynes et al., 1991), readers are referred to Chapter 3 in this volume for further discussion of that important soil health indicator.

For decades, researchers, extension agents, and producers realized the importance of maintaining soil structural stability to enhance land sustainability and soil health (Kibblewhite et al., 2008; Baldock and Skjemstad, 2000; USDA-NRCS Soil Health Team, 2015). Aggregate size distribution is therefore a key indicator for soil health and structural stability. Quantifying the amount and size classes of macro- and micro-aggregates within a specific soil type is important for soil management in order to reduce soil and nutrient loss, increase water infiltration and retention, decrease runoff, facilitate soil air exchange, support biodiversity, and thus enhance sustainability. Several methods for evaluating aggregate size distribution have been proposed, but most can be traced to a multiple sieve method adapted from Yoder (1936). That procedure simulates, to a large extent, the influence of soil and crop management as well as water slaking on the: (1) size of water-stable aggregates (*i.e.*, macroaggregates), and (2) quantity of both macro- and micro-aggregates within an individual soil. Our objective for this chapter is to present a modified multiple sieve method adapted from the Yoder wet-sieving apparatus. A schematic diagram to guide construction of the apparatus is included for easy implementation.

## Soil sampling and preparation

Soil samples should be collected when soil moisture is near field capacity (−0.033 MPa) to avoid aggregate deformation or destruction if the soil is too wet. If the soil is too dry, clusters or clods formed during sampling will need to be broken during processing, but that may lead to breaking soil aggregates and result in higher microaggregates or individual particle fractions.

The tools or probes used to collect the soil sample are also important, because probe edges can break large soil aggregates into smaller aggregate size fractions. Within a specific soil sampling volume, the potential of soil aggregate destruction due to probe size and edge effects are reduced as the probe diameter increased (Fig. 4.1). Using a spade (Kemper and Rosenau, 1986) for soil sampling, can further reduce aggregate destruction during sampling. Care should be taken when

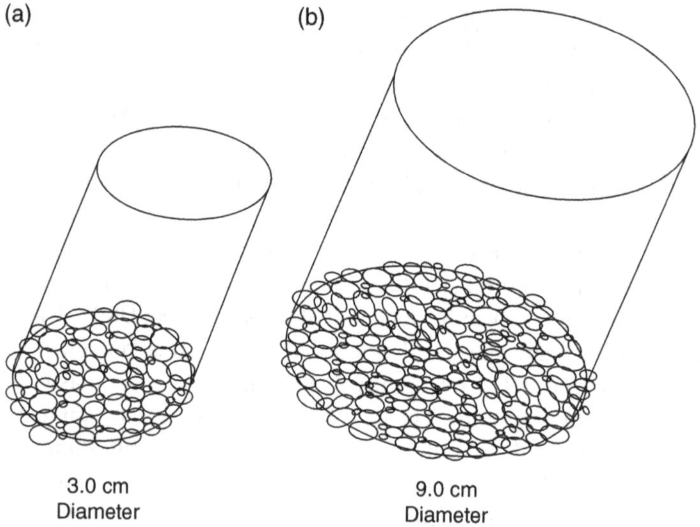


**Figure 4.1** Soil sampling probes for aggregate size distribution. (A) Represents a small diameter probe (approximately 3.0 cm) and (B) represents a large diameter probe (approximately 9.0 cm).

using a spade to ensure that the sample is neither compressed nor loosened by applying excessive force or abrupt movement if the soil is to wet or dry. The spade should also be pointed vertically ($90^0$ angle) to the soil surface to ensure the desired sampling depth is achieved.

The preferred sampling period is spring after the ground thaws (cold environments) but before any field operations (fertilization, tillage, planting, etc.) commence. Spring sampling also eliminates the influence of crop roots and/or wetting-drying cycles on soil aggregate fractions. The most important sampling guideline, however, is that time of sampling should be kept constant throughout each study and/or across study sites for accurate comparisons. Soil cores should also be taken to the same depth(s) throughout or across sites. After collecting the samples, they should be transferred in crush resistant containers to eliminate aggregate destruction and/or compression as they are shipped or transported to the laboratory (Nimmo and Perkins, 2002). During transportation and before processing, soil samples should be kept cool to prevent accelerated SOM decomposition. Again, this is extremely important because it has been well documented that SOM plays a major role in binding soil microaggregates into macroaggregates (Tisdall and Oades, 1982; Miller and Jastrow, 1990; Haynes et al., 1991; Jastrow and Miller, 1997; Six et al., 1999, 2000).

In the laboratory, (1) pass the field moist samples through a 6- to 8-mm sieve to establish the largest aggregate size, (2) homogenize the soil sample, (3) remove stone and plant debris and (4) air dry the samples at 35–38 °C (95–100 °F) for approximately 24 to 48 h until constant water content is achieved. There should then be no additional disturbance before aggregate sizes are evaluated.

## Water-stable aggregation (WSA)

Soil aggregate fractions consist of two size classifications: microaggregates (<250 µm diameter) and macroaggregates (>250 µm diameter). Microaggregates consist of soil particle clusters bound together by degraded aromatic humic material derived from plant roots, fungal hyphae, and bacterial cells that is stabilized by chemical bonds with amorphous iron, aluminum, and aluminosilicates. Microaggregates are bound into macroaggregates (>250 µm diameter) by two types of bonding agents: (i) temporary “glues” consisting of fine roots, hyphae, and microbial cells; and (ii) transient, rapidly decomposable organic materials including microbial byproducts, added organic materials, and plant-derived polysaccharides in the rhizosphere (Edwards and Bremner, 1967; Tisdall and Oades, 1982; Jastrow and Miller, 1997). The SOM associated with macroaggregates was found to be the primary source of nutrients released by cultivation (Elliott, 1986), and macroaggregate stability was strongly influenced by management practices and rapid wetting (Edwards and Bremner, 1967; Tisdall and Oades, 1982; Jastrow and Miller, 1997; Jiao et al., 2006; Zibilske and Bradford, 2007). Furthermore, since macroaggregates consist of microaggregate clusters (Jastrow and Miller, 1997; Six et al., 2000a; Denef et al., 2004), their disintegration results in an increase in the quantity of microaggregates (Mikha et al., 2010).

This chapter provides guidelines to quantify water-stable aggregates into different size fractions. The largest macroaggregate fraction, with aggregates greater than 2-mm (> 2000 µm) in diameter, will be found only if SOM levels are high and soil texture (*i.e.*, % sand, silt, and clay) is clay or clay loam. In contrast, soils with low clay content and/or low SOM levels will genreally have aggregates that are equivalent or less than 1-mm (< 1000 µm) in diameter. The largest aggregates will generally be found in prairie and forest soils where they may be 4 to 5 mm (4,000 to 5,000 µm) in size, compared to agricultural land where aggregates are often no larger than 1 to 2 mm (1,000 to 2,000 µm).

Recent studies confirming these general size ranges include a moderately well-drained Kennebec silt loam soil near Manhattan, KS, where Mikha and Rice (2004) had macroaggregates that were 2,000 µm in size. Conversely, for a Tripp sandy loam soil in the North Platte River Valley near Sidney, NE, Mikha et al. (2015) found the largest macroaggregate fraction to be 1,000 µm in size. These results, coupled with many other published studies, confirm there is no specific macroaggregates size fraction that is appropriate for all applications. What is important is that for regional- to national-scale studies, a standard macroaggregate sieve size should be designated for making meaningful comparisons among study sites.

Our guidelines for water-stable aggregate evaluation in this chapter are based on the studies cited and many others for which we used soil that was sieved through 6 to 8 mm screens under field moist conditions and then air dried. The amount of air-dry soil (ADS) used ultimately needs to be adjusted on an oven-dry soil (ODS) weight basis to account for differences in soil moisture, also known as gravimetric water content. To adjust ADS weight to ODS weight we suggest the following protocol.

## Materials and Procedures

1) Label and weigh small aluminum tins (62 mm; 75 mL) using an analytical balance with 4 digits or decimal places (0.1 mg). Using an analytical balance is important because soil moisture differences between ADS and ODS samples is small and difficult to detect with 2-digit scales.
2) Weigh approximately 5 to 10 g of ADS using an analytical balance, place the soil in the tin, and record the total weight (ADS + tin).

3) Place the tin plus soil into an oven for approximately 24 h at 105 °C.
4) When the tins are cool, reweigh and record the ODS plus tin weight using the same analytical balance and calculate ODS by subtracting the tin weight.
5) Calculate the soil moisture content (g) using the following equation after subtracting the tin weight from both ADS and ODS measurements:

$$\text{Soil moisture content}\left(\text{SMC, g}\right) = \frac{\text{ADS} - \text{ODS}}{\text{ODS}} \quad (1)$$

Adjust the 50 g of ADS used for aggregate size distribution measurements to the ODS basis using the following equation:

$$\text{ODS weight}\left(\text{g}\right) = \frac{\text{ADS}}{1 + \text{SMC}} \quad (2)$$

## Aggregate Sieving Apparatus

The suggested water-stable aggregation apparatus used is similar to the Yoder wet-sieve apparatus (Yoder, 1936) as modified by Mikha and Rice (2004). The modifications were made to accommodate a stacked nest of sieves, thus enabling complete recovery of all particle fractions from individual samples. The modified wet sieving apparatus, with specified dimensions, is shown in Fig. 4.2.

Macroaggregates are typically defined as aggregates ≥250 μm, whereas microaggregates are defined as those <250 μm in size (Tisdall and Oades, 1982; Miller and Jastrow, 1990; Mikha and Rice, 2004). Within the stack, sieves with mesh opening ≥250 μm diameters were contained inside the oscillation cylinders. Sieves with mesh opening <250 μm diameter were placed outside the oscillation cylinders, to prevent soil particles from blocking the mesh of the <250 μm diameter sieves and thus preventing water from passing through the nest of sieves.

The apparatus shown in Figure 4.2 is designed to handle four nested sieves, but it is also possible to only use two nested sieves at a time. The example for macroaggregates presented was taken form Mikha and Rice (2004) and Mikha et al. (2005, 2015). Soil collected on the 250-μm sieve represents macroaggregates and that collected on the 53 μm sieve represents microaggregates. Both measures can ultimately be used for macro- and micro-aggregate evaluations, such as those used for the Soil Management Assessment Framework (Stott, 2019), and the two-sieve approach does not provide sufficient information to calculate aggregates size distribution or mean weight diameter (MWD). Those indicators require the use of multiple sieves, usually consisting of four mesh sizes: 1000 μm, 500 μm, 250 μm (macroaggregates) and 53 μm (microaggregates), but a 2000-μm sieve can be

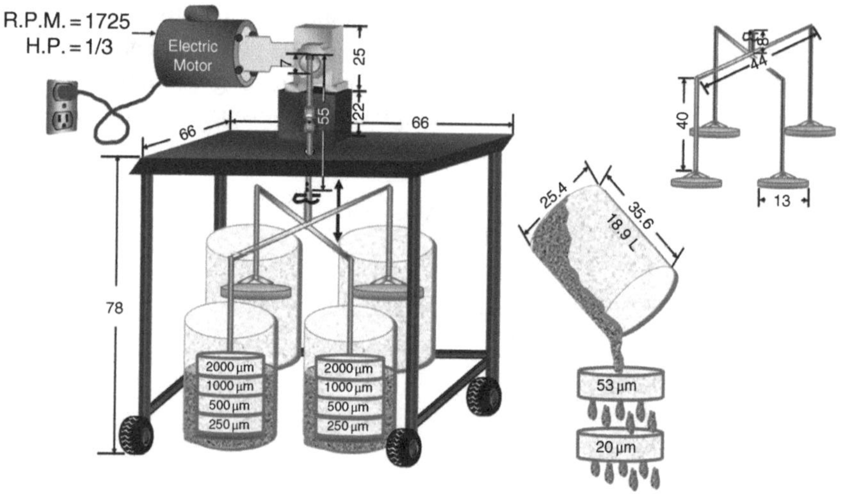


**Figure 4.2** Dimensional diagram of the modified Yoder (1936) wet-sieving apparatus. The sieves <250 μm are outside the oscillation cylinders. The apparatus allows for complete recovery of all particle fractions from individual samples. The ground fault interrupts (GFI) device needed to be added to prevent electrocution. All dimensions are given in centimeters.

added if desired. This was done for the examples below (i.e., the largest macroaggregate size was 2000 μm).

## Construction Supplies

The dimensions and specifications given below are for the apparatus shown in Figure 4.2.

1) Electric switch. It is advisable to have the electric switch a safe distance from the apparatus and above ground level to prevent electrocution due to water spills.
2) Electric motor capable of producing the required immersion cycle of sieves (1725 rotations per minute and 1/3 horsepower).
3) Four buckets to immerse the nest of sieve, about 18.9 L (25.4 cm diameters and 35.6 cm height).
4) Two or four sets of stainless-steel sieves of about 12.7-cm diameter for macroaggregate evaluation.
5) One set of stainless-steel sieves at of 53 μm mesh size needs to be used for microaggregate evaluation. The sieve will be used outside the oscillation

cylinders. Preferably using the sieve's size of approximately 20 cm diameter for microaggregate evaluation.

6) The apparatus specifications of oscillation time are 10 min, stroke length of 3.5 cm, and stroke frequency of 30-cycle $min^{-1}$, all of which should be held constant.

## Water-stable aggregate size distribution with the slaking method

1) Label and weigh the aluminum pans (approximately 19-cm base diameter) using a 2-digit balance.
2) Label crush-resistant containers corresponding to each aggregate size class associated with each sample.
3) Place the nest of sieves in the oscillation cylinders inside the bucket in order with the sieve with the smallest mesh on bottom and the largest mesh size at the top of the nest of sieves (Fig. 4.2).
4) The oscillation cylinders need to be at the highest end and the nest of sieves (Fig. 4.2).
5) Weigh 50 g of air-dried (AD) soil using a 2-digit balance.
6) Fill the buckets with distilled water where the bottom sieves will be completely submerged in water but does not reach the top sieve of the nest of sieves (Fig. 4.2).
7) Watch for the air bubbles that might be trapped between the nests of sieves during water addition. Tilt the nest of sieves slightly and the air bubbles will float on the surface.
8) Evenly distribute the air-dried soil sample on the top sieve of the nest of each set.

   **Note:** the amount of soil used should be 0.4 g of air-dried soil per $cm^2$ of sieve area.
9) Add enough distilled water to each bucket to cover the soil and top sieve's mesh with water, approximately 1 L, to begin the slaking process (Fig. 4.3).
10) No more than four sieves should be used in this design.
11) Submerge the nest of sieves in water for 10 min. The oscillation cylinders need to be at the lowest end and the nest of sieves should be at the lowest point where the water is filling the top sieve, but not overflow (Fig. 4.4).
12) Start of the wet-sieving apparatus for 10 minutes at a stroke length of 3.5 cm and frequency of 30-cycle $min^{-1}$ that should be held constant.
13) Remove the sieves from the bucket and each individual sieve into individual round aluminum pans (approximately 22.7 cm inner top diameter, volume of 1,301 mL) and backwash the aggregates completely from the sieve (Fig. 4.5).

**Figure 4.3** The water level on the top sieve of the nest of sieves that containing soils. This figure also shows the highest point of the oscillation cylinders during the soil slaking process.

**Note**: some of the aggregates are broken due to the backwashing procedure, but this is the individual size aggregate amount that was originally associated with the specific mesh size before the backwashing procedure.

14) If desired, the aggregates can also be transferred to an aluminum pan (10.9 cm inner top diameter, volume of 200 mL) to conserve room in the drying oven. These aluminum pans need to be labeled and weighed using a balance with 2 decimal places.
15) Floating organic matter (density <1 g $cm^{-3}$, which is water density) needs to be removed from the >2000 or 1000 μm aggregate size class (top sieve) because it is mostly plant debris.
16) Organic matter associated with the other aggregate size classes is considered as the organic matter associated with that specific size class and will not be removed.

**Figure 4.4** The placement of the submerged nest sieves containing soils and the water level in the top sieve of the nest.

17) Pour the soil and the water remaining in the bucket onto the finer mesh sieves of 53 μm and 20 μm in diameter (the 20 μm size fraction is optional) for microaggregates evaluation (Fig. 4.2).
18) Each sieve needs to be shaken by hand horizontally for approximately 1 or 2 minutes to allow water and particle fractions smaller than the sieve size to pass through.
19) Using distilled water, backwash the soil and the sieve several times to be sure that all the small soil aggregates and particles have passed through and the water going through the sieve mesh is clear.
20) Repeat step 13 and 14 by backwashing the microaggregates fraction from the individual sieve into the aluminum pan.
21) Place the aluminum pan with the aggregates in the oven to dry at 105 °C for 24 h (Fig. 4.6) to evaluate aggregates on oven-dry basis.
22) Weigh the dry aggregates with the pan from each size class using analytical balance with 2- or 4-digits and store the aggregates in crush-resistant containers at room temperature.
23) Calculate the mass of aggregate size at all size fractions as follow:

$$\text{Aggregate fraction weight}\left(g\right)=\left(\text{pan weight}+\text{Aggregate}\right)-\left(\text{pan weight}\right) \quad (3)$$

24) Soil materials <53 or 20 μm diameter (the smallest mesh size used) need to be discarded and the quantity of soil lost needs to be calculated.

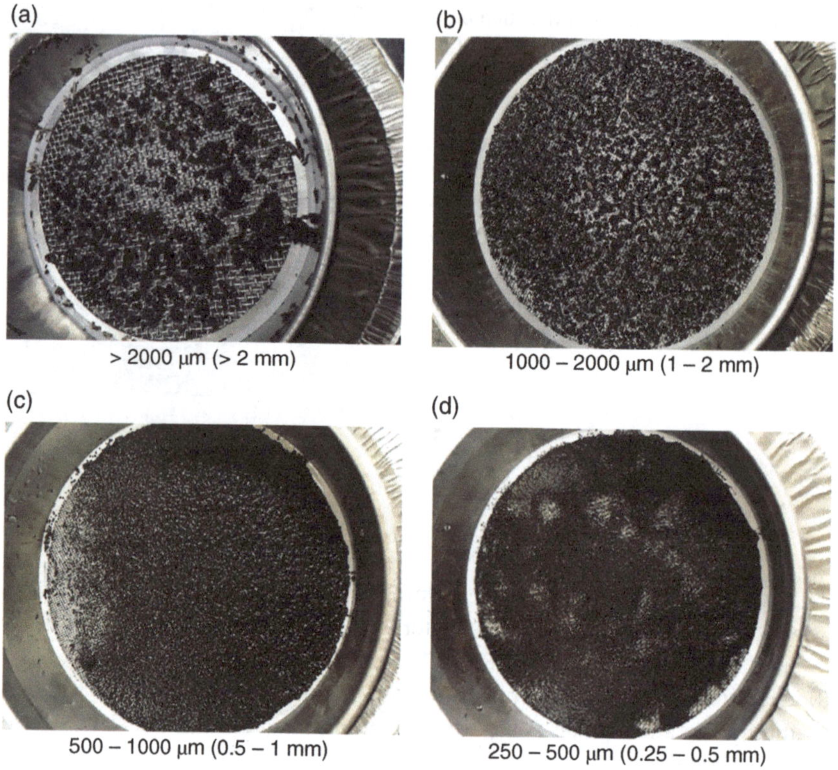


**Figure 4.5** Soil macroaggregates on an individual sieve places in the aluminum pan (approximately 22.7 cm inner top diameter, volume of 1,301 mL) before backwashed completely from the sieve.

25) The recovery will be calculated as the sum of aggregate masses associated with soil sample used as shown in Equation 4:

$$\text{Aggregate recovery}\left(\text{g}\right) = \text{Soil oven-dry weight} - \Sigma \text{ aggregate fractions weight} \quad (4)$$

## Sand-Free Water-Stable Soil Aggregates

It is important to report soil aggregates on a sand-free basis because the sand fraction is not part of soil aggregate formation and stabilization (Tisdall and Oades, 1982; Miller and Jastrow, 1990; Jastrow and Miller, 1997). The different sizes of sand, very fine, fine, coarse, very coarse, and stone, are associated with different size aggregates depending upon the fraction being studied. Therefore, the entire

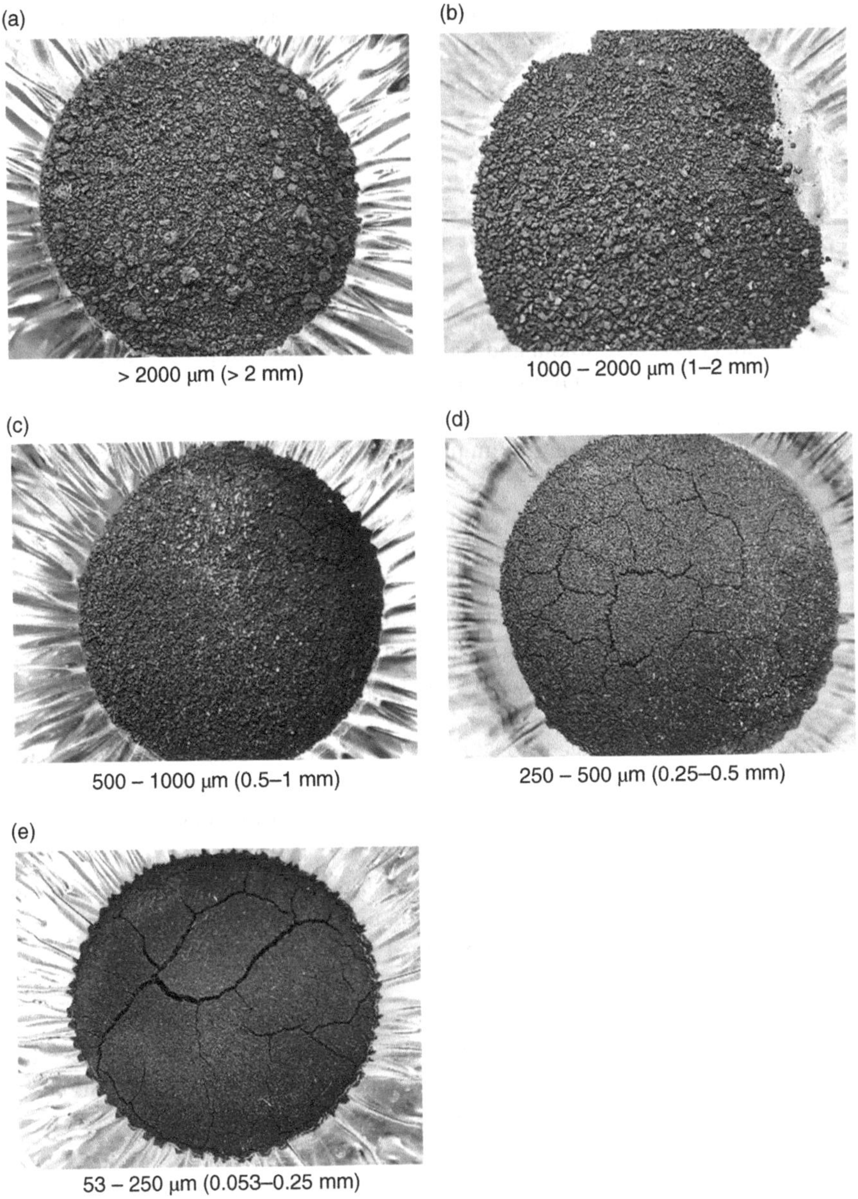


**Figure 4.6** Dried soil macroaggregates and microaggregates at 105 °C for 24 h on an individual aluminum pan.

sand size fraction (≥50 to 53 μm) associated with all aggregate size classes needs to be quantified so that the sand weight can be subtracted from individual aggregate size fraction weights as reported by Nimmo and Perkins (2002) and Mikha and Rice (2004).

## Supplies and procedure for sand content determination

1) Prepare 5 g $L^{-1}$ sodium hexametaphosphate $(NaPO_3)_6$ as previously reported in Six et al. (1999) and Mikha and Rice (2004) for soil with pH > 7 or 2 g $L^{-1}$ of sodium hydroxide (NaOH) for soil with pH < 7 as reported in Kemper and Rosenau (1986).
2) Label and weigh small aluminum tins (62 mm; 75 mL) using an analytical balance (0.1 mg).
3) Label and weigh glass flasks (250 mL) using analytical balance (0.1 mg).
4) Weigh a subsample of 2 to 5 g of each individual oven-dried soil aggregate size classes that are ≥ 53 μm, using an analytical balance (0.1 mg). Using the analytical balance capable of weighing to 0.1 mg is important because of the small amount (g) of aggregate used for this procedure.
   **Note:** the entire weight of the individual aggregate size classes generated from the previous step can be dispersed, if desire. However, the weight of the aggregate used in this step needs to be recorded accurately using an analytical balance (0.1 mg).
5) Add approximately five-fold volume (10 to 25 mL) of the previously prepared sodium hexametaphosphate.
   **Note:** the entire weighed aggregate sample needs to be completely covered with sodium hexametaphosphate solution to insure accurate dispersion.
6) Leave the aggregate samples submerged with sodium hexametaphosphate solution overnight.
7) Shake the samples on an orbital shaker at approximately 350 RPM for 4 h.
8) The dispersed soil aggregates that contained organic matter and sand need to be washed well with distilled water and collected on a 50 μm or 53 μm mesh sieve. All silt and clay fractions will pass through, and only sand greater than the 50 μm or 53 μm mesh size will remain on the sieve.
   **Note:** (1) Wash the sample thoroughly with distilled water to eliminate any remaining sodium hexametaphosphate that could be dried out and overestimate the sand weight; (2) Try to backwash the sand and avoid forcing the sand particles through the sieve's mesh.
9) Transfer the collected materials associated with each dispersed aggregate size class to a small aluminum tin (Labeled and weighted previously).

10) Place the sand in the tin to dry at 105 °C for 24 h.
11) The oven-dried material and tin need to be weighed using analytical balance (0.1 mg).
12) Calculate the sand-free water stable aggregate (WSA) quantity for each aggregate size fraction *i* as follows:

$$\%WSA_i = \frac{(\text{Aggregate fraction weight}) - (\text{Weight of sand})}{\text{Total soil used as OD basis}} \times 100 \quad (5)$$

where *i* represents each aggregate size fraction, 1, 2, 3, ..., n.

## Aggregate mean weight diameter

Previously, Van Bavel (1950) reported the importance of mean weight diameter (MWD) evaluation as a sensitive index of aggregation status of the soil, soil structure, and estimation of average aggregate size. The MWD is calculated by adding the upper and lower limits of each aggregate size class and then dividing by 2 as reported by Hurisso et al. (2013) and explained by Equation 5 as follow.

$$\text{Mean diameter}(x_i) = \frac{\text{upper size fraction} + \text{lower size fraction}}{2} \quad (6)$$

For example, the mean diameter for aggregates > 2000 μm is calculated by adding the upper limit (6 mm) and lower limit (2 mm) and then dividing by 2. The MWD values can be calculated for any size range of aggregates and then multiplied by the weight of aggregates in that that size range as a fraction of the total dry weight of soil used, as reported by Grant and Groenevelt (2007) and using Equation 7:

$$MWD = \Sigma_{i=1}^{n} x_i WSA_i \quad (7)$$

where *i* represents each aggregate size fraction, 1, 2, 3,..., n; and $x_i$ represents the sieves size mean diameter of each aggregate size fraction Equation 6.

## Summary

Soil aggregate stability is one of the important soil health parameters because it mediates many soil biological, chemical, and physical processes. The size and stability of soil aggregates, specifically macroaggregates, control pore size distribution, which can influence soil water and air relationships, crop production, and soil biological process. Soil structural stability depends on macroaggregates being able to remain intact when subjected to stresses such as tillage, compaction, freeze-thaw cycles, and rapid wetting. Unstable macroaggregates can slake due to rapid wetting and the detached microaggregates may then be transported into pores making them narrower or discontinuous. Disruption of surface aggregates can lead to crust formation (an undesirable soil structure). They then inhibit water infiltration, increase runoff and soil erosion, and reduce air movement into the soil which can affect seed germination. Finally, soil macroaggregates are sensitive to management practices such as tillage practices where mechanical destruction may expose SOM that was protected within the macroaggregates. This increases the microaggregates fraction which can then make the soil more susceptible to wind erosion. For these and undoubtedly many other reasons, soil aggregate stability and size distribution are important measurements and should be seriously considered when assessing soil health.

## References

Baldock, J.A., and Skjemstad, J.O. (2000). Role of the soil matrix and minerals in protecting natural organic materials against biological attack. *Organic Geochem.* 31, 697–710.

Cambardella, C.A., and Elliott, E.T. (1993). Carbon and nitrogen distribution in aggregates from cultivated and native grassland soils. *Soil Sci. Soc. Am. J.* 57, 1071–1076.

Chaney, K., and Swift, R.S. (1984). The influence of organic matter on aggregate stability in some British soils. *J. Soil Sci.* 35, 223–230.

Denef, K., Six, J., Merckx, R., and Paustian, K. (2004). Carbon sequestration in microaggregates of no-tillage soils with different mineralogy. *Soil Sci. Soc. Am. J.* 68, 1935–1944.

Dormaar, J.R. (1983). Chemical properties of soil and water-stable aggregation after sixty-seven years of cropping to spring wheat. *Plant Soil* 75, 51–61.

Edwards, A.P., and Bremner, J.M. (1967). Microaggregates in soils. *J. Soil Sci.* 18, 64–73.

Elliott, E.T. (1986). Aggregate structure and carbon, nitrogen, and phosphorus in native and cultivated soils. *Soil Sci. Soc. Am. J.* 50, 627–633.

Elliott, E.T., and Coleman, D.C. (1988). Let the soil work for us. *Ecol. Bull.* 39, 23–32.

Gäth, S., and Frede, H.G. (1995). Mechanisms of air slaking. In Hartge K.H., and B.A. Stewart (Ed.), *Soil structure: Its development and function* (p. 159–173). Boca Raton, FL: CRC Press.

Grant, C.D. and Groenevelt, P.H. (2007). Aggregate stability to water. In Carter M.R., and E.G. Gregorich (Eds.) *Soil sampling and methods of analysis* (p. 811–821). 2nd edition. Boca Raton, FL: CRC Press.

Haynes, R.J., Swift, R.S., and Stephen, R.C. (1991). Influence of mixed cropping rotations (pasture-arable) on organic matter content, stable aggregation and clod porosity in a group of soils. *Soil Tillage Res.* 19, 77–87.

Hurisso, T.T., Davis, J.G., Brummer, J.E., Stromberger, M.E., Mikha, M.M., Haddix, M.L., Booher, M.R., and Paul, E.A. (2013). Rapid changes in microbial biomass and aggregate size distribution in response to changes in organic matter management in grass pasture. *Geoderma* 193-194, 68–75.

Jastrow, J.D., and Miller, R.M. (1997). Soil aggregate stabilization and carbon sequestration: Feedbacks through organomineral associations. In Lal, R., J.M. Kimble, R.F. Follett, B.A. Stewart (Ed.), *Soil processes and carbon cycle* (p. 209–223). Boca Raton, FL: CRC Press.

Jiao, Y., Whalen, J.K., and Hendershot, W.H. (2006). No-tillage and manure application increase aggregation and improve nutrient retention in a sandy-loam soil. *Geoderma* 134, 24–33.

Kemper, W.D., and Chepil, W.S. (1965). Size distribution of aggregates. In C.A. Black (Ed.) *Method of soil analysis. Part 1. Physical and mineralogical properties, including statistics of measurement and sampling* (p. 499–510). Madison, WI: SSSA.

Kemper, W.D., and Rosenau, R.C. (1986). Aggregate stability and size distribution. In A.S. Campbell (Ed.) *Method of soil analysis. Part 4. Physical methods* (p. 425–442). Madison, WI: SSSA.

Kibblewhite, M.G., Ritz, K., and Swift, M.J. (2008). Soil health in agricultural systems. *Phil. Trans. R. Soc. B* 363, 685–701.

LeBissonnais, Y. (1996). Aggregate stability and assessment of soil crustability and erodibility. Part 1. Theory and methodology. *Eur. J. Soil Sci.* 47, 425–437.

Lynch, J.M., and Bragg, E. (1985). Microorganisms and soil aggregate stability. *Adv. Soil Sci.* 2, 133–171.

Martin, J.P., Martin, W.P., Page, J.B., Raney, W.A., and De Ment, J.G. (1955). Soil aggregation. *Adv. Agron.* 7, 1–37.

Mimmo, J.R., and Perkins, K.S. (2002). Aggregate stability and size distribution. In A. Klute (Ed.) *Method of soil analysis. Part 1* (p. 317–328). 2nd edition. Madison, WI: ASA and SSSA.

Mikha, M.M., Hergret, G.W., Benjamin, J.G., Jabro, J.D., and Nielsen, R.A. (2015). Long-term manure impacts on soil aggregates and aggregate-associated carbon and nitrogen. *Soil Sci. Soc. Am. J.* 79, 626–636.

Mikha, M.M., and Rice, C.W. (2004). Tillage and manure effects on soil and aggregate-associated carbon and nitrogen. *Soil Sci. Soc. Am. J.* 68, 809–816.

Mikha, M.M., Rice, C.W., and Milliken, G.A. (2005). Carbon and nitrogen mineralization as affected by drying and wetting cycles. *Soil Biol. Biochem.* 37, 339–347.

Mikha, M.M., Benjamin, J.G., Vigil, M.F., and Nielsen, C. (2010). Cropping intensity impacts on soil aggregation and carbon sequestration in the central Great Plains. *Soil Sci. Soc. Am. J.* 74, 1712–1719.

Miller, R.M., and Jastrow, J.D. (1990). Hierarchy of root and mycorrhizal fungal interactions with soil aggregation. *Soil Biol. Biochem.* 22, 570–584.

Miller, R.M., and Jastrow, J.D. (2000). Mycorrhizal fungi influence soil structure. In Kapulnik, Y., D.D. Douds, Jr. (Eds.), *Arbuscular mycorrhizas: Physiology and function* (p. 3–18). Netherlands: Kluwer Academic Publishers.

Oades, J.M. (1988). The retention of organic matter in soils. *Biogeochemistry* 5, 35–70.

Rathore, T.R., Ghildyal, B.P., and Sachan, R.S. (1982). Germination and emergence of soybean under crusted soil conditions. Part 2. Seed environment and varietal differences. *Plant Soil* 65, 73–77.

Six, J., Elliott, E.T., and Paustian, K. (1999). Aggregate and soil organic matter dynamic under conventional and no-tillage systems. *Soil Sci. Soc. Am. J.* 63, 1350–1358.

Six, J., Elliott, E.T., and Paustian, K. (2000a). Soil macroaggregate turnover and microaggregate formation: a mechanism for C sequestration under no-tillage agriculture. *Soil Bio. Biochem.* 32, 2099–2103.

Six, J., Paustian, K., Elliott, E.T., and Combrink, C. (2000b). Soil structure and organic matter: Part 1. Distribution of aggregate-size classes and aggregate-associated carbon. *Soil Sci. Soc. Am. J.* 64, 681–689.

Stott, D.E. (2019). *Recommended soil health indicators and associated laboratory procedures.* Soil Health Technical Note No. 450-03. Washington, DC: USDA Natural Resources Conservation Service. Retrieved from https://directives.sc.egov.usda.gov/OpenNonWebContent.aspx?content=43754.wba

Tisdall, J.M. (1996). Formation of soil aggregates and accumulation of soil organic matter. In M.R. Carter and B.A. Stewart (Eds.), *Structure and soil organic matter storage in agricultural soils* (p. 57–96). Boca Raton, FL: Adv. Soil Sci., Lewis Publ.

Tisdall, J.M., and Oades, J.M. (1982). Organic matter and water stable aggregates in soil. *J. Soil Sci.* 33, 141–161.

USDA-NRCS Soil Health Team. (2015) *Healthy soil for life.* Washington, DC: USDA Natural Resources Conservation Service. Retrieved from http://www.nrcs.usda.gov/wps/portal/nrcs/main/soils/health/

Van Bavel, C.H.M. (1950). Mean weight-diameter of soil aggregates as a statistical index of aggregation. *Soil Sci. Soc. Am. J.* 14, 24–28.

Van Veen, J.A., and Kuikman, P.J. (1990). Soil structural aspects of decomposition of organic matter by microorganisms. *Biogeochemistry* 11, 213–223.

Yoder, R.E. (1936). A direct method of aggregate analysis of soil and a study of the physical nature of soil erosion losses. *J. Am. Soc. Agron.* 28, 337–351.

Zibilske, L.M., Bradford, J.M. (2007). Soil aggregation, aggregate carbon and nitrogen, and moisture retention induced by conservation tillage. *Soil Sci. Soc. Am. J.* 71, 793–802.

# 5

# Determination of Infiltration Rate and Bulk Density in Soils

*Jalal D. Jabro and Maysoon M. Mikha*

Numerous field methods have been developed for measuring infiltration rate and soil bulk density. We discuss those methods as two critical indicators reflecting soil compaction and soil health under field conditions. First we examine three infiltration methods and then examine bulk density measurement, with the emphasis on both being ease of measurement in terms of time and labor.

## Soil Infiltration

When rainfall or irrigation water is applied to a land surface, it enters (infiltrates) and seeps (percolates) through the soil profile under the influence of gravity and capillary forces (Hanks & Ashcroft, 1980; Hillel, 1971). Soil infiltration is one of the most complex processes within the hydrologic cycle because it is affected not only by the physical, chemical, and biological properties of every soil but also by the vegetation cover, rainfall intensity, irrigation method, and other management practices such as tillage or crop rotation (Hillel, 1998; Jury, Gardner, & Gardner, 1991; Massimo et al., 2017).

The infiltration rate (or flux) is defined as the rate at which water enters the soil surface [Equation 5.1]. It is expressed as the volume of water flowing into the soil profile per unit soil surface area per unit time and has the dimensions of velocity (length per time, i.e., inches/h or cm/h).

$$i = \frac{Q}{At} \tag{5.1}$$

*Soil Health Series: Volume 2 Laboratory Methods for Soil Health Analysis*, First Edition.
Edited by Douglas L. Karlen, Diane E. Stott, and Maysoon M. Mikha.

where $i$ is the soil infiltration rate or flux ($L\ T^{-1}$), $Q$ is the quantity or volume of water applied ($L^3$), $A$ is the cross-sectional area of the ring ($L^2$), and $t$ is time (Hanks & Ashcroft, 1980; Jury et al., 1991).

Measurement of the infiltration rate is very important in soil health studies relating to the soil water budget, hydrology, runoff, erosion, irrigation, drainage, and water conservation. Several types of infiltrometers and techniques have been developed and used to measure infiltration. In this chapter, we discuss single-ring, double-ring, and constant-head well infiltrometer methods for measuring in situ soil infiltration rates.

## Single-Ring Infiltrometer (Bouwer, 1986; Reynolds, Elrick, Young, & Amoozegar, 2002)

Materials needed

- metal ring (recommend 15 cm [6 inches] in diameter or larger and 30 cm [12 inches] in height)—size of ring and height may vary
- a hammer and 2 by 4 piece of wood or a block of wood
- a bucket or a jug of water
- a meter or yard stick
- a paper tablet and pen
- a stopwatch
- an operator

Summarized procedure (Figure 5.1)

1) Select a flat area in the field and make sure there are no holes or cracks.
2) Drive the ring about 8 cm (3 inches) into the soil.
3) Pour water in the ring to about a 15-cm (6-inch) depth.
4) Record the time and the level of water in the ring.
5) Measure changes in the water level per unit time in the ring until a steady state is reached.
6) Refill the ring if needed when the level of water has dropped to about 5 cm (2 inches), noting the water level before and after refilling each time.
7) Continue measurements of the water level in the ring per unit time until a steady state is reached.
8) The average of the last three readings is the final infiltration rate in centimeters or inches per hour.

Note: Sometimes, the single ring is lined with plastic wrap prior to water application to prevent soil disturbance inside the ring.

**Figure 5.1** Clockwise steps for measuring soil infiltration rate using a single-ring infiltrometer.

## Double-Ring Infiltrometer (Bouwer, 1986; Jabro, Toth, & Jemison, 1996; Reynolds et al., 2002)

Materials needed

- metal rings (recommend 15-cm [6-inch] diameter inner ring and 30-cm [12-inch] diameter outer ring or larger and 30 cm in height)—size of rings and height may vary
- a hammer and 2 by 4 piece of wood or a block of wood
- a bucket or a jug of water
- a meter or yard stick
- a paper tablet and a pen
- a stopwatch
- an operator

Summarized procedure

1) Select a flat area in the field and make sure there are no holes or cracks.
2) Drive metal rings about 8 cm (3 inches) into the soil so the inner ring is centered in the outer ring.
3) Pour water in both rings to about a 15-cm (6-inch) depth.
4) Record the time and the initial level of water in the inner ring.
5) Measure changes in the water level per unit time in the inner ring until a steady state is reached.
6) Refill the ring when the level of water has dropped to about 5 cm (2 inches), noting the water level before and after refilling each time.
7) Maintain equal water levels in both the inner ring and the zone between the rings.
8) Continue the measurements in the water level per unit time in the inner ring until a steady state is reached.
9) The average of the last three readings is the final infiltration rate in centimeters or inches per hour.

## Constant-Head Pressure Infiltrometer (Figure 5.2)

Materials

- a constant-head pressure infiltrometer
- a hammer and 2 by 4 piece of wood or a block of wood
- a bucket or a jug of water
- a paper tablet and a pen
- a stopwatch
- an operator

Summarized procedure

1) Select a flat area in the field and make sure there are no holes or cracks.
2) Drive the steel ring about 5 cm (2 inches) into the soil surface.
3) Attach the end cap to the Mariotte reservoir.
4) Close the water outlet of the infiltrometer by pushing the air inlet tube down into the outlet tip.
5) Fill the infiltrometer's reservoir to the top with water.
6) Raise the air inlet tube of the outlet tip to maintain the desired pressure head of water on the soil surface in the ring.
7) The infiltrometer is operating properly when air bubbles rise up regularly into the reservoir.
8) Measure changes in the water level per unit time in the infiltrometer reservoir until a steady state is reached, which can be converted to the rate of water flow out of the infiltrometer into the soil.

**Figure 5.2** Constant-head pressure infiltrometer in sugarbeet plots.

9) Equations [5.2] and [5.3] are used to calculate the infiltration rate:

$$I_r = \frac{GQ}{G\pi a^2 + a\left(H + \alpha^{-1}\right)} \tag{5.2}$$

where $I_r$ (L $T^{-1}$) is the infiltration rate at the soil surface, $Q$ is a steady-state flow rate out of the infiltrometer and into the soil ($L^3$ $T^{-1}$), $H$ is the pressure head of water on the soil surface in the steel ring, $\alpha$ is a soil texture–structure parameter ($L^{-1}$), $a$ is the radius of the stainless steel ring (L), and $G$ is a dimensionless shape parameter:

$$G = 0.316\left(\frac{d}{a}\right) + 0.184 \tag{5.3}$$

where $d$ is the depth of ring insertion into the soil (L). More details are given in the user's manual (Reynolds, 1993; Reynolds et al., 2002).

## Soil Bulk Density

Quantifying soil bulk density, defined as the ratio of oven-dried soil mass per unit volume, is important for evaluating land management and farming practice effects on soil health (Blake & Hartge, 1986; Jabro et al., 1996). Bulk density is also used to convert soil water content measurements from a gravimetric to a volumetric basis and to calculate total soil porosity. Bulk density is an important soil characteristic that is often used as an indicator of soil compaction, which affects soil porosity, water movement, rooting depth, and water holding capacity. Factors influencing bulk density include soil texture, total porosity, organic matter content, moisture content, and land management (Blake & Hartge, 1986; Culley, 1993; Hillel, 1971). For undisturbed and uncompacted soils, typical bulk density values range 1.26–1.44 g $cm^{-3}$ for clayey soils and 1.44–1.63 g $cm^{-3}$ for sandy soils (Campbell, 1985).

Several field techniques, including core or cylinder, excavation, clod, and radiation methods, have been developed for measuring soil bulk density. The core method is the most common, so for soil health assessments, this chapter focuses on that method for measuring soil bulk density (Blake & Hartge, 1986).

### Soil Core Method

Materials needed

- a steel core or cylinder with an internal diameter of 5 cm (2 inches) and 5 cm in length. Diameter and length of core may vary
- a soil core sampler
- weighing tin cans for soil core samples
- a spatula
- a balance
- a drying oven of 105–110 °C

Summarized procedure (Blake & Hartge, 1986; Culley, 1993; Grossman & Reinsch, 2002; Jabro et al., 1996; Thien & Graveel, 1997)

1) Select an area, remove about 2–3 cm (an inch or so) of surface soil, clean away all plant residues where soil samples will be taken, and level the area (Figure 5.3a).
2) Push or drive a 5-cm-long by 5-cm-diameter core of known mass (M1, g) and volume ($V$, $cm^3$) into the soil using a core sampler hammer or mallet (dimensions of core may vary) (Figures 5.3b and 5.3c).

**Figure 5.3** Clockwise soil core sampling processes for measuring soil bulk density.

3) Dig out the soil from around the core, remove the core, cut the soil beneath the core bottom, and trim excess soil from the core ends. This process can also be accomplished by wiggling the core sampler and loosening the soil around the core (Figures 5.3d, 5.3e, and 5.3f).
4) The volume of the core ($cm^3$) is calculated as: $V = \pi r^2 h$, where $r$ is the radius of the core (cm) and $h$ is the length of the core (cm).

5) Place the soil core in a tin can of known mass (M2; g)
6) Place the tin can with the soil core in the oven and dry the soil sample at 105–110 °C for 8–24 h.
7) Weigh the oven-dry soil core including the tin can (M3; g).
8) Calculate soil bulk density ($B_d$, g cm$^{-3}$) using

$$B_{\mathrm{d}} = \frac{\mathrm{M3} - \mathrm{M1} - \mathrm{M2}}{V} \tag{5.4}$$

Soils with coarse fragments can be corrected by subtracting the mass and volume of the fragments from the total mass and volume of the soil core. If the soil core is emptied into the tin can, oven dried, and weighed with the can (M4; g), Equation [5.4] can be written as

$$B_{\mathrm{d}} = \frac{\mathrm{M4} - \mathrm{M2}}{V} \tag{5.5}$$

Soil core sampling processes are illustrated in Figure 5.3.

## Acknowledgments

The use of trade, firm, or corporation names in this publication is for the information and convenience of the reader. Such use does not constitute an official endorsement or approval by the USDA or the Agricultural Research Service of any product or service to the exclusion of others that may be suitable. The USDA is an equal opportunity provider and employer.

## References

Blake, G. R., & Hartge, K. H. (1986). Bulk density. In A. Klute (Ed.), *Methods of soil analysis. Part 1. Physical and mineralogical methods* (2nd ed., pp. 363–375). Madison, WI: ASA & SSSA. https://doi.org/10.2136/sssabookser5.1.2ed.c13

Bouwer, H. (1986). Intake rate: Cylinder infiltrometer. In A. Klute (Ed.), *Methods of soil analysis. Part 1. Physical and mineralogical methods* (2nd ed., pp. 825–844). Madison, WI: ASA & SSSA. https://doi.org/10.2136/sssabookser5.1.2ed.c32

Campbell, G. S. (1985). *Soil physics with Basic: Transport models for soil–plant systems.* Amsterdam: Elsevier.

Culley, L. L. B. 1993. Density and compressibility. In M. R. Carter (Ed.), *Soil sampling and methods of analysis* (pp. 529–539). Ann Arbor, MI: Lewis Publishers.

Grossman, R. B., & Reinsch, T. G. 2002. The solid phase. In J. H. Dane & G. C. Topp (Eds.), *Methods of soil analysis. Part 4. Physical methods* (pp. 201–228). Madison, WI: SSSA. https://doi.org/10.2136/sssabookser5.4.c9

Hanks, R. J., & Ashcroft, G. L. (1980). *Applied soil physics*. Berlin: Springer. https://doi.org/10.1007/978-1-4684-0184-4

Hillel, D. (1971). *Soil and water: Physical principles and processes*. New York: Academic Press.

Hillel, D. (1998). *Environmental soil physics*. San Diego, CA: Academic Press.

Jabro, J. D., Toth, J. D., & Jemison, J. M., Jr. (1996). *Physical and hydraulic characteristics of a Hagerstown silt loam soil* (Agronomy Series 138). University Park, PA: Department of Agronomy, Pennsylvania State University.

Jury, W. A., Gardner, W. R., & Gardner, W. H. (1991). *Soil physics* (5th ed.). New York: John Wiley & Sons.

Massimo, I., Angulo-Jaramillo, R., Bagarello, V., Gerke, H. H., Jabro, J., & Lassabatere, L. 2017. Thematic issue on soil water infiltration. *Journal of Hydrology and Hydromechanics* 63, 205–208. https://doi.org/10.1515/johh-2017-0036

Reynolds, W. D. 1993. Saturated hydraulic conductivity: Field measurement. In M. R. Carter (Ed.), *Soil sampling and methods of analysis* (pp. 599–613). Ann Arbor, MI: Lewis Publishers.

Reynolds, W. D., Elrick, D. E., Young, E. G., & Amoozegar, A. (2002). The soil solution phase. In J. H. Dane & G. C. Topp (Eds.), *Methods of soil analysis. Part 4. Physical methods* (pp. 817–858). Madison, WI: SSSA. https://doi.org/10.2136/sssabookser5.4.c32

Thien, S. J., & Graveel, J. G. 1997. *Laboratory manual for soil science: Agricultural & environmental principles*. Dubuque, IA: Wm. C. Brown Publishers.

# 6

# Chemical Reactivity: pH, Salinity and Sodicity Effects on Soil Health

*Yaakov Anker, Vladimir Mirlas, Michael Zilberbrand, and Adi Oren*

## Introduction

Soils are heterogeneous three-dimensional entities, located in site-specific physical settings. They reflect different mineralogical compositions that support above- and below-ground life by mediating energy inputs through biological, chemical, and physical processes. Soils provide many ecosystem services including water purification, carbon sequestration, nutrient cycling and provision of habitat for soil biota. In contrast to natural ecosystems, agroecosystems are regularly disturbed through physical disruption (e.g., tillage), plant biomass removal, fertilizer input, and biocidal applications (Neher, 1999). In response to those disturbances, soil organic C content (SOC) in U.S. Corn Belt soils declined by an average of 47% in just 40 years (1907–1947) following conversion from forest or prairie to agricultural vegetation (Matson, Parton, Power, & Swift, 1997). Similarly, based on long-term data sets, Paul et al. (1997) reported an exponential decrease in soil organic matter (SOM) after just two or three decades of harvesting crops. One consequence of reduced SOM is decreased soil biodiversity which impacts nutrient cycling through food web decomposition processes (Hendrix et al., 1986). Beginning in the 1950s, conservation agriculture, with practices such as crop residue retention and reduced tillage, was initiated to offset detrimental effects of intensive agriculture and restore soil quality.

Variation in soil chemical reactivity, induced by shifts in acidity or salinity, can affect ecological food web dynamics and resultant biotic functions. Under natural

*Soil Health Series: Volume 2 Laboratory Methods for Soil Health Analysis*, First Edition.
Edited by Douglas L. Karlen, Diane E. Stott, and Maysoon M. Mikha.

conditions, significant changes, beyond normal spatial-temporal patterns, occur only after severe disturbances associated with fire, drought, flood and subsequent changes in vegetative cover (Coyle et al., 2017). In contrast, significant ecosystem changes can quickly occur with agricultural management, through its continuous application of chemicals (e.g., fertilizers, biocides, treated wastewater and other soil amendments) and practices that modify soil structure and chemistry (e.g., tillage, agricultural traffic, grazing). As a result, agricultural management studies often focus on short-term soil chemical effects on soil biota. However, natural gradients in soil chemical properties due to geographic position, climatic regions and other edaphic factors must also be considered when assessing soil health effects of acidity, salinity, or sodicity.

Irrigation is a well-recognized soil salinization factor in semi-arid and arid regions because of its effect on groundwater flow dynamics and solute transport (Hanson, Grattan, & Fulton, 1999). For example, the application of chemical fertilizers with high-salinity recycled irrigation water can impair soil health (Tanji, 2002), but not all salinity sources have similar severity. Elements such as boron cause plant toxicity by root injury and seed germination inhibition, while increasing exchangeable sodium percentage (ESP) results in soil particle dispersion, thus decreasing permeability for both water and air, and decreasing structural stability (Sparks, 2002). Regulating anthropogenic soil salinization is a rather straight forward action based on appropriate management, but terrain and climate must also be considered. Precipitation-to-evaporation ratios of less than one, a characteristic of arid conditions, encourage soluble salts to rise to upper soil horizons and the surface (Wua et al., 2019). In temperate regions, salt accumulation is related to soil freezing–thawing processes (Korolyuk, 2014). As a result of those two processes, 6% of terrestrial land is naturally salinized (Nachshon, 2018). This makes planning and maintaining optimal agricultural management complicated and challenging.

## Natural Soil Salinization Processes

Soil salinity accumulates under natural conditions primarily by: (a) the rise of water through the capillaries, followed by evaporation, and precipitation of salts near the soil surface; (b) dissolution and transfer of salts from sediments composed of interlayers containing high initial salt concentrations; (c) aerial salt deposition of eroded soil or geologic parent material, and (d) salt infusion from saline groundwater. The dominant mechanism depends upon physicochemical processes occurring during salt transport. Physically-based water flow processes impose a convective solute transport with resulting solute dispersion at several different rates (Wagenet & Jury, 1984). Salt transport by water has two mechanisms: first by diffusion with a tendency to level-off any salt accumulations except in porous materials where it is a rather slow process; and second by advection which is much faster and within porous material, will readily compete with diffusion (Sawdy, Heritage, & Pel, 2008). Solute mixing of ions in a solution or with soil surface materials (e.g., SOM) modifies the chemical composition resulting in either dissolution or precipitation. Salt precipitation in soil pores or at the surface decreases permeability and alters water distribution patterns (Weisbrod, Nachshon, Dragila, & Grader, 2014).

## Salt Accumulation and Solute Transport Mechanisms

Soil salinization is a process in which dissolved salts concentrate on the soil surface and in upper soil layers. The main causes are high dissolved salt concentrations in soil or groundwater and a high groundwater table. Groundwater flow in the soil causes salinization as dissolved salts are transported by capillary movement from groundwater to upper soil layers. Soil salinization intensity increases as the groundwater table rises because the hydraulic head increases to a level where evapotranspiration starts and downward water flow is reduced. This decreases salt transport by infiltration and percolation of precipitation or irrigation water (Mirlas, Anker, Aizenkod, & Goldshleger, 2020). Buried riverbed channels (Goldshlger, Mirlas, Ben Dor, & Eshel, 2006; Liu et al., 2014) are one of the predominant natural factors causing high soil salinity because of abnormally high hydrostatic pressure heads and variability in groundwater salinity (Figure 6.1).

Geochemical weathering of the earth's upper crust is the initial source of salt to soil and water solutions, although the primary salinization mechanisms may involve atmospheric deposition, seawater intrusion and saline aquifers groundwater intrusion (Corwin & Scudiero, 2019). Owing to capillary forces, upward water movement can carry soluble salts toward the surface and following evaporation, the salts accumulate in upper soil horizons (Wua et al., 2019). In higher-latitude regions, low temperatures result in longer periods of frozen soil, causing salt accumulation to be closely related to freeze–thaw processes (Korolyuk, 2014). As soils thaw in the spring, high surface temperatures can lead to high evaporation rates and further increase salt accumulation. When frozen, temperature differences between lower and upper soil layers induce upward water flow and the transfer of soluble salts to the soil crust (Li, 2010).

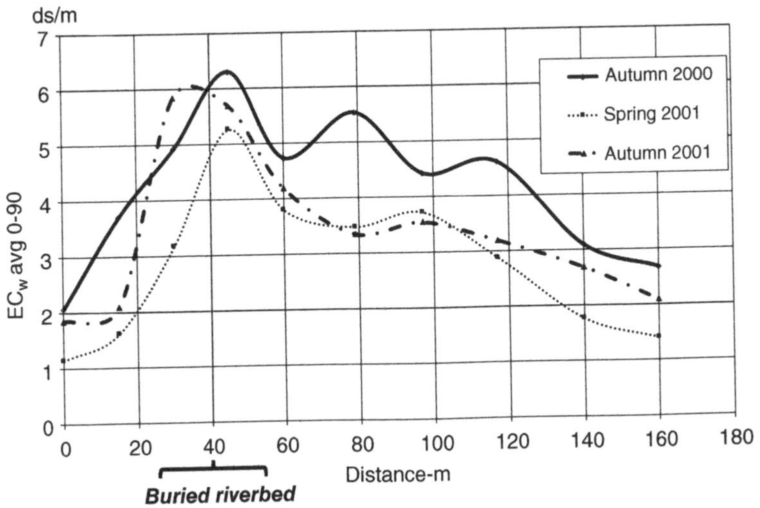


**Figure 6.1** Salinity accumulation in the upper soil layer (0–90 cm) under the buried riverbed of the Mizra creek in Jezre'el Valley, Israel.

# Anthropogenic Soil Salinization Processes

Anthropogenic activities are secondary soil salinity processes. They include over-fertilization, mismanaged irrigation, excessive application of reclaimed wastewater and/or sewage sludge (Tanji, 2002), road salt, and industrial pollution (Litalien & Zeeb, 2020). Infiltration or leakage from constructed reservoirs or dams can also cause watertables to rise and carry stored salts to waterways and the soil surface (Seeboonruang, 2012). Reservoirs were found to induce groundwater rise to about 1 m from the surface, in areas located 400 to 2000 m from reservoir boundaries (Cui & Shao, 2005), thus illustrating the land loss potential due to salinization (Figure 6.2).

Salinity distribution to soil depths of up to 1 m in the land surrounding a reservoir is completely different from typical salt distribution in soils. For example, in irrigated heavy soils (such as in Kfar Yehoshua, Israel, Figure 6.3), salinity within the accumulation horizon (60–90 cm) is higher than in upper soil layers, but adjacent to water reservoirs salinity is greatest in the topsoil (0–30 cm). Figure 6.3 illustrates the inversion point where surface and subsurface salinity concentrations cross, thus defining the salinization boundary area associated with the reservoir. The same process is why land along main irrigation channels is often salinized and abandoned.

Irrigation-induced salinization can be caused by poor soil water management, manifested by high watertables, poor drainage conditions, and the use

**Figure 6.2** Soil salinization around Kfar Baruch reservoir, Israel (Google Earth, Orion-ME, Maxar Technologies, 2021) and the phenomenon manifestation, as salt crusts in peanut field furrows.

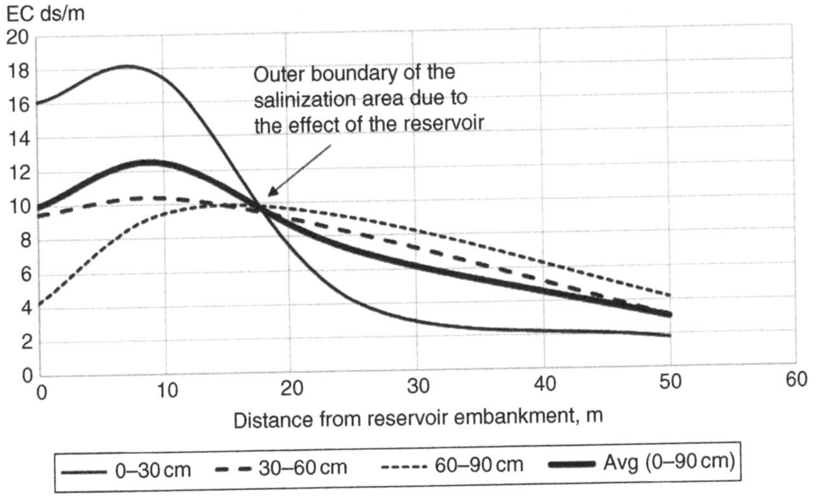


**Figure 6.3** Soil salinity (EC) distribution with depth around the Kfar Yehoshua water reservoir in Israel.

of saline-brackish water for irrigation with less emphasis on leaching fraction (Shahid, 2012). Soil salinization can also be caused by saline water irrigation, fertilizer salt input, and intensification of evapotranspiration processes. The latter is enhanced by irrigation and resultant water-balance shift toward an upward groundwater flow regime that prevents salt removal from the plant root zone (Hanson et al., 1999; Houk, Frasier, & Schuck, 2006; Kaman, Kirda, Cetin, & Topcu, 2005). Irrigation promotes salt transfer by increasing water flow gradients compared with non-irrigated conditions. Salinization is also accelerated when fields are underlain by shallow, semi-confined aquifers that exert upward hydraulic pressure, and impede drainage of overlying soil layers (Christen & Skehan, 2001; Gupta, 2002; Mirlas, 2012; Purkey & Wallender, 2001).

Many numerical models have been developed to quantify evaporation from saline water-saturated soils based on liquid water, water vapor, salt solute, and soil heat (Zhang, Li, & Lockington, 2014). These simulation models incorporate salt precipitation, transport processes, and evaporation to illustrate salt accumulation and precipitation due to evaporation under different watertable conditions. The most popular models are the Richards equation for variably saturated flow and Fickian-based convection–dispersion equation for solute transport (Ramos et al., 2011; Šimůnek, van Genuchten, & Šejna, 2012). Further discussion of these models per se is beyond the scope of this book.

## Chemical Characterization of Salinity Sources

The chemistry and mineralogy of soil salts are defined by their source. Figure 6.4 shows representative vertical profiles for major ions in brackish/saline aquifer pore water under arid (Crimea, Black Sea shore) and extremely arid (Dead Sea shore) conditions. The typical trends are an upward increase in pore solution ion concentration and a general decrease in calcium and sulfate percentage among the total cations and anions, respectively. This decrease is connected to calcite and gypsum precipitation in accordance with calculated positive saturation indices of these minerals in the pore water (Shvets, 1988; Yechieli, 1993). At the Dead Sea site (Figure 6.4b), the ion percentage change points to halite precipitation in the uppermost 10-cm layer.

Capillary rise of shallow groundwater encourages salt precipitation with composition defined by chemical concentrations in the water. Most precipitates are dominated by calcite, chlorides, and sulfates, but when groundwater is enriched

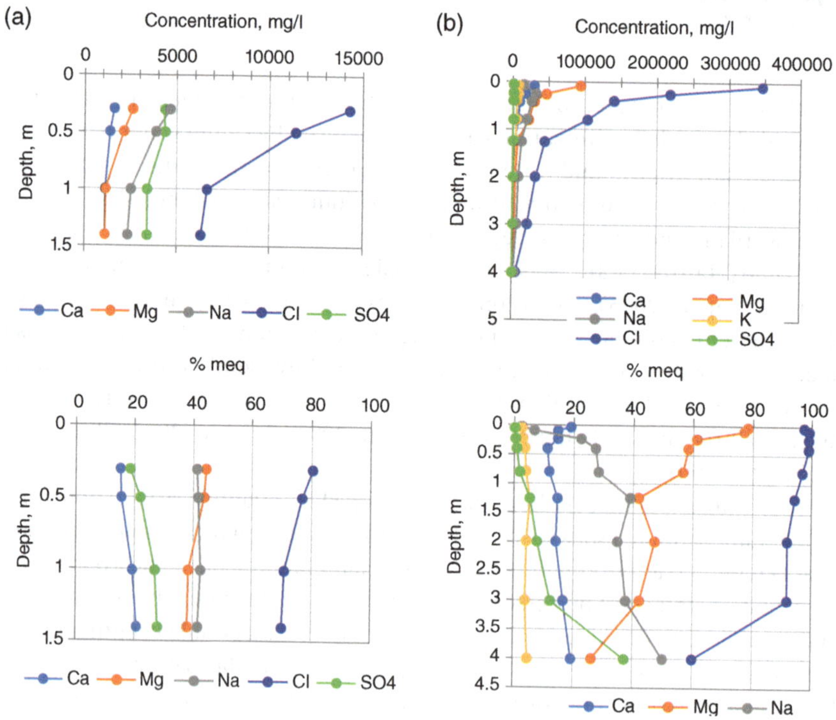


**Figure 6.4** Major ions profiles in the shallow unsaturated zone due to calcite and gypsum precipitation following capillary rise at (a) Black Sea shore, Ogni village, Crimea Peninsula (Shvets, 1988), and (b) a newly exposed Dead Sea shore (Yechieli, 1993).

with bicarbonate soda ($Na_2CO_3$), can also precipitate. When bicarbonate-rich water rises by capillary forces, $Ca(HCO_3)_2$ and $Mg(HCO_3)_2$ release hydrogen and precipitate as carbonates, whereas $NaHCO_3$ that reaches the upper soil layer precipitates as soda (Whittig & Janitzky, 1963). Bicarbonate-rich groundwater can be formed by several processes (Plummer, Prestemon, & Parkhurst, 1994) including (a) oxidation of organic materials; (b) iron reduction; (c) sulfate reduction; (d) methanogenesis; and (e) inorganic reactions involving proton exchange between sodium and calcium.

Evaporation of solutions entering soil is the main mechanism for salt precipitation and soil salinization, with calcite ($CaCO_3$), gypsum ($CaSO_4$:$2H_2O$) and halite (NaCl) being among the first minerals to precipitate (Sonnenfeld, 1984). Calcite deposition due to evapotranspiration normally occurs in zones of maximum water uptake by roots. Under arid conditions, calcite precipitation forms slightly permeable calcrete (caliche) layers in B horizons and crusts at the soil surface (Milnes, 1992; Reeves, 1976). Formation of these water movement barriers exacerbates salinization and negative consequences for soil health.

Gypsum layers may form under hyper-arid conditions, and in addition to halite precipitation in soil crusts, more soluble hygroscopic salts such as Trona ($Na_2CO_3 \cdot NaHCO_3 \cdot 2H_2O$), mirabilite ($Na_2SO_4 \cdot 10H_2O$), glauberite ($Na_2Ca(SO_4)_2$), bloedite ($Na_2Mg(SO_4)_2$:$4H_2O$) and epsomite ($MgSO_4$:$7H_2O$) may form (Anker, Rosenthal, Schulman, & Flexer, 2009). Under semi-arid conditions, soil solution calcium concentrations typically decrease and root zone calcite precipitation is nearly inhibited, thus increasing the role of sodium and magnesium. The descending water flux with prevailing sodium and magnesium concentrations can cause cation exchange in the C horizon with an increase in exchangeable sodium ratio (ESR) considerably beyond 15% (Zilberbrand, 1995), thus causing the horizon to become sodic.

Dry fallout is another calcite input to soil that may prevail. Analysis of dust storm samples collected at different locations in Israel showed an average calcite content of 30–40% (Ganor, 1975; Ganor, Stupp, & Alpert, 2009), with annual dust deposition ranging from 50 to 200 $g/m^2$ (Frumkin & Stein, 2004; Ganor & Foner, 2001). Such figures suggest annual calcite deposition from dust of 15 to 80 $g/m^2$. Subsequent dissolution of calcite in rainwater and re-precipitation in soil pores can further decrease permeability (Ganor et al., 1991).

The second mineral that precipitates from soil solution via evaporation is gypsum. Primary sulfate sources for gypsum precipitation are irrigation water and sulfate-enriched shallow groundwater resulting from pyrite oxidation in underlying bedrock (Nachshon, 2018). Dry deposition of gypsum may also be a notable input to soil. Calcium and sulfate in rain and dry fallout within the Mediterranean region originate primarily in sabkhas of North Africa, Red Sea and Sinai or as evaporated sea spray (Herut, 1992). These dust sources, especially the Sahara Desert in North Africa, account for more than 60% of global dust fallout (Tanaka & Chiba, 2006). During dust storms, gypsum particles constitute ~10% of the total dust mass (Frumkin & Stein, 2004; Ganor et al., 2009; Ganor & Foner, 2001; Kishcha, Volpov, Starobinets, Alpert, & Nickovic, 2020).

Among all ions, increases in sodium and chloride concentrations in pore solution cause the maximum soil fertility damage by decreasing soil water activity, increasing osmotic pressure, and thus decreasing water availability to plant roots. Direct toxicity of sodium and chloride to plants is also a concern (Table 6.1). Furthermore, sodium prevalence in pore water after calcium-bearing minerals (e.g., calcite and gypsum) precipitate causes soil sodification (Mandal et al., 2008). Halite is the third most common mineral to precipitate from soil solution via evaporation. Most halite precipitates in the uppermost soil layers due to evaporation following rain or runoff events, but it is generally redissolved by subsequent rains, creating an annual dissolution-precipitation cycle (Anker, 2007). Progressive halite accumulation in soils may also be related to capillary rise from shallow groundwater, causing an irreversible loss of soil fertility (Nachshon, 2018).

To prevent irrigation-induced soil salinization and sodification, water quality should meet certain criteria (http://www.fao.org/3/t0234e/t0234e01.htm) (Table 6.1). Despite the availability of good guidelines, 20% of total cropland and 33% of irrigated cropland worldwide are salinized due to poor agricultural practices. For example, some irrigation-introduced fertilizers contain high levels of potentially harmful salts, such as potassium chloride or ammonium sulfate. Excessive use of those fertilizers with insufficient flushing can exacerbate soil salinization processes.

## Soil Chemistry, Biota and Ecosystem Services

The capacity of soil to function, within ecosystem boundaries (i.e., sustain plant and animal productivity and maintain water and air quality) constitutes the "soil quality" concept (Doran, 2002). "Soil health" is determined primarily by ecological characteristics (Karlen et al., 1997). The provision of soil ecosystem services reflects interactions between soil biota and their chemical and physical environment. The soil food web consists of microflora (bacteria, fungi, algae, and actinomycetes), microfauna (protozoa), mesofauna (nematodes, arthropods, enchytraeids, mites, and springtails), macrofauna (insects) and megafauna (earthworms). The microflora (mainly bacteria and fungi) comprise a labile pool of nutrients (C, N, P, S) that has a pivotal role in nutrient immobilization, mineralization and plant supply (Yu et al., 2011). The release of nutrients from the microbial biomass is partly regulated through grazing by the soil fauna (Crossley, Coleman, & Hendrix, 1989).

**Table 6.1** Guidelines for Interpretations of Water Quality for Irrigation.

| Potential irrigation problem | Units | Degree of restriction on use: None | Slight to Moderate | Severe |
|---|---|---|---|---|
| Salinity (affects crop water availability)[a] | | | | |
| $EC_w$ | dS/m | < 0.7 | 0.7–3.0 | >3.0 |
| (or) | | | | |
| TDS | mg/l | < 450 | 450–2000 | >2000 |
| Infiltration (affects infiltration rate of water into the soil. Evaluate using ECw and SAR together)[b] | | | | |
| SAR =0–3 and ECw = | | > 0.7 | 0.7–0.2 | <0.2 |
| =3–6 = | | > 1.2 | 1.2–0.3 | <0.3 |
| =6–12 = | | > 1.9 | 1.9–0.5 | <0.5 |
| =12–20 = | | > 2.9 | 2.9–1.3 | <1.3 |
| =20–40 = | | > 5.0 | 5.0–2.9 | <2.9 |
| Specific ion toxicity *(affects sensitive crops)* | | | | |
| **Sodium (Na)** | | | | |
| surface irrigation | SAR | <3 | 3–9 | >9 |
| sprinkler irrigation | meq/l | <3 | > 3 | |
| **Chloride (Cl)** | | | | |
| surface irrigation | meq/l | <4 | 4–10 | >10 |
| sprinkler irrigation | meq/l | <3 | >3 | |
| **Boron (B)** | mg/l | <0.7 | 0.7–3.0 | >3.0 |
| Miscellaneous effects *(affects susceptible crops)* | | | | |
| Nitrogen ($NO_3$—N)[c] | mg/l | <5 | 5–30 | >30 |
| Bicarbonate ($HCO_3$) | | | | |
| (overhead sprinkling only) | meq/l | <1.5 | 1.5–8.5 | >8.5 |
| pH | | Normal Range 6.5–8.4 | | |

[a] $EC_W$—electrical conductivity, a measure of the water salinity, reported in units of deciSiemens per meter at 25 °C (dS/m) or millimhos per centimeter (mmho/cm). Both are equivalent. TDS means total dissolved solids, reported in milligrams per liter (mg/l).

[b] SAR—sodium adsorption ratio, a measure of the amount of sodium relative to calcium and magnesium in the water extract from saturated soil paste. It is the ratio of the Na concentration (in meq/l) divided by the square root of one-half of the Ca + Mg concentration (in meq/l). At a given SAR, the infiltration rate increases as water salinity increases.

[c] $NO_3$—N—nitrate-nitrogen reported in terms of elemental nitrogen ($NH_4$—N and Organic—N should be included when wastewater is being tested).

*Source:* Adapted from the University of California Committee of Consultants (1974).

Soil food webs tend to have longer food chains, greater incidences of omnivory, and greater complexity than other food webs (Neher, 1999). Practically, the closely linked taxonomic and functional diversity of soil food webs (Fierer, Leff et al., 2012) account for ecosystem integrity and functioning (Wittebolle et al., 2009), as well as resistance and resilience to soil chemistry fluctuations. The more complex food web structures are, the greater the system's capacity for adjustment and regulation of changing chemical (or ecological) conditions (Maggiotto et al., 2019; Sánchez-Moreno & Ferris, 2007). Therefore, soil buffering capacity to fluctuations in essential habitat conditions is an essential soil health status determinant (Thomsen, Faber, & Sorensen, 2012; Van Bruggen & Semenov, 2000). Conversely, the resistance of biological diversity to chemical stress impacts, although a complex matter difficult to quantify, has also become a key soil health indicator (Nielsen, Wall, & Six, 2015; Van Bruggen & Semenov, 2000).

## Chemistry Effects on Soil Biota

Currently, soils and their biota are being threatened by degradation, due to land use, climate change, chemical pollution, and invasions of new species that have the potential to cause widespread global impacts on Earth's ecosystems (Nielsen et al., 2015). Changes in soil solution chemical properties (e.g., redox potential, pH, ionic strength, specific ion concentrations), beyond threshold limits for biological viability, might have an immediate influence on soil organisms (Bünemann, Schwenke, & Van Zwieten, 2006), but furthermore, chemistry-mediated effects on soil physical properties can also cause indirect effects on soil biology. Modification in the balance between $Ca^{2+}$ and $Na^{+}$ ion inputs to soil (i.e., sodification) is the most prominent example. This is most often caused by changing irrigation water quality or applications of exogenous materials (Mandal et al., 2008) that in clayey soils, can degrade soil structure, intensify erosion and soil loss (Goldberg, Nachshon, Argaman, & Ben-Hur, 2020) and impair hydraulic properties (Assouline & Narkis, 2011). All of those processes eventually deplete SOM pools, reducing substrate availability, and the value of the soil as a plant growth medium (Setia, Rengasamy, & Marschner, 2014; Singh, 2016).

A review by Tecon and Or (2017) highlights several controlling factors determining soil microbe biogeography at various spatial scales, as well as management-related physico-chemical factors that shape microbial abundance and diversity in agricultural systems. Factors expected to create environmental and chemical variation relevant to soil microbes include plant cover (Berg & Smalla, 2009), animal excreta (Singh, Nunan, & Millard, 2009), wetness (Or, Smets, Wraith, Dechesne, & Friedman, 2007a), fertilizer application (Fierer, Lauber et al., 2012), pH (Rousk et al., 2010), and salinity (Rajaniemi & Allison, 2009).

Microbial biomass varies within and across biomes by up to three orders of magnitude and is positively associated with moisture availability, SOC, essential nutrients, and near-neutral pH, regardless of climate conditions or plant biomass (Tecon & Or, 2017). Soil pH was the best predictor of bacterial community composition. As fungi are generally more tolerant of acidic conditions than bacteria (Rousk et al., 2010), fungal to bacterial biomass ratios tend to correlate negatively with soil pH (Fierer, Strickland, Liptzin, Bradford, & Cleveland, 2009; Orgiazzi, Bardgett, & Barrios, 2016). Furthermore, since fungal hyphal structure influences soil structure formation (Beare, Cabrera, Hendrix, & Coleman, 1994; Dighton, 2003; Eash, Karlen, & Parkin, 1994), the filamentous fungal element is considered an essential soil quality determinant (Adl, Coleman, & Read, 2006).

Soil acidity can indirectly influence solubility and bioavailability of chemical compounds (nutrients or pollutants) to plant roots and soil microbes (Troeh & Thompson, 1993). For instance, the characteristic increase in pH of naturally-acidic grazed grasslands due to sheep urine input may lead to different phylogenetic responses of bacterial versus fungal communities, with the bacterial community variation related to the change in soil pH, and the fungal community variation more responsive to the dissolution of SOC induced by increasing pH (Singh et al., 2009). Measurement of $CO_2$ respiration in soil, often used as a surrogate for microbial biomass (Anderson & Domsch, 1978), is also affected by soil pH, with exponentially increasing underestimates above pH 6, mostly in calcareous soils due to partitioning of respired $CO_2$ between soil gaseous and solution phases (Oren & Steinberger, 2008).

Soil pH is also an important determinant of microbially mediated processes related to nutrient cycling (e.g., nitrification, denitrification, and others) with the optimum pH for nitrification generally being between 6.5 and 8. Above pH 8, $NH_4^+$ is present mainly as $NH_3$ gas, which can be lost from soil and also inhibits the activity of $NO_2^-$ oxidizing bacteria. Below pH 5.5–6, bacterial nitrification is greatly reduced in most soils (Smith & Doran, 1996). Soil acidification is a natural process influenced by parent material (e.g., buffered in calcareous soils) and favored in humid regions (forest soils) by rainfall and cation leaching. In contrast, most arid and semi-arid soils are alkaline and buffered (Brady & Weil, 2014).

Salinity can vary spatially and temporally in soils and affects plants and microbes by changing water (osmotic) potential. An increased soil solution osmotic pressure is defined by decreased water activity (Sparks, 2002) compared to cell sap, which makes moisture extraction by plants or soil fauna more difficult. If the salt concentration in soil is greater than that of plant or microbial cells, water moves from cells into soil potentially causing plasmolysis that often leads to wilting and death (Choat et al., 2018; Hillel, 2003). High salt concentrations can also result in ion-specific toxicities (Rath et al., 2016) and inhibition of metabolic processes (Or, Phutane, & Dechesne, 2007b).

Overall, high soil salinity results in lower microbial biomass, lower metabolic activity (Sardinha, Müller, Schmeisky, & Joergensen, 2003; Yan & Marschner, 2012) and decreased microbial biodiversity (Zhang et al., 2019). In addition to direct negative impacts on microbial function, those reductions may also be associated with global decreases in SOM stocks due to salinization (Setia et al., 2013). In saline soils, drought severity for the soil microbial community is exacerbated by increasing salt concentrations during drying (Rath, Maheshwari, & Rousk, 2017). Soil salinity gradient studies have shown that bacteria are more sensitive than fungi (Rath, Murphy, & Rousk, 2019); yet, salinization can also lower fungal biomass indirectly by reducing vegetation and thus input of root and litter debris, which is a very important energy source for saprotrophic fungi (Sardinha et al., 2003; Setia et al., 2011).

Soil redox potential which decreases with soil saturation (Tecon & Or, 2017), can be highly variable in both space and time (Hinsinger, Bengough, Vetterlein, & Young, 2009; Standing & Killham, 2006) due to changes in water content and associated oxygen availability. Under anaerobic conditions (widespread in wetland soils), the availability of electron acceptors ($NO_3^-$, $Mn^{4+}$, $Fe^{3+}$, $SO_4^{2-}$ in decreasing order of redox potential), determines what type of anaerobic metabolism can take place (Ben-Noah & Friedman, 2018). Reduced $O_2$ and elevated $CO_2$ concentrations negatively affect plant growth and productivity (Harris & van Bavel, 1957), as well as microbial activity (Sierra & Renault, 1995). Excluding wetlands, those conditions are best correlated with wet and warm soils, such as intensively irrigated fields with fine-textured soils (high water retention), especially during the summer (Ben-Noah & Friedman, 2018). Several active aeration methods have been proposed and evaluated, such as adding air, $O_2$ bubbles, or $H_2O_2$ to irrigation water and air injection into the soil, resulting in improved yield, quality, and water-use efficiency for a variety of crops (Ben-Noah & Friedman, 2018). While soil-aeration procedures are widely used as an environmental practice for soil and groundwater remediation, future agricultural use of those practices depends on design simplicity (Ben-Noah & Friedman, 2019).

## Habitat Heterogeneity and Soil Microbial Diversity

While soil might be portrayed as an unfavorable habitat for microbial life due to its harsh and fluctuating environmental conditions, evidence suggests a higher microbial density and unparalleled diversity compared with other biosphere components (Bahram et al., 2018; Stotzky, 1997). Genetic diversity in soils may exceed diversity in aboveground or aquatic communities by three orders of magnitude (Tecon & Or, 2017 and references within). Heterogeneity, coupled with spatial and temporal microhabitat fragmentation are often cited as key factors promoting immense soil microbial diversity (Bardgett, 2002; Curtis & Sloan, 2005; Dion, 2008; Zhou et al., 2002).

A fundamental trait promoting soil functioning is a high specific surface area (reaching 100 $m^2/g^1$ in clayey soils) and, particularly, the structural arrangement of variable-size particles, often glued by SOM. This trait creates complex pore spaces inhabited by microbial life, retaining water and nutrients, and providing gas transport and cell dispersion pathways (Crawford, Harris, Ritz, & Young, 2005). Conversely, microbial communities often moderate heterogeneity of their surroundings, altering not only the chemical environment but also physical structure through hydrophobic films and aggregate formation (Totsche et al., 2010). While bacteria and fungi both typically demonstrate co-occurrence with clay and fine silt-sized particles, soil textural heterogeneity was found to positively influence bacterial diversity while having little impact upon fungal diversity (Seaton et al., 2020).

Patchy microbial biomass distributions and diversity can be observed at the meter scale, often in association with so-called "islands of fertility" (Schlesinger, Raikes, Hartley, & Cross, 1996) or "hot spot" phenomena (Bundt, Widmer, Pesaro, Zeyer, & Blaser, 2001; Kuzyakov & Blagodatskaya, 2015), typically close to plant roots (rhizosphere) or associated with decaying organic material. "Hot moments" are short-term events inducing accelerated process rates controlled by water availability that are vital for biochemical function (Kuzyakov & Blagodatskaya, 2015). The microscale arrangement of water retained in soil pores defines the aqueous microbial habitat size distribution and connectedness that, in turn, affects substrate diffusion rates, cell motility (Kim & Or, 2016; Wang & Or, 2010) and timing of cell-to-cell interactions (Tecon, Ebrahimi, Kleyer, Levi, & Or, 2018). Correspondingly, small microbes-fed fauna (e.g., nematodes, protozoa) may regulate microbial distribution (Tecon et al., 2018) through their dependence on water films that enable them to live and move through the soil (Griffiths, 1994; Lavelle, Lattaud, Trigo, & Barois, 1995). As a result, microbial diversity is enhanced at intermediate water contents where the aqueous phase forms numerous disconnected habitats.

Relationships between microbial biomass and diversity are positive only when the aqueous phase is fragmented and spatial isolation suppresses dominance of a few species (Bickel & Or, 2020). This could be one reason why overall microbial abundance is higher, but diversity is lower, in arable (tilled) compared to pasture soils. Generally, moisture fluctuations are less pronounced in the former, regularly irrigated, and more structurally homogenized soil systems (Torsvik, Øvreås, & Thingstad, 2002). An additional "hidden" factor, often preserving microbial diversity and potentially offsetting factors related to water or substrate availability, is the greater prevalence of microbial dormancy in soils compared to other environments. Thus, under unfavorable conditions, microbial "seed banks" may be developed and then revived when environmental conditions improve (Blagodatskaya & Kuzyakov, 2013; Lennon & Jones, 2011).

## Agricultural Ecosystems: Degradation Prevention and Conservation

As agricultural intensification occurs, regulation of functions through soil biodiversity is progressively replaced by regulation through chemical and mechanical inputs (Giller, Beare, Lavelle, Izac, & Swift, 1997). For instance, soil oxidation state, typically enhanced by tillage, was correlated positively with nitrate leaching in conventionally tilled soils (Doran, 1980). In no-till soils, both nitrification and denitrification were greater in surface soil (0–5 cm) compared to conventionally tilled soils, but the reverse was true at greater depths (5–21 cm) (Beare, 1997). These trends demonstrate the preservative regulatory power of a less disturbed soil food web. Likewise, high use of inorganic fertilizers in modern agriculture is likely to have ecological costs due to the loss of natural cycles among organisms (Crossley et al., 1989) and potential pollution by excess nutrients. In contrast, healthy soils often have increased detrital food webs containing more groups of organisms and with faster decomposition, nutrient turnover rates (Setälä, Tyynismaa, Martikainen, & Huhta, 1991), and increased primary production (Setälä & Huhta, 1991). Balanced with organisms that immobilize nutrients, the net effect is regulation of nutrient availability to plants and other organisms (Neher, 1999). The routine application of inorganic N to agricultural soils may well modify microbial community structure, favoring a more N-demanding and active copiotrophic microbial community, characterized by lower C:N biomass ratios (Fierer, Lauber et al., 2012). Fungal to bacterial biomass ratios might decrease as well, as the C:N ratio is higher in fungal relative to bacterial cells (Paul & Clark, 1996). On the other hand, in non-buffered soils, N fertilization might cause soil acidification (Bünemann et al., 2006), selectively depressing the bacterial component. Furthermore, higher trophic groups in soil food webs often respond to agricultural management and soil nutrient status according to their feeding habits (e.g., nematode guilds feeding on bacteria, fungi, or plant parasites) as discussed by Ferris and Bongers (2006).

A significant increase in species diversity of soil fauna, including nematodes, microarthropods, amoebae, flagellates, ciliates, and protozoans, often occurs with an increase in years of no-tillage management due to an increase in organic matter content and profile stratification (Adl et al., 2006). Similarly, faunal diversity was suggested to be driven mainly by plant productivity (van der Wal et al., 2009) which enhances microbial growth, thereby increasing resources for soil fauna (Cole, Buckland, & Bardgett, 2005). Finally, according to a global meta-analysis of 96 recent publications, no-tillage with residue retention increased microbial C, microbial N and the microbial-to-organic C ratio by 25, 64, and 57%, respectively, regardless of the soil condition (e.g., soil pH and texture), experimental duration, and climate (Li, Chang, Tian, & Zhang, 2018).

Crop rotations, especially those that include cover crops, sustain soil quality and productivity by enhancing soil C, N, and microbial biomass, making them a cornerstone for sustainable agroecosystems (McDaniel, Tiemann, & Grandy, 2014). Organic amendments such as animal manures and their composted products can increase SOM content, replenish nutrient pools, and generally promote soil biota activity (Enwall et al., 2007; Ruppel, Torsvik, Daae, Øvreås, & Ruhlmann, 2007). However, the full impact of organic amendments and their characteristics on microbial and faunal community composition as well as structure and diversity remains unclear (Jangid et al., 2008).

Spatial and temporal scales are significant factors affecting soil pH, salinity, and sodicity measurements. For illustration, Terrat et al. (2017) reported that only 48% of the observed variance in soil biodiversity within samples collected across France and analyzed using the most advanced molecular tools available, could be accounted for with traditional soil measurements. This presumably reflects the fact that the volume of soil in a typical soil sample is huge, compared to actual habitats of individual microbes. Furthermore, virtually all spatial information regarding microbial individuals, relative to each other and their resources, is lost during sample processing (Vos et al., 2013).

What is needed to improve soil health assessment with regard to the overviewed factors is more quantitative information on the spatial heterogeneity manifested at the micron scale. This type of information can now be obtained with advanced spectroscopic and microscopic technologies that can provide detailed observations with sharp spatial heterogeneity and thus help quantify soil chemical makeup over minute distances (Veum et al., 2021). Coupled with current advances in soil microbiology, specific microbes can be identified in soils, and their spatial distribution determined at scales of 3 μm or less. Three-dimensional distributions can then be computed using statistical algorithms. Transport and physicochemical processes occurring in soil pores can also be modeled to describe the dynamics of microbes (Baveye et al., 2018). Finally, when combined with traditional pH, salinity, sodicity, and other measurements it will be possible to better elucidate many more factors influencing microbial diversity and abundance within natural or managed soil ecosystems.

## Soil Health Indices

Soil health assessment is important for understanding potential agricultural practice effects on crop yield and other soil ecosystem services (e.g., environmental quality and human health). A practical approach for comparing and monitoring soil health is to integrate selected soil quality attributes, with or without weighting, into a soil health index (SHI) as described by Bi, Zhen-Lou, Wang, and Zhou (2013). Several soil parameters have been identified as soil quality indicators for use in minimum datasets used to compute an SHI (Haberern, 1992). The indicators should be: (a) representative of chemical, biological, and physical soil properties and processes; (b) responsive to soil management, land use, and climate change; and (c) measured in accurate and cost-effective ways (Doran, 2002; Andrews et al., 2004). Several soil parameters including pH, TOC, SOC, cation exchange capacity (CEC), SAR, and concentrations of essential macro- and micro-nutrients, as well as soil biological traits such as soil microbial biomass, respiration and enzyme activities (Acosta-Martinez, Cano, & Johnson, 2018; Xue et al., 2019), have been evaluated. Other soil parameters, attainable using remote sensing (Xiong, Chen, Xia, Ye, & Anker, 2020) or other spectral analysis techniques for soil samples (Ben-Dor et al., 2009) have become common as well.

Using appropriate, site-specific scoring curves, each soil parameter receives a value between 0 and 1 that can be combined (integrated) to compute an overall SHI or individual SHI values for soil biological, chemical, and physical properties and processes (Amacher et al., 2007). The most common approach is "additive" with the equal weight assigned to each parameter (Svoray, Hassid, Atkinson, Moebius-Clune, & van Es, 2015). Relative contributions (i.e., weighted, additive values) based on literature values or expert opinion can be used if there are known or logical reasons for assigning different levels of importance of each parameter based on targeted ecosystem services (Mukherjee & Lal, 2014; Stott et al., 2021).

## Technological Advances in the Study of Soil Biogeochemical Interfaces

Soil deterioration can be monitored by solid and liquid phase analysis. Recent technological advancements have made suitable spectroscopic and microscopic equipment available to soil researchers, allowing the quantification and modeling of the physical, chemical, and microbiological properties of soils, thus facilitating the knowledge of the interactions between microbial habitats and soil spatial structure, essential for understanding soil processes and management in the context of soil function and health.

Soil alkalinity/acidity is measured by using commercially available electronic pH meters, in which a glass or solid-state electrode is inserted into moistened soil or a mixture (suspension) of soil and water (Table 6.2). Soil salinity is generally measured by the electrical conductivity (EC) of a soil extract. A soil is considered saline if the EC of the saturation extract exceeds 4 dS/m at 25 °C. Soil sodicity

**Table 6.2** USDA-NRCS Soil pH Classifications (Soil Survey Manual, 1993).

| Classification | pH Range | Classification | pH Range |
|---|---|---|---|
| Ultra acidic | <3.5 | Neutral | 6.6–7.3 |
| Extremely acidic | 3.5–4.4 | Slightly alkaline | 7.4–7.8 |
| Very strongly acidic | 4.5–5.0 | Moderately alkaline | 7.9–8.4 |
| Strongly acidic | 5.1–5.5 | Strongly alkaline | 8.5–9.0 |
| Moderately acidic | 5.6–6.0 | Very strongly alkaline | >9.0 |
| Slightly acidic | 6.1–6.5 | | |

is determined by measuring the ESP basing on results of displacement of exchangeable cations using concentrated solutions of $SrCl_2$ or $NH_4Cl$; or more commonly, estimated from the sodium adsorption ratio (SAR) of a soil–water extract. If the SAR of the saturation extract exceeds 13, the soil is considered sodic (Carter & Gregorich, 2008).

Direct soil solutions analysis is a very promising way of characterization soil chemistry changes. Extraction of these solutions can be performed by vacuum pumping (in situ and the lab), and by pressing out, centrifugation and exclusion by an immiscible liquid in the lab (Rhoades & Oster, 1986; Zilberbrand, 1995). All these methods are effective only at water content above field capacity.

X-ray computed tomography (CT) systems provide increasingly reliable information about soil pores and solid geometry at resolutions as small as 0.05 mm as well as for three-dimensional salt distribution and its effect on gas and water permeability (Weisbrod et al., 2014) as well as microbial function (Tecon & Or, 2017). Progress in synchrotron-based, near-edge X-ray spectromicroscopy (NEXAFS), scanning transmission X-ray microscopy (STXM), X-ray absorption spectroscopy (XAS), X-ray micro-fluorescence spectroscopy (μXRF) and nanoscale secondary ion mass spectrometry (Nano-SIMS), when applied to soil thin sections, has also led to observations of sharp spatial heterogeneity in elemental composition of soils over very small distances. For a review of those technological advances, see Baveye et al. (2018).

The integrative application of various non-invasive, analytical SOC speciation methods, including NEXAFS and nuclear magnetic resonance (NMR) and Fourier transform infrared (FT-IR) spectroscopies, is useful for understanding speciation and turnover of SOM (Mitchell et al., 2013; Prietzel, Müller, Kögel-Knabner, & Thieme, 2018). Fluorescence spectroscopy is another particularly sensitive tool for characterizing dissolved SOM dynamics, which are important for characterizing abiotic and/or biotic decomposition processes and agricultural management (Rinot, Rotbart, Borisover, Bar-Tal, & Oren, 2021; Sharma et al., 2017).

Transmission and scanning electron microscopes (TEM and SEM, respectively) and confocal laser microscopes (CLM) have also provided a wealth of qualitative information regarding the state of bacteria and fungi (i.e., intact living organisms, spores, residues) within organo-mineral complexes. Furthermore, analytical TEM characterization of microaggregates by electron energy loss spectroscopy (EELS) or energy dispersive X-rays spectroscopy (EDX) provides in situ identification of microbial involvement in biogeochemical cycles of elements. For micrometer characterization associated with other methodologies such as Nano-SIMS or soil fractionation, EDX enables monitoring both incorporation of biodegraded litter within soil aggregates and impacts of microbial dynamics on soil aggregation, particularly due to production of extracellular polymeric substances (Watteau & Villemin, 2018).

Fluorescence in situ hybridization (FISH) techniques are also promising, because they allow specific classes of microorganisms to be localized in soils using oligonucleotide probes that provide reduced background interference due to soil fluorescence. Combining FISH and CLM techniques provides a powerful tool to visualize and quantify distribution of soil microorganisms with satisfactory resolution (Eickhorst & Tippkötter, 2008a, 2008b; Baveye et al., 2018). DNA and RNA-based methods for soil microbial community characterization can provide a huge volume of data relating microbial community composition to its functioning. However, use of "-omics"-based approaches in biogeochemical models is an extreme challenge (Blankinship et al., 2018).

Finally, there is an urgent need to integrate various disciplinary perspectives (i.e., physical, chemical, and biological) to describe the processes by which micro-scale heterogeneity influences and generates macroscopic soil behavior (Baveye et al., 2018). This could be promoted by combining various biogeochemical imaging methods (i.e., correlative imaging or correlative microscopy; Schlüter, Eickhorst, & Mueller, 2019) with current soil health assessment protocols.

## Conclusions

Soil salinity is a dynamic soil property with potentially dramatic effects on soil health due to direct ion toxicities or osmotic stress to plants and fauna. It can also adversely affect soil structure and adversely affect soil as a habitat. Under irrigation, non-saline soils can be easily become saline, depending on water composition and irrigation rate, as well as intrinsic soil properties and climatic conditions. Those circumstances necessitate constant soil health monitoring of salinization, especially for irrigated land.

The basic processes affecting salt movement in soil involve a complex physiochemical system. Solute transport models have been developed to estimate a large number of simultaneous physical, chemical, and biological processes, including sorption–desorption, volatilization, photolysis, and biodegradation, as well as their kinetics. The models are designed to identify equilibrium conditions between chemical reactions of soluble salt components including complexation, cation exchange and precipitation-dissolution.

This chapter provides global perspectives as well as optional strategies for pH, salinity, and sodicity analyses and modeling of dynamic agricultural soil processes relevant to soil health assessment. Monitoring salinity dynamics is relatively simple, but must not be overlooked, especially for irrigated agricultural production systems.

## References


Acosta-Martinez, V., Cano, A., Johnson, J. 2018. Simultaneous determination of multiple soil enzyme activities for soil health-biogeochemical indices. *Applied Soil Ecology*;126:121–8. doi: 10.1016/j.apsoil.2017.11.024.

Adl, S.M., Coleman, D.C., Read, F. 2006. Slow recovery of soil biodiversity in sandy loam soils of Georgia after 25 years of no-tillage management. *Agriculture, Ecosystems and Environment* 114:323–334.

Amacher, M.C., O'Neill, K.P., Perry, C.H. 2007. Soil vital signs: A new soil quality index (SQI) for assessing forest soil health. USDA Forest Service Research Paper. Washington, DC: USDA. doi:10.2737/RMRS-RP-65

Anderson, J.P.E., Domsch, K.H. 1978. A physiological method for the quantitative measurement of microbial biomass in soils. *Soil Biology and Biochemistry* 10:215–221.

Andrews, S.S., Karlen, D.L., Cambardella, C.A. 2004. The soil management assessment framework. *Soil Science Society of America Journal* 68:1945–62. doi: 10.2136/sssaj2004.1945.

Anker, Y. 2007. Hydrogeochemistry of the Central Jordan Valley. Thesis. Tel Aviv University.

Anker, Y., Rosenthal, E., Schulman, H., Flexer, A. 2009. Geochemical evolution of runoff in a hyper-saline watershed (Lower Jordan Valley). *Israel Journal of Earth Sciences* 59:41–61.

Assouline, S., Narkis, K. 2011. Effects of long-term irrigation with treated wastewater on the hydraulic properties of a clayey soil. *Water Resources Research* 47: W08530.

Bahram, M., Hildebrand, F., Forslund, S.K., Anderson, J.L., Soudzilovskaia, N.A., Bodegom, P.M., Bengtsson-Palme, J., Anslan, S., Coelho, L.P., Harend, H., Huerta-Cepas, J., Medema, M.H., Maltz, M.R., Mundra, S., Olsson, P.A., Pent, M., Põlme, S., Sunagawa, S., Ryberg, M., Tedersoo, L., Bork, P. 2018. Structure and function of the global topsoil microbiome. *Nature* 560:233–237.

Bardgett, R.D. 2002. Causes and consequences of biological diversity in soil. *Zoology* 105:367–374.

Baveye, P.C., Otten, W., Kravchenko, A., Balseiro-Romero, M., Beckers, E., Chalhoub, M., Darnault, C., Eickhorst, T., Garnier, P., Hapca, S., Kiranyaz, S., Monga, O., Mueller, C.W., Nunan, N., Schlüter, V. Pot, S., Schmidt, H., Vogel, H.J. 2018. Emergent properties of microbial activity in heterogeneous soil microenvironments: Different research approaches are slowly converging, yet major challenges remain. *Frontiers in Microbiology* 9: 1929.

Beare, M.H. 1997. Fungal and bacterial pathways of organic matter decomposition and nitrogen mineralization in arable soil. In: *Soil ecology in sustainable agricultural systems*. Brussaard, L., & R. Ferrera-Cerrato, eds., p. 37–70. Lewis Publishers, Boca Raton, LA.

Beare, M.H., Cabrera, M.L., Hendrix, P.F., Coleman, D.C. 1994. Aggregate-protected and unprotected pools of organic matter in conventional and no-tillage Ultisols. *Soil Science Society of America Journal* 58:787–795.

Ben-Dor, Y., Goldshleger, N., Mor, E., Mirlas, V., Basson, U. 2009. Combined active and passive remote sensing methods for assessing soil salinity: A case study from Jezre'el Valley, Northern Israel. In: *Remote sensing of soil salinization. Impact on land management*. G. Metternicht & J.A. Zinck, eds. pp. 236–253. CRC Press, Taylor & Francis Group,LLC, New York.

Ben-Noah, I., Friedman, S.P. 2018. Review and evaluation of root respiration and of natural and agricultural processes of soil aeration. *Vadose Zone Journal* 17:170119.

Ben-Noah, I., Friedman, S.P. 2019. Bounds to air-flow patterns during cyclic air injection into partially saturated soils inferred from extremum states. *Vadose Zone Journal* 18:180023.

Berg, G., Smalla, K. 2009. Plant species and soil type cooperatively shape the structure and function of microbial communities in the rhizosphere. *FEMS Microbiology Ecology* 68:1–13.

Bi CJ, Zhen-Lou C, Wang J, Zhou, D. 2013. Quantitative assessment of soil health under different planting patterns and soil types. *Pedosphere* 23:194–204. doi: 10.1016/S1002-0160(13)60007-7.

Bickel S., Or, D. 2020. Soil bacterial diversity mediated by microscale aqueous-phase processes across biomes. *Nature Communications* 11:116.

Blagodatskaya, E., Kuzyakov, Y. 2013. Active microorganisms in soil: Critical review of estimation criteria and approaches. *Soil Biology and Biochemistry* 67:192–211.

Blankinship, J.C., Berhe, A.A., Crow, S.E., Druhan, J.L., Heckman, K.A., Keiluweit, M., Lawrence, C.R., Marín-Spiotta, E., Plante, A.F., Rasmussen, C., Schädel, C., Schimel, J.P., Sierra, C.A., Thompson, A., Wagai, R., Wieder, W.R. 2018. Improving understanding of soil organic matter dynamics by triangulating theories, measurements, and models. *Biogeochemistry* 140:1–13.

Brady, N.C., Weil, R.R. 2014 .*The nature and properties of soils*. Harlow: Pearson Education Limited.

Bundt, M., Widmer, F., Pesaro, M., Zeyer, J., Blaser, P. 2001. Preferential flow paths: Biological "hot spots" in soils. *Soil Biology and Biochemistry* 33:729–738.

Bünemann, E.K., Schwenke, G.D., Van Zwieten, L. 2006. Impact of agricultural inputs on soil organisms—A review. *Soil Research* 44:379–406.

Carter M.R. and E.G. Gregorich 2008. *Soil sampling and methods of analysis*. Boca Raton, FL: CRC Press. 198 p.

Choat, B., Munns, R., McCully, M., Passioura, J.B., Tyerman, S.D., Bramley, H., Canny, M. 2018. Water movement in plants. Chapter 3. In: Plants in Action. Australian Society of Plant Scientists. pp. 1–56. http://plantsinaction.science.uq.edu.au/content/chapter-3-water-movement-plants

Christen, E., Skehan, D. 2001. Design and management of subsurface horizontal drainage to reduce salt loads. *Journal of Irrigation and Drainage Engineering* 127:148–155.

Cole, L., Buckland, S.M., Bardgett. R.D. 2005. Relating microarthropod community structure and diversity to soil fertility manipulations in temperate grassland. *Soil Biology and Biochemistry* 37:1707–1717.

Corwin, D., Scudiero, E. 2019. Review of soil salinity assessment for agriculture across multiple scales using proximal and/or remote sensors. *Advances in Agronomy* 158: 1–130. doi: 10.1016/bs.agron.2019.07.001

Coyle, D.R, Nagendra, U.J., Taylor, M.K., Campbell, J.H., Cunard, C.E., Joslin, A.H., Mundepi, A., Phillips, C.A., Callaham, M.A. 2017. Soil fauna responses to natural disturbances, invasive species, and global climate change: Current state of the science and a call to action. *Soil Biology and Biochemistry* 110:116–133.

Crawford, J.W., Harris, J.A., Ritz, K., Young, I.M. 2005. Towards an evolutionary ecology of life in soil. *Trends in Ecology and Evolution* 20:81–87.

Crossley Jr., D.A., Coleman, D.C., Hendrix. P.F. 1989. The importance of the fauna in agricultural soils: Research approaches and perspectives. *Agriculture, Ecosystems and Environment* 27:47–55.

Cui, Y., Shao J. 2005. The role of ground water in arid/semiarid ecosystems, Northwest China. *Groundwater* 43: 471–477. doi: 10.1111/j.1745-6584.2005.0063.x

Curtis, T.P., Sloan, W.T. 2005. Exploring microbial diversity—A vast below. *Science* 309:1331–1333.

Dighton, J. 2003. *Fungi in ecosystem processes*. Dekker, New York.

Dion, P. 2008. Extreme views on prokaryote evolution. In: *Microbiology of extreme soils*. Dion, P., and Nautiyal, C.S., editors, p. 45–70. Berlin, Heidelberg: Springer.

Doran, J.W. 1980. Soil microbial and biochemical changes associated with reduced tillage. *Soil Science Society of America Journal* 44:765–741.

Doran, J.W. 2002. Soil health and global sustainability: Translating science into practice. *Agriculture, Ecosystems and Environment* 88:119–127.

Eash, N.S., Karlen, D.L., Parkin, T.B. 1994. Fungal contributions to soil aggregation and soil quality. In: *Defining soil quality for a sustainable environment*, Vol. 35. Doran, J.W., D.C. Coleman, D.F. Bezdicek, and B.A. Stewart, eds., p. 221–228. Madison WI: SSSA.

Eickhorst, T., Tippkötter, R. 2008a. Improved detection of soil microorganisms using fluorescence in situ hybridization (FISH) and catalyzed reporter deposition (CARD-FISH). *Soil Biology and Biochemistry* 40:1883–1891.

Eickhorst, T., Tippkötter, R. 2008b. Detection of microorganisms in undisturbed soil by combining fluorescence in situ hybridization (FISH) and micropedological methods. *Soil Biology and Biochemistry* 40:1284–1293.

Enwall, K., Nyberg, K., Bertilsson, S., Cederlund, H., Stenstrom, H.J., Hallin, S. 2007. Long-term impact of fertilization on activity and composition of bacterial communities and metabolic guilds in agricultural soil. *Soil Biology and Biochemistry* 39:106–115.

Ferris, H., Bongers, T. 2006. Nematode indicators of organic enrichment. *Journal of Nematology* 38:3–12.

Fierer, N., Lauber, C.L., Ramirez, K.S., Zaneveld, J., Bradford M.A., Knight, R. 2012. Comparative metagenomic, phylogenetic and physiological analyses of soil microbial communities across nitrogen gradients. *ISME Journal* 6:1007–1017.

Fierer, N., Leff, J.W., Adams, B.J., Nielsen, U.N., Bates, S.T., Lauber, C.L., Owens, S., Gilbert, J.A., Wall, D.H., Caporaso, J.G. 2012. Cross-biome metagenomic analyses of soil microbial communities and their functional attributes. *Proceedings of the National Academy of Sciences in the United States of America* 109:21390–21395.

Fierer, N., Strickland, M.S., Liptzin, D., Bradford, M.A., Cleveland, C.C. 2009. Global patterns in belowground communities. *Ecology Letters* 12:1238–1249.

Frumkin, A., Stein, M. 2004. The Sahara—East Mediterranean dust and climate connection revealed by strontium and uranium isotopes I a Jerusalem speleothem. *Earth and Planetary Science Letters* 217: 451–464.

Ganor, E. 1975. Atmospheric dust in Israel. Sedimentological and meteorological analysis of dust deposition. Thesis. Hebrew University, Jerusalem.

Ganor, E., Foner, H.A. 2001. Mineral dust concentrations, deposition fluxes and deposition velocities in dust episodes over Israel, *Journal of Geophysical Research* 106: 18431–18437.

Ganor, E., Foner, H. A., Brenner, S., Neeman, E., Lavi, N. 1991. The chemical composition of aerosols settling in Israel following dust storms. *Atmospheric Environment* 25A: 2665–2670.

Ganor, E., Stupp, A. Alpert, P. 2009. A method to determine the effect of mineral dust aerosols on air quality. *Atmospheric Environment* 43: 5463–5468.

Giller, K.E., Beare, M.H., Lavelle, P., Izac, A.M.N., Swift, M.J. 1997. Agricultural intensification, soil biodiversity and agroecosystem function. *Applied Soil Ecology* 6:3–16.

Goldberg, N., Nachshon, U., Argaman, E., Ben-Hur, M. 2020. Short term effects of livestock manures on soil structure stability, runoff and soil erosion in semi-arid soils under simulated rainfall. *Geosciences* 10:213.

Goldshlger, N., Mirlas, V., Ben Dor, E., Eshel, M. 2006. New approach for localization, prediction, and management of saline-infected soils. In 46th Congress of the European Regional Science Association: Enlargement, Southern Europe and the Mediterranean. August 30 to September 3rd, 2006. Volos, Greece. http://hdl.handle.net/10419/118166

Griffiths, B.S. 1994. Soil nutrient flow. In: *Soil Protozoa*. Darbyshire, J., ed., p. 65–91. CAB International, Wallingford, Oxon, UK.

Gupta, S.K. 2002. A century of subsurface drainage research in India. *Irrigation and Drainage Systems* 16, 69–84

Haberern, J. 1992. Viewpoint: A soil health index. *Journal of Soil and Water Conservation* 47: 6.

Hanson, B., Grattan, SR, Fulton, A. 1999. Agricultural salinity and drainage. Published in the United States of America by the Department of Land, Air and Water Resources, University of California, Davis, CA.

Harris, D.G., van Bavel. C.H.M. 1957. Root respiration of tobacco, corn and cotton plants. *Agronomy Journal* 49:182–184.

Hendrix, P.F., Parmelee, R.W., Crossley Jr., D.A., Coleman, C.D., Odum, E.P., Groffman, P. 1986. Detritus food webs in conventional and no-tillage agroecosystems. *Bioscience* 36:374–380.

Herut, B. 1992. The chemical composition and sources of dissolved salts in rainwater in Israel. Thesis. Hebrew Univ., Jerusalem.

Hillel, D. 2003. *Introduction to environmental soil physics*. Elsevier Academic Press London.

Hinsinger, P., Bengough, G., Vetterlein, D., Young. I.M. 2009. Rhizosphere: Biophysics, biogeochemistry and ecological relevance. *Plant and Soil* 321:117–152.

Houk, E., Frasier, M., Schuck, E. 2006. The agricultural impacts of irrigation induced waterlogging and soil salinity in the Arkansas Basin. *Journal of Agricultural Water Management* 85: 175–183. doi: 10.1016/j.agwat.2006.04.007.

Jangid, K., Williams, M.A., Franzluebbers, A.J., Sanderlin, J.S., Reeves, J.H., Jenkins, M.B., Endale, D.M., Coleman, D.C., Whitman. W.B. 2008. Relative impacts of land-use, management intensity and fertilization upon soil microbial community structure in agricultural systems. *Soil Biology and Biochemistry* 40:2843–2853.

Kaman, H., Kirda, C., Cetin, M., Topcu, S. 2005. Salt accumulation in the root zones of tomato and cotton irrigated with partial root-drying technique. *Journal of Irrigation and Drainage Engineering* 55: 533–544. doi: 10.1002/ird.276

Karlen, D.L., Mausbach, M.J., Doran, J.W., Cline, R.G., Harris, R.F., Schuman, G.E. 1997. Soil quality: A concept, definition, and framework for evaluation (a guest editorial). *Soil Science Society of America Journal* 61:4–10.

Kim, M., Or, D. 2016. Individual-based model of microbial life on hydrated rough soil surfaces. *PLoS One* 11: e0147394.

Kishcha, P., Volpov, E., Starobinets, B., Alpert, P. and S. Nickovic. 2020. Dust dry deposition over Israel. *Atmosphere* 11: 197–214.

Korolyuk, T. V. 2014. Specific features of the dynamics of salts in salt affected soils subjected to long term seasonal freezing in the south Transbaikal Region. *Journal of Eurasian Soil Science* 47: 339–352 doi: 10.1134/S1064229314050093.

Kuzyakov, K., Blagodatskaya, E. 2015. Microbial hotspots and hot moments in soil: Concept & review. *Soil Biology and Biochemistry* 83:184–199.

Lavelle P., Lattaud, C., Trigo, D., Barois, I. 1995. Mutualism and biodiversity in soils. *Plant and Soil* 170:23–33.

Lennon, J.T., Jones, S.E. 2011. Microbial seed banks: The ecological and evolutionary implications of dormancy. *Nature Reviews Microbiology* 9:119–130.

Li, B. 2010. Soil salinization. In: *Desertification and its control in China*. Ci, L. and X. Yang, eds., Springer, Berlin, Heidelberg. 263-298. doi: 10.1007/978-3-642-01869-5

Li, Y., Chang, S.X., Tian, L,. Zhang, Q. 2018. Conservation agriculture practices increase soil microbial biomass carbon and nitrogen in agricultural soils: A global meta-analysis. *Soil Biology and Biochemistry* 121:50–58.

Litalien, A., Zeeb, B. 2020. Curing the earth: A review of anthropogenic soil salinization and plant-based strategies for sustainable mitigation. *Science of the Total Environment* 698. doi: 10.1016/j.scitotenv.2019.134235.

Liu, S., Feng, A., Du, J., Xia, D., Li, P., Xue, Z., Hu, W., Yu, X. 2014. Evolution of the buried channel systems under the modern Yellow River Delta since the Last Glacial Maximum. *Journal of Quaternary International* 349: 327–338. doi: 10.1016/j.quaint.2014.06.061.

Maggiotto, G., Sabatté, L., Marina, T.I., Fueyo-Sánchez, L., Londoño, A.M.R., Porres, M.D., Rionda, M., Domínguez, M., Perelli, R., Momo, F.R. 2019. Soil fauna community and ecosystem's resilience: A food web approach. *Acta Oecologica* 99:103445.

Mandal, U.K., Warrington, D.N., Bhardwaj, A.K., Bar-Tal, A.. Kautsky, L., Minz, D., Levy, G.J. 2008. Evaluating impact of irrigation water quality on a calcareous clay soil using principal component analysis. *Geoderma* 144:189–197.

Matson, P.A., Parton, W.J., Power, A.G., Swift, M.J., 1997. Agricultural intensification and ecosystem properties. *Science* 277:504–509.

McDaniel, M.D., Tiemann, L.K., Grandy, A.S. 2014. Does agricultural crop diversity enhance soil microbial biomass and organic matter dynamics? A meta-analysis. *Ecological Applications* 24:560–570.

Milnes, A.R. 1992. Calcrete. In: I.P. Martini, W. Chesworth, editors, *Developments in earth surface processes*, Vol. 2. Elsevier. pp. 309–347. https://doi.org/10.1016/B978-0-444-89198-3.50018-0.

Mirlas, V. 2012. Assessing soil salinity hazard in cultivated areas using MODFLOW model and GIS tools: A case study from the Jezre'el valley, Israel. *Journal of Agriculture Water Management* 109: 144–154. doi: 10.1016/j.agwat.2012.03.003.

Mirlas, V., Anker, Y., Aizenkod, A., Goldshleger, N. 2020. Soil salinization risk assessment owing to poor water quality drip irrigation: A case study from an olive plantation at the arid to semi-arid Beit She'an Valley, Israel, Geoscientific Model Development Discussions. p. 1–31. doi: 10.5194/gmd-2020-231.

Mitchell, P.J., Simpson, A.J., Soong, R., Oren, A., Chefetz, B. Simpson, M.J. 2013. Solution-state NMR investigation of the sorptive fractionation of dissolved organic matter by alkaline mineral soils. *Environmental Chemistry* 10: 333–340.

Mukherjee, A., & Lal, R. (2014). Comparison of soil quality index using three methods. *PLoS ONE,* 9(8), e105981. https://doi.org/10.1371/journal.pone.0105981.

Nachshon, U. 2018. Cropland soil Saliniza tion and associated hydrology: Trends, processes and examples. *Water* 10: 1–21. https://www.mdpi.com/2073-4441/10/8.

Neher, D. A. (1999). Soil community composition and ecosystem processes: Comparing agricultural ecosystems with natural ecosystems. *Agroforestry Systems*, 45, 159–185.

Nielsen, U.N., Wall, D.H., Six, J. 2015. Soil biodiversity and the environment. *Annual Review of Environment and Resources* 40:63–90.

Or, D., Phutane, S., Dechesne, A. 2007b. Extracellular polymeric substances affecting pore-scale hydrologic conditions for bacterial activity in unsaturated soils. *Vadose Zone Journal* 6:298–305.

Or, D., Smets, B.F., Wraith, J.M., Dechesne, A., Friedman, S.P. 2007a. Physical constraints affecting bacterial habitats and activity in unsaturated porous media—A review. *Advances in Water Resources* 30:1505–1527.

Oren, A., and Steinberger, Y. 2008. Coping with artifacts induced by $CaCO_3$–$CO_2$–$H_2O$ equilibria in substrate utilization profiling of calcareous soils. *Soil Biology and Biochemistry* 40:2569–2577.

Orgiazzi, A., Bardgett, R.D. Barrios, E. 2016.*Global soil biodiversity atlas.* Luxembourg: European Union.

Paul, E.A., Clark, F.E. 1996. *Soil microbiology and biochemistry.* Academic Press, San Diego.

Paul, E. A., Follett, R. F., Leavitt, S. W., Halvorson, A., Peterson, G. A., Lyon, D. J. (1997). Radiocarbon dating for determination of soil organic matter pool sizes and dynamics. *Soil Science Society of America Journal*, 61, 1058–1067.

Plummer, L.N., Prestemon, E.C. and D. L. Parkhurst. 1994. Origin of sodium bicarbonate waters in the Atlantic Coastal Plain. In: An interactive code NETPATH for modeling net geochemical reactions along a flow path. U.S. Geological Survey. Water-Resources Investigation Report 94-4169. p. 58-71.

Prietzel, J., Müller, S., Kögel-Knabner, I., Thieme, J. 2018. Comparison of soil organic carbon speciation using C NEXAFS and CPMAS 13C NMR spectroscopy. *Science of the Total Environment* 628–629:906–918.

Purkey, D.R., Wallender, W.W. 2001. Drainage reduction under land retirement over a shallow water table. *Journal of Irrigation and Drainage Engineering* 127: 1–7. doi: 10.1061/(ASCE)0733-9437(2001)127:1(1).

Rajaniemi, T.K., Allison, V.J. 2009. Abiotic conditions and plant cover differentially affect microbial biomass and community composition on dune gradients. *Soil Biology and Biochemistry* 41:102–109.

Ramos, T.B., Šimunek, J., Gonçalves, M.C., Martins, J.C., Prazeres, A., Castanheira, N.L., Pereira, L.S. 2011. Field evaluation of a multicomponent solute transport model in soils irrigated with saline waters. *Journal of Hydrology* 407: 129–144. doi: 10.1016/j.jhydrol.2011.07.016.

Rath, K.M., Maheshwari, A., Bengtson, P., Rousk, J. 2016. Comparative toxicities of salts on microbial processes in soil. *Applied and Environmental Microbiology* 82:2012–2020.

Rath, K.M., Maheshwari, A., Rousk, J. 2017. The impact of salinity on the microbial response to drying and rewetting in soil. *Soil Biology and Biochemistry* 108:17–26.

Rath, K.M., Murphy, D.N., Rousk, J. 2019. The microbial community size, structure, and process rates along natural gradients of soil salinity. *Soil Biology and Biochemistry* 138:107607.

Reeves, C.C., Jr. 1976. *Caliche: Origin, classification, morphology and uses.* Lubbock, Texas: Estacado Books. 233 pp.

Rhoades, J.D. and J.D. Oster 1986. Solute content. In A. Klute, Ed. *Methods of soil analysis*, 2. Agronomy Monograph 9, ASA and SSSA, Madison, WI, p. 985–1006.

Rinot, O., Rotbart, N., Borisover, M., Bar-Tal, A., Oren, A. 2021. Proteinaceous and humic fluorescent components in chloroform-fumigated soil extracts: Implication for microbial biomass estimation. Soil Research 59, 373–382. (in press: doi.org/10.1071/SR20205).

Rousk, J., Baath, E., Brookes, P.C., Lauber, C.L., Lozupone, C., Caporaso, J.G., Knight, R., Fierer, N. 2010. Soil bacterial and fungal communities across a pH gradient in an arable soil. *ISME Journal* 4:1340–1351.

Ruppel, S., Torsvik, V., Daae, F.L., Øvreås, L., Ruhlmann, J. 2007. Nitrogen availability decreases prokaryotic diversity in sandy soils. *Biology and Fertility of Soils* 43:449–459.

Sánchez-Moreno, S., Ferris, H. 2007. Suppressive service of the soil food web: Effects of environmental management. *Agriculture, Ecosystems and Environment* 119:75–87.

Sardinha, M., Müller, T., Schmeisky, H., Joergensen, R.G. 2003. Microbial performance in soils along a salinity gradient under acidic conditions. *Applied Soil Ecology* 23:237–244.

Sawdy, A., Heritage, A., Pel, L. 2008. *A review of salt transport in porous media, assessment methods and salt reduction treatments.* SWBSS, Copenhagen. https://www.researchgate.net/publication/254875494.

Schlesinger, W.H., Raikes, J.A., Hartley, A.E., Cross, A.F. 1996. On the spatial pattern of soil nutrients in desert ecosystems. *Ecology* 77:364–374.

Schlüter, S., Eickhorst, T., Mueller, C.W. 2019. Correlative imaging reveals holistic view of soil microenvironments. *Environmental Science and Technology* 53:829–837.

Seaton, F.M., George, P.B.L., Lebron, I., Jones, D.L., Creer, S., Robinson, D.A. 2020. Soil textural heterogeneity impacts bacterial but not fungal diversity. *Soil Biology and Biochemistry* 144:107766.

Seeboonruang, U. 2012. Impacts of reservoir on groundwater level and quality in a saline area, Nakhon Panom Province, Thailand, *APCBEE Procedia* 4: 16–21. doi: 10.1016/j.apcbee.2012.11.004.

Setälä, H., Huhta, V. 1991. Soil fauna increase Betula pendula growth: Laboratory experiments with coniferous forest floor. *Ecology* 72:665–671.

Setälä, H., Tyynismaa, M., Martikainen, E., Huhta, V. 1991. Mineralisation of C, N and P in relation to decomposer community structure in coniferous forest soil. *Pedobiologia* 95:285–296.

Setia, R., Gottschalk, P., Smith, P., Marschner, P., Baldock, J., Setia, D., Smith, J. 2013. Soil salinity decreases global soil organic carbon stocks. *Science of the Total Environment* 465:267–272.

Setia, R., Rengasamy, P., Marschner, P. 2014. Effect of mono- and divalent cations on sorption of water-extractable organic carbon and microbial activity. *Biology and Fertility of Soils* 50:727–734.

Setia, R., Smith, P., Marschner, P., Gottschalk, P., Baldock, J., Verma, V., Setia, D., Smith, J. 2011. Simulation of salinity effects on past, present, and future soil organic carbon stocks. *Environmental Science and Technology* 46:1624–1631.

Shahid, S. 2012. Developments in soil salinity assessment, modeling, mapping, and monitoring from regional to submicroscopic scales. In: *Developments in soil salinity assessment and reclamation. Innovative thinking and use of marginal soil and water resources in irrigated agriculture*. S.A. Shahid, M.A. Abdelfattah, F.K. Taha, eds. Springer Dordrecht Heidelberg New York London, p. 3–41. doi: 10.1007/978-94-007-5684-7.

Sharma, P., Laor, Y., Raviv, M., Medina, S., Saadi, I., Krasnovsky, A., Vager, M., Levy, G.J., Bar-Tal, A., Borisover, M. 2017. Compositional characteristics of organic matter and its water-extractable components across a profile of organically managed soil. *Geoderma* 86:73–82.

Shvets, A.P. 1988. The regularities of water transfer in the unsaturated zone of the Crimea Plain. Thesis. Institute of Geological Sciences, Kiev, Ukraine.

Sierra, J., Renault, P. 1995. Oxygen consumption by soil microorganisms as affected by oxygen and carbon dioxide levels. *Applied Soil Ecology* 2:175–184.

Šimůnek, J., van Genuchten M. Th., Šejna, M. 2012. The HYDRUS Software Package for Simulating Two- and Three Dimensional Movement of Water, Heat, and Multiple Solutes in Variably Saturated Porous Media, Technical Manual, Version 2.0, PC Progress, Prague, Czech Republic.

Singh, B.K., Nunan, N., Millard, P. 2009. Response of fungal, bacterial and ureolytic communities to synthetic sheep urine deposition in a grassland soil. *FEMS Microbiology Ecology* 70:109–117.

Singh, K. 2016. Microbial and enzyme activities of saline and sodic soils. *Land Degradation and Development* 27:706–718.

Smith, J.L., Doran, J.W. 1996. Measurement and use of pH and electrical conductivity for soil quality analysis. In: J. W. Doran and A. J. Jones (eds.) *Methods for assessing soil quality (p. 169–185)*. Soil Science Society of America, Special Publication No. 49. SSSA, Madison, WI.

Soil Survey Division Staff. Soil survey manual. 1993. Chapter 3. Soil Conservation Service. U.S. Department of Agriculture Handbook 18.

Sonnenfeld, P. 1984. *Brines and Evaporites*. Elsevier. 614p.

Sparks, D.L. 2002. *Environmental soil chemistry*. Elsevier. Academic Press. 352p.

Standing, D. and K. Killham. 2006. The soil environment. pp. 1–22. In: J. D. Van Elsas, J.T. Trevors, and J.K. Jansson, eds., *Modern soil microbiology*. CRC Press, Boca Raton, FL.

Stotzky, G. 1997. Soil as an environment for microbial life. pp. 1–20. In: J. D. Van Elsas, J.T. Trevors, and E. Wellington, eds., *Modern soil microbiology*. Marcel Deeker, New York City, NY.

Svoray, T., Hassid, I., Atkinson, P.M., Moebius-Clune, B.N., van Es, H.M. 2015. Mapping soil health over large agriculturally important areas. *Soil Science Society of America Journal* 79:1420–1434.

Tanaka, T.Y., Chiba, M. 2006. A numerical study of the contributions of dust source regions to the global dust budget. *Global and Planet Change* 52:88–104. doi: 10.1016/j.gloplacha.2006.02.002.

Tanji, K. 2002. Salinity in the soil environment. In: "*Salinity: Environment—plants—molecules*", A. Läuchli and U. Lüttge, eds., 21–51. Kluwer Academic Publishers. doi: 10.1007/0-306-48155-3_2.

Tecon, R., Ebrahimi, A., Kleyer, H., Levi, S.E., Or, D. 2018. Cell-to-cell bacterial interactions promoted by drier conditions on soil surfaces. *Proceedings of the National Academy of Sciences in United States of America* 115:9791–9796.

Tecon, R, Or, D. 2017. Biophysical processes supporting the diversity of microbial life in soil. *FEMS Microbiology Reviews* 41:599–623.

Terrat, S., W. Horrigue, S. Dequietd, N.P.A. Saby, M. Lelièvre, V. Nowak, J. Tripied, T. Régnier, Jolivet, C., Arrouays, D., Wincker, P., Cruaud, C., Karimi, B., Bispo, A., Maron, P.A., Prévost-Bouré, N.C., Ranjard, L. 2017. Mapping and predictive variations of soil bacterial richness across France. *PLoS One* 12: e0186766.

Thomsen, N., Faber, J.H., Sorensen, P.B. 2012. Soil ecosystem health and services—Evaluation of ecological indicators susceptible to chemical stressors. *Ecological Indicators* 16:67–75.

Torsvik, V., Øvreås, L., Thingstad, T.F. 2002. Prokaryotic diversity—Magnitude, dynamics, and controlling factors. *Science* 296:1064–1066.

Totsche, K.U., Rennert, T., Gerzabek, M.H., Kögel-Knabner, I., Smalla, K., Spiteller, M., Vogel, H.-J. 2010. Biogeochemical interfaces in soil: The interdisciplinary challenge for soil science. *Journal of Plant Nutrition and Soil Science* 173:88–99.

Troeh, F.R., Thompson, L.M. 1993. *Soils and soil fertility*. Oxford University Press, New York.

Van Bruggen, A.H.C., Semenov, A.M. 2000. In search of biological indicators for soil health and disease suppression. *Applied Soil Ecology* 15:13–24.

Van der Wal, A., Geerts, R.H.E.M., Korevaar, H., Schouten, A.J., Jagersop Akkerhuis, G.A.J.M., Rutgers, M., Mulder, C. 2009. Dissimilar response of plant and soil biota communities to long-term nutrient addition in grasslands. *Biology and Fertility of Soils* 45:663–667.

Vos, M., Wolf, A. B., Jennings, S. J., & Kowalchuk, G. A. (2013). Micro-scale determinants of bacterial diversity in soil. *FEMS Microbiology Reviews*, 37, 936–954.

Wagenet, R.J., Jury, W.A. 1984. Movement and accumulation of salts in soils. In *Soil salinity under irrigation. Processes and management*. Ed. I. Shainberg and J. Shalhevet. Springer-Verlag Berlin Heidelberg. P. 100–130. doi: 10.1007/978-3-642-69836-1.

Wang, G., Or, D. 2010. Aqueous films limit bacterial cell motility and colony expansion on partially saturated rough surfaces. *Environmental Microbiology* 12: 1363–1373.

Watteau, F., Villemin, G. 2018. Soil microstructures examined through transmission electron microscopy reveal soil-microorganisms interactions. *Frontiers in Environmental Science* 6:1–10.

Weisbrod N., Nachshon U., Dragila M., Grader A. 2014. Micro-CT analysis to explore salt precipitation impact on porous media permeability. In: Mercury L., Tas N., Zilberbrand M., eds., *Transport and reactivity of solutions in confined Hydrosystems*. NATO Science for Peace and Security Series C: Environmental Security. Springer, Dordrecht. P. 231–241. doi: 10.1007/978-94-007-7534-3_20.

Whittig, L.D. and Janitzky, P. 1963. Mechanisma of formation of sodium carbonate in soil. *European Journal of Soil Science* 14: 322–333.

Wittebolle, L., Marzorati, M., Clement, L., Balloi, A., Daffonchio, D., Heylen, K., De Vos, P., Verstraete, W., Boon, N. 2009. Initial community evenness favours functionality under selective stress. *Nature* 458:623–626.

Wua, M., Wua, J., Tana, X., Huanga, J., Jansson P-E., Zhang, W. 2019. Simulation of dynamical interactions between soil freezing/thawing and salinization for improving water management in cold/arid agricultural region. *Geoderma* 338: 325–342. doi: 10.1016/j.geoderma.2018.12.022.

Xiong B, Chen R, Xia Z, Ye C, Anker Y. 2020. Large-scale deforestation of mountainous areas during the 21st century in Zhejiang Province. *Land Degradation & Development* 2015:1–14. doi: 10.1002/ldr.3563.

Xue R, Wang C, Liu M, Zhang D, Li K, Li N. 2019. A new method for soil health assessment based on analytic hierarchy process and meta-analysis. *Science of the Total Environment* 650:2771–2777. doi: 10.1016/j.scitotenv.2018.10.049.

Yan, N., Marschner. P. 2012. Response of microbial activity and biomass to increasing salinity depends on the final salinity, not the original salinity. *Soil Biology and Biochemistry* 53:50–55.

Yechieli, Y. 1993. The effects of water level changes in closed lakes (Dead Sea) on the surrounding groundwater and country rocks. Thesis. Weizmann Institute of Science. https://weizmann.primo.exlibrisgroup.com/discovery/delivery/972WIS_INST:-972WIS_V1/1229661480003596.

Yu, W., Brookes, P.C., Ma, Q., Zhou, H., Xu, Y., Shen, S. 2011. Extraction of soil nitrogen by chloroform fumigation—A new index for the evaluation of soil nitrogen supply. *Soil Biology and Biochemistry* 43:2423–2426.

Zhang, C., Li, L., Lockington, D. 2014. Numerical study of evaporation-induced salt accumulation and precipitation in bare saline soils: Mechanism and feedback. *Water Resources Research* 50: 8084–8106. doi: 10.1002/2013WR015127.

Zhang, K., Shi, Y., Cui, X., Yue, P., Kaihui, L., Liu, X., Tripathi, B.M., Chu, H. 2019. Salinity is a key determinant for soil microbial communities in a desert ecosystem. *mSystems* 4: e00225-18.

Zhou, J., Xia, B., Treves, D.S., Wu, L.Y., Marsh, T.L., O'Neill, R.V., Palumbo, A.V., Tiedje, J.M. 2002. Spatial and resource factors influencing high microbial diversity in soil. *Applied and Environmental Microbiology* 68:326–334.

Zilberbrand, M. 1995. The effect of carbonates and gypsum precipitation in the root zone on the chemical composition of groundwater. *Journal of Hydrology* 171: 5–22.

# 7

# Nutrient Availability: Macro- and Micronutrients in Soil Quality and Health

*James A. Ippolito*

## Introduction

### Macronutrients

Macronutrients (C, H, O, N, P, K, Ca, Mg, and S) are required by plants in relatively large quantities. The average concentrations of C, H, O, N, P, K, Ca, Mg, and S in plants are 45, 6, 45, 1.5, 0.2, 1.0, 0.5, 0.2, and 0.2%, respectively (dry weight basis; Havlin et al., 2014). When these elements are lacking in soils, this leads to decreased plant or crop yields with that yield proportional to the most limiting nutrient; this is the basis of Justus von Liebig's Law of the Minimum. Thus, it is important to not only understand soil macronutrient concentrations, but how to interpret those concentrations from the standpoint of a soil quality and health perspective. It is relatively easy to imagine that the assumption of "less is better" is the incorrect approach for interpreting soil macronutrient concentrations for soil quality and health determination. Instead, one might be under the assumption that a "more is better" approach is correct for interpreting soil macronutrient concentrations in lieu of soil quality and health determinations. For example, The Soil Management Assessment Framework uses a "more is better" approach for extractable K, and this is likely not entirely the correct approach, as soil systems that contain excessive K could lead to plants with N or Cl deficiencies. The Framework also uses the "more is better" approach for soil C, yet if excessive manure applications have led to substantial increases in soil C, excessive manure

*Soil Health Series: Volume 2 Laboratory Methods for Soil Health Analysis*, First Edition.
Edited by Douglas L. Karlen, Diane E. Stott, and Maysoon M. Mikha.

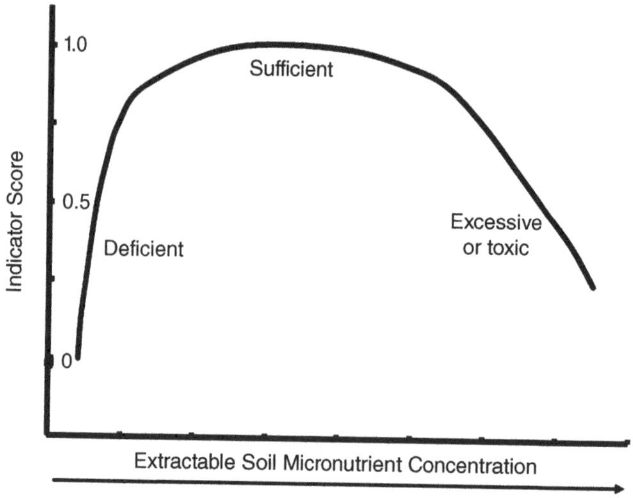


**Figure 7.1** Hypothetical function describing increasing soil-extractable macro- and micronutrient concentrations as related to indicator score.

application more than likely will lead to detrimental effects with regards to other nutrients (e.g., excessive N, P, K).

In all likelihood, the approach of "somewhere in the middle is better" is the correct approach for soil macronutrient concentrations and their interpretation into soil quality and health functions (a pseudo-Gaussian-type function; Fig. 7.1). A perfect example found in the Soil Management Assessment Framework is that of soil extractable P. The soil extractable P concentration interpretation follows the "somewhere in the middle is best" approach, relying on the fact that too little P limits plant growth while too much P is detrimental to the environment.

With the above stated, the main issue with the "more macronutrient(s) is better" approach is the fact that "more" is a relative term, and excess of any nutrient leads to either plant toxicities due to that particular nutrient, or the excessive nutrient triggers plant deficiencies with regards to other nutrients (e.g., excessive P may lead to Zn deficiency symptoms). Furthermore, the main issue with the "more is better" approach is pinpointing exactly where excessive nutrient concentrations limit plant productivity and thus have a detrimental effect on soil quality and health. For certain macronutrients (e.g., P) this seems relatively straightforward; for others (e.g., C), secondary consequential effects on soil quality need to be considered. Pinpointing where detrimental macronutrient effects occur is no easy task, as it varies by climatic conditions, soil conditions, crops and varieties grown, amount of macronutrient present, etc. Finally, if one is working with "normal", arable soils (e.g., soils that are not contaminated with excess nutrients), then for most macronutrients a "more is better" approach is likely warranted. If one is working with soils that have been contaminated with excess macronutrients (e.g., due to excess manure applications, fertilizer spills, etc.), the "somewhere in the middle" approach for interpreting soil quality/health should be considered.

## Micronutrients

Micronutrients are as necessary for plant growth as macronutrients, and as with macronutrients, micronutrient deficiencies or toxicities can limit plant productivity. On an average dry weight basis, plants require micronutrients at concentrations less than 100 mg $kg^{-1}$ (Table 7.1). In soils, total micronutrient concentrations range from 0.2 to over 500,000 mg $kg^{-1}$, yet total micronutrient concentrations do not indicate micronutrient availability. Micronutrient concentrations found in the soil solution, from which plants attain micronutrients, range from $< 0.0001$ to 10 mg $L^{-1}$ (Table 7.1). Soil solution micronutrient concentrations and availability are affected by several factors including soil pH, clay and organic matter content, soil (oxyhydr)oxide and carbonate content, soil water, temperature, etc. Because these factors all play a role in binding, sorbing, precipitating, and desorbing micronutrients, a fine line exists between micronutrient deficiency and sufficiency, and in some cases toxicity. Thus, it is imperative to quantify soil-extractable micronutrient concentrations to not only generate fertilizer recommendations, but also to quantify changes in soil quality and health due to land management practices. Although they have yet to be developed, it is easy to envision the functions describing changes in soil quality associated with changing micronutrient concentrations as following a "somewhere in the middle is better" approach (Fig. 7.1).

As soil and plant scientists, we are aware that the line between micronutrient sufficiency and toxicity to plants is ill-defined. Issues arise when targeting the area between sufficiency and toxicity in Fig. 7.1, and as with macronutrients, this portion of the curve is a function of plant species, variety, soil type, etc. For example, Ippolito et al. (2010, 2011) grew either corn or alfalfa in the same soil (Typic Calciaquoll) with increasing Cu applications. Based on plant growth as well as effects on the soil microbial community, thresholds between sufficiency and toxicity were recommended to be 130 and 63 mg extractable-Cu $kg^{-1}$ soil, respectively (quite a range). Moore et al. (2013) grew potatoes in a sandy or silt loam soil, affected by increasing Cu applications. Based on plant growth, thresholds between sufficiency and toxicity were recommended at 25 and 90 mg extractable Cu $kg^{-1}$ soil, respectively (again, quite a range). Kabata–Pendias (2011) presented a range of the greatest plant Cu, Ni, and Zn concentrations that would suppress growth

**Table 7.1** Average micronutrient concentrations in plants and the range of total micronutrient concentrations in soils (adapted from Havlin et al., 2014; Adriano, 2001; Lindsay and Norvell, 1978).

| Micronutrient | Average Plant Concentration | Common Range for Total Soil Micronutrient Concentrations | Common Range for Soil Solution Micronutrient Concentrations | Comments and Concerns |
|---|---|---|---|---|
| | (mg $kg^{-1}$; dry wt. basis) | (mg $kg^{-1}$) | (mg $L^{-1}$) | |
| Cl | 100 | 20 to 900 | Very low (see comments and concerns) | Highly mobile in soils. |
| Fe | 100 | 7000 to 550,000 | < 0.0001 to 0.06 | Phytotoxicity in acidic soils. |
| Mn | 50 | 20 to 3000 | 0.01 to 1 | Wide margin for plant growth. Phytotoxicity in acidic soils. |
| B | 20 | 2 to 100 | ~ 0.1 | Narrow margin for plant deficiency/sufficiency/toxicity. Phytotoxicity more prone in arid environments. |
| Zn | 20 | 60 to 2000 | 0.002 to 0.07 | Wide margin for plant growth. Easily complexed and relatively immobile in soils. |
| Cu | 6 | 2 to 100 | 0.0006 to 0.06 | Narrow margin for plant growth. Easily complexed in soils by clays, organic matter. Relatively immobile in soils. |
| Ni | 0.1 | 5 to 500 | Very low (see comments and concerns) | Highly mobile in soils and plants. |
| Mo | 0.1 | 0.2 to 5 | < 0.0001 to 10 | Can be enriched in plants, but toxicities are typically associated with feeding animals plants containing excess Mo. |

(15–20, 20–30, and 150–200 mg $kg^{-1}$, respectively) or reduce yield by 10% (10–30, 10–30, and 100–500 mg $kg^{-1}$, respectively). It is quite clear that pinpointing exactly where toxicity effects occur in plants is not straightforward.

Unlike differences between sufficiency and toxicity, the fine line between micronutrient deficiency and sufficiency is reported in numerous U.S. state fertilizer or plant tissue test recommendations. Thus, targeting the area between deficiency and sufficiency in Fig. 7.1 should be relatively easy in terms of quantifying micronutrients and soil quality and health. A more comprehensive literature review of the fine line between micronutrient deficiency and sufficiency, and the extractants used to identify micronutrient concentrations, are presented in Table 7.2. One could envision Table 7.2 information being used as the starting point for creating a soil micronutrient–soil quality and health framework.

## Soil Nutrient Availability Measurements

Some soil extractants used are relatively universal for determining nutrient concentrations. For example, the use of 2M KCl as an extractant for determining soil $NO_3$–N and $NH_4$–N concentrations are used globally and without regard for soil pH. Also, soil B availability is typically measured using a hot water extraction, regardless of soil pH. However, soil pH plays a major role in determining mineral species present (e.g., soils dominated by (hydr)oxides versus carbonates) and thus which form(s) of macro- and micronutrients are present and available for plants. Scientists and testing facilities typically do not use one extractant for determining most macro- and micronutrient concentrations in soils, regardless of pH. An example of using various extractants for various soils is illustrated in Table 7.2. The following sections below attempt to briefly describe extractants utilized and why they are used based on soil pH.

### Acid to Neutral, Non-Calcareous Soils

Soil macro- and micronutrient concentrations determined in acid to neutral, noncalcareous soils (up to ~ pH 7.2) typically use multielement or single extractants developed specifically for these soils. For example, the Mehlich-1 and Mehlich-3 extractants were developed to estimate both macro- and micronutrient soil concentrations, and both are currently used by eastern and southeastern U.S. states. These extractants use weak acids (e.g., Mehlich-1, 0.05M HCl + 0.0125M $H_2SO_4$; Mehlich-3, 0.2M $CH_3COOH$ + 0.013M $HNO_3$; Reed and Martens, 1996) to dissolve a fraction of the nutrients associated with oxide and hydroxide soil mineral phases (e.g., Al-P, Fe-P). In addition, some extractants utilize additional chemicals to both aid in nutrient release from soil and prevent those nutrients from re-reacting with soil particles and re-precipitating. Specifically, the Mehlich-3 utilizes weak acids to partially dissolve Al–P phases, and to prevent re-precipitation, 0.015M $NH_4F$ is added to the extractant.

**Table 7.2** US state by state soil extractants used to determine soil micronutrient deficiency concentration or state by state micronutrient fertilizer recommendations for various crops (N/A = not available).

| State | Soil Extractant | Soil Micronutrient Concentration Below Which is Considered Deficient | Link (all verified January 26, 2019) |
|---|---|---|---|
| Alabama | Mehlich-1 | For reseeding clover or clover seed harvest: apply 1 to 1.5 lb B $acre^{-1}$. | http://www.aces.edu/anr/soillab/forms/documents/ay-324B.pdf |
| | Mehlich-1 | For cotton: apply 0.3 lbs B $acre^{-1}$ in the fertilizer or in the insecticide spray. | " |
| | Mehlich-1 | for (non)irrigated (sweet) corn: on sandy soils apply 3 lb B $acre^{-1}$ in fertilizer or after liming where pH is above 6.0. | " |
| | Mehlich-1 | For peanuts: apply 0.3 to 0.5 lb B $acre^{-1}$ in the macronutrient fertilizer, gypsum, or disease control spray or dust. | " |
| | Mehlich-1 | For alfalfa: apply 3 lb B $acre^{-1}$ annually. | " |
| | Mehlich-1 | For soybean: on all soils of northern Alabama and on fine-textured, acid soils in other areas of Alabama, apply the equivalent of 1 ounce per acre of sodium molybdate or ammonium molybdate to the seed at planting. | " |
| | | | The state may also follow guidelines found here: http://aesl.ces.uga.edu/sera6/PUB/MethodsManualFinalSERA6.pdf |
| Alaska | N/A | For vegetables: apply 1 lb B $acre^{-1}$ | https://www.uaf.edu/ces/agriculture/soil/ |
| | N/A | For vegetables: 1–2 lb Cu $acre^{-1}$ banded or 4–8 lb Cu $acre^{-1}$ broadcast. | " |
| | N/A | For vegetables: 3 lb Mn $acre^{-1}$ as $MnSO_4$ or 0.5 lb Mn $acre^{-1}$ in chelated form if banded. | " |
| | N/A | For vegetables: 0.5–1.0 oz. Mo $acre^{-1}$ if banded or 0.5–5.0 oz. Mo $acre^{-1}$ if broadcast. | " |

| | | | |
|---|---|---|---|
| | N/A | For vegetables: 0.5–1.0 lb Zn $acre^{-1}$ in chelated form or 2–4 lb Zn $acre^{-1}$ as $ZnSO_4$ if banded, or 1–2 lb Zn $acre^{-1}$ in chelated form or 4–8 lb Zn $acre^{-1}$ as $ZnSO_4$ if broadcast. | " |
| Arizona | Saturated Paste Extract | For Alfalfa: B < 0.2 mg $kg^{-1}$ | http://www.ipni.net/publication/bettercrops.nsf/0/B3D37D5A6F4A11C2852581D00057AA17/$FILE/BC-2017–4%20p21.pdf |
| | N/A | For vegetable gardens: A soil test reading < 1.0 mg $kg^{-1}$ zinc may also indicate the need for additional zinc for these crops. When applying zinc to vegetable gardens, broadcast about 1 pound of zinc sulfate per 1000 sq. ft. | https://delange.org/Arizona_Vegetable_Garden_Fertilizer/Arizona_Vegetable_Garden_Fertilizer.htm |
| | N/A | For vegetable gardens: A soil test reading < 4 mg $kg^{-1}$ iron may indicate a low level of available iron in the soil. Applications of granular iron sulfate materials before planting or liquid iron products after planting can be helpful in supplying iron for plant growth. | " |
| Arkansas | N/A | for corn: apply 10 lb of Zn as a granular when Zn levels are < 4 mg $kg^{-1}$ and pH is > 6.0 | https://www.corn-sorghum.org/fertilizer-recommendations.html |
| | Mehlich-3 | For most row crops: Zn < 4.0 mg $kg^{-1}$; Cu < 1.0 mg $kg^{-1}$; Mn < 40 mg $kg^{-1}$. | https://www.uaex.edu/publications/pdf/FSA-2118.pdf |
| California | Saturated Paste Extract | For alfalfa: < 0.4 mg B $kg^{-1}$ | http://www.ipni.net/publication/bettercrops.nsf/0/B3D37D5A6F4A11C2852581D00057AA17/$FILE/BC-2017–4%20p21.pdf |
| | DTPA | For onion, corn, potatoes, and canning tomatoes: < 0.3 mg Zn $kg^{-1}$;<br>For other warm season vegetables: < 0.2 mg Zn $kg^{-1}$;<br>For other cool season vegetables: < 0.5 mg Zn $kg^{-1}$. | http://sfp.ucdavis.edu/pubs/Family_Farm_Series/Veg/Fertilizing/tests/ |

(Continued)

**Table 7.2** (Continued)

| State | Soil Extractant | Soil Micronutrient Concentration Below Which is Considered Deficient | Link (all verified January 26, 2019) |
|---|---|---|---|
| Colorado | AB-DTPA | For Corn: < 1.5 mg Zn $kg^{-1}$, and if pH > 7.8 and soil is calcareous, add 50–100 lb ferrous sulfate $acre^{-1}$ band application at planting | (Davis and Westfall, 2014) https://extension.colostate.edu/topic-areas/agriculture/fertilizing-corn-0–538/ |
| Connecticut | | | May follow Vermont micronutrient fertilizer recommendations. |
| Delaware | Mehlich-3 | Closely monitor grain corn when < 1.7 mg Mn $acre^{-1}$. Grain Corn and Barley and Grain Sorghum and Soybean and Wheat: Mn < 25 ppm and pH (1:1); Zn < 1.9 lb $acre^{-1}$ OR < 3.1 and pH > 7 OR | https://cdn.extension.udel.edu/wp-content/uploads/2017/01/12045621/AGR-Grain-Corn_Final.pdf |
| | " | < 1.0 mg Zn $kg^{-1}$, or < 1.6 mg Zn $kg^{-1}$ and pH > 7.0 | " |
| Florida | Mehlich-1 | 0.1–0.3 mg Cu $kg^{-1}$ and pH 5.5–5.9, or 0.3–0.5 mg Cu $kg^{-1}$ and pH 6.0–6.4, or 0.5 mg Cu $kg^{-1}$ and pH 6.5–7.0 | http://edis.ifas.ufl.edu/cv101 |
| | " | 3–5 mg Mn $kg^{-1}$ and pH 5.5–5.9, or 5–7 mg Mn $kg^{-1}$ and pH 6.0–6.4, or 7–9 mg Mg $kg^{-1}$ and pH 6.5–7.0 | " |
| Georgia | Hot Water | < 0.5 mg B $kg^{-1}$ | The state may also follow guidelines found here: http://aesl.ces.uga.edu/sera6/PUB/MethodsManualFinalSERA6.pdf |
| Hawaii | N/A | N/A | N/A |
| Idaho | DTPA | For Corn silage and grain: < 0.6 mg Zn $kg^{-1}$. Apply 10 lb Zn $acre^{-1}$ when soil test is < 0.6 ppm (Same for eastern OR and WA) | (Brown, Hart, Horneck, and Moore, 2010) http://www.cals.uidaho.edu/edComm/pdf/PNW/PNW0615.pdf |
| | DTPA | For Alfalfa: < 1.0 mg Zn $kg^{-1}$; < 0.2 mg Cu $kg^{-1}$ | (Koenig et al., 2009) http://cru.cahe.wsu.edu/CEPublications/PNW0611/PNW0611.pdf |

| | | | |
|---|---|---|---|
| Illinois | Hot Water | For Alfalfa: < 0.4–0.8 mg B $kg^{-1}$ depending on soil texture. | " |
| | Hot Water | < 1.0 mg B $kg^{-1}$ | http://extension.cropsciences.illinois.edu/handbook/pdfs/chapter08.pdf |
| Indiana | DTPA | < 4.0 mg Fe $kg^{-1}$; < 2.0 mg Mn $kg^{-1}$; < 1.0 mg Zn $kg^{-1}$. | " |
| | 0.1 N HCl | Varies between 2 and 24 mg Mn $kg^{-1}$ and pH between 6.3–7.5; varies between 1 and 12 mg Zn $kg^{-1}$ and pH 6.6–7.6. | https://www.agry.purdue.edu/ext/forages/publications/ay9–32.htm#Table 23. |
| | 1.0 N HCl | < 16 mg Cu $kg^{-1}$ | " |
| Iowa | DTPA | For corn: < 0.9 mg Zn $kg^{-1}$ | https://store.extension.iastate.edu/product/A-General-Guide-for-Crop-Nutrient-and-Limestone-Recommendations-in-Iowa |
| Kansas | DTPA | For corn, wheat and sorghum: < 1.0 mg Zn or B $kg^{-1}$, and < 45 mg Cl $kg^{-1}$. | https://www.bookstore.ksre.ksu.edu/pubs/MF2586.pdf |
| Kentucky | Hot Water | < 1.0 mg B $kg^{-1}$ | (Leikam, Lamond, and Mengel, 2003) http://aesl.ces.uga.edu/sera6/PUB/MethodsManualFinalSERA6.pdf |
| Louisiana | N/A | For sugarcane: < 2.25 mg Zn $kg^{-1}$ | https://www.lsuagcenter.com/profiles/lblack/articles/page1487777401591 |
| | Mehlich-3 | For rice: < 1.5 mg Zn $kg^{-1}$ | https://www.lsuagcenter.com/~/media/system/9/0/e/9/90e93160aba5daccea90c6d955299f74/5—chapter-3-soils-plant-fertilization.pdf |
| Maine | N/A | N/A | N/A |
| Maryland | Mehlich-1 | N/A | https://extension.umd.edu/sites/extension.umd.edu/files/_images/programs/anmp/SFM-1.pdf |

(*Continued*)

**Table 7.2** (Continued)

| State | Soil Extractant | Soil Micronutrient Concentration Below Which is Considered Deficient | Link (all verified January 26, 2019) |
|---|---|---|---|
| Massachusetts | N/A | Only lists tolerance to B, but not concentrations are presented. | https://nevegetable.org/cultural-practices/micronutrients |
| Michigan | Hot Water | $< 1.0$ mg B $kg^{-1}$ | https://soil.msu.edu/wp-content/uploads/2014/06/MSU-Nutrient-recomdns-field-crops-E-2904.pdf |
| | 0.1 N HCl | $< 6.0$ mg Mn $kg^{-1}$ at pH 6.3 and $< 12$ mg Mn $kg^{-1}$ at pH 6.7 | " |
| | 0.1 N HCl | $< 2$ mg Zn $kg^{-1}$ at pH 6.6 and $< 7$ mg Zn $kg^{-1}$ at pH 7.0 | " |
| | 1.0 M HCl | $< 0.5$ mg Cu $kg^{-1}$ | " |
| Minnesota | DTPA | For corn and edible beans: $< 0.75$ mg Zn $kg^{-1}$ | https://extension.umn.edu/micro-and-secondary-macronutrients/zinc-crop-production |
| | " | For crops grown in high organic soils, no soil test value is listed but soils with pH > 7.5 should be monitored closely. | https://extension.umn.edu/micro-and-secondary-macronutrients/copper-crop-production |
| | Hot Water | $< 1.0$ mg B $kg^{-1}$ | https://extension.umn.edu/micro-and-secondary-macronutrients/boron-minnesota-soils |
| Mississippi | N/A | $< 1.0$ mg Zn $kg^{-1}$ | https://www.mssoy.org/uploads/files/pub-2647-msu-ext.pdf The state may also follow guidelines found here: http://aesl.ces.uga.edu/sera6/PUB/MethodsManualFinalSERA6.pdf |
| Missouri | DTPA | For corn and sorghum: $< 1.0$ mg Zn $kg^{-1}$ | http://aes.missouri.edu/pfcs/soiltest.pdf |

| | | | |
|---|---|---|---|
| Montana | DTPA | < 5.0 mg Fe $kg^{-1}$; < 1.0 mg Mn $kg^{-1}$; < 0.5 mg Zn or Cu $kg^{-1}$ | http://landresources.montana.edu/soilfertility/documents/PDF/pub/FertGuidelMTCropsEB161.pdf (And Dr. Clain Jones, personal communication) |
| | Hot Water | < 1.0 mg B $kg^{-1}$ | " |
| Nebraska | DTPA | For corn, field beans, sweet corn: < 0.8 mg Zn $kg^{-1}$ | http://extensionpublications.unl.edu/assets/pdf/ec155.pdf |
| | DTPA | For most crops: < 4.5 mg Fe $kg^{-1}$ | " |
| | Hot Water | For most crops: < 0.25 mg B $kg^{-1}$ | " |
| | DTPA | For soybean: < 0.4 mg Zn $kg^{-1}$ | https://digitalcommons.unl.edu/cgi/viewcontent.cgi?referer = &httpsredir = 1&article = 2712&context = extensionhist |
| | DTPA | for sugarbeet: < 0.5 mg Zn $kg^{-1}$ | http://extensionpublications.unl.edu/assets/html/g1459/build/g1459.htm |
| Nevada | N/A | N/A | N/A |
| New Hampshire | N/A | N/A | N/A |
| New Jersey | Mehlich-3 | < 1.0 mg Zn $kg^{-1}$ | https://njaes.rutgers.edu/pubs/publication.php?pid = FS174 |
| | Hot Water | < 0.75 mg B $kg^{-1}$ | " |
| New Mexico | DTPA | < 4.5 mg Fe $kg^{-1}$; < 2.5 mg Mn $kg^{-1}$; < 1.0 mg Zn or Cu $kg^{-1}$ | https://aces.nmsu.edu/ces/mastergardeners/manual/docs/chap_1/chap1.f.pdf |
| | Hot Water | B, although not soil test concentration is listed. | " |

(*Continued*)

**Table 7.2** (Continued)

| State | Soil Extractant | Soil Micronutrient Concentration Below Which is Considered Deficient | Link (all verified January 26, 2019) |
|---|---|---|---|
| New York | Morgan | $< 0.5$ mg Zn kg$^{-1}$ | http://nmsp.cals.cornell.edu/publications/factsheets/factsheet32.pdf |
| | Hot Water | $< 0.38$ mg B kg$^{-1}$ | http://nmsp.cals.cornell.edu/publications/factsheets/factsheet47.pdf |
| North Carolina | N/A | Uses an availability index. | http://www.ncagr.gov/agronomi/obpart2.htm |
| North Dakota | DTPA | $< 0.75$ mg Zn kg$^{-1}$; $< 0.3$ mg Cu kg$^{-1}$ | https://www.ag.ndsu.edu/publications/crops/north-dakota-fertilizer-recommendation-tables-and-equations#section-9 |
| | Hot Water | B, although not soil test concentration is listed. | " |
| Ohio | 0.1 N HCl | $< 2$ mg Zn kg$^{-1}$ at pH 6.6 and $< 7$ mg Zn kg$^{-1}$ at pH 7.0 | https://ohioline.osu.edu/tags/fertilizer-recommendations (same as Michigan) |
| | 1.0 N HCl | $< 0.5$ mg Cu kg$^{-1}$ | |
| Oklahoma | Mehlich-3 (?) | $< 4.5$ mg Fe kg$^{-1}$; $< 0.3$ mg Zn kg$^{-1}$; $< 0.5$ mg B kg$^{-1}$ | http://factsheets.okstate.edu/documents/pss-2225-osu-soil-test-interpretations/ |
| Oregon | DTPA | For Corn silage and grain: $< 0.6$ mg Zn kg$^{-1}$ | (Brown, Hart, Horneck, and Moore, 2010) http://www.cals.uidaho.edu/edComm/pdf/PNW/PNW0615.pdf |
| | DTPA | For Alfalfa: $< 1.0$ mg Zn kg$^{-1}$; $< 0.2$ mg Cu kg$^{-1}$ | (Koenig et al., 2009) http://cru.cahe.wsu.edu/CEPublications/PNW0611/PNW0611.pdf |
| | Hot Water | For Alfalfa: B $< 0.4$ to 0.8 depending on soil texture. | |

| | | | |
|---|---|---|---|
| Pennsylvania | Mehlich-3 | Citation states that tests are not reliable for Zn or Cu. | https://agsci.psu.edu/aasl/soil-testing/soil-fertility-testing/handbooks/agronomic/forms/st4-interpreting-soil-test-for-agronomic-crops |
| Rhode Island | N/A | N/A | N/A |
| South Carolina | Hot Water (assumed) | < 0.05 mg B $kg^{-1}$ | https://www.clemson.edu/public/regulatory/ag-srvc-lab/soil-testing/pdf/micronutrients.pdf |
| | Mehlich-3 | < 1.0 or < 1.25 mg Zn $kg^{-1}$ when pH is less than 6 or greater than 6, respectively; Mn is dependent on pH and extractable Mn content (ranges from 2.0 mg Mn $kg^{-1}$ at pH < 5.6 and 2 ppm, to < 8.5 mg Mn $kg^{-1}$ at pH > 6.8 | http://aesl.ces.uga.edu/sera6/PUB/MethodsManualFinalSERA6.pdf |
| South Dakota | N/A | N/A | Might follow North Dakota Guidelines |
| Tennessee | Mehlich-1 | < 0.6 mg B $kg^{-1}$; < 8 mg Mn $kg^{-1}$ when pH > 7; < 1.0 mg Zn $kg^{-1}$ | https://ag.tennessee.edu/spp/SPP%20Publications/SP645.pdf |
| Texas | N/A | N/A | http://aesl.ces.uga.edu/sera6/PUB/MethodsManualFinalSERA6.pdf (general guidelines for the southeastern US) |
| Utah | DTPA | < 0.8 mg Zn $kg^{-1}$ | https://extension.usu.edu/waterquality/files-ou/Agriculture-and-Water-Quality/Fertilizer/AG-431.pdf |
| Vermont | Modified Morgan | < 1.0 mg Zn $kg^{-1}$ | https://pss.uvm.edu/vtcrops/articles/VT_Nutrient_Rec_Field_Crops_1390.pdf |
| Virginia | Mehlich-1 | < 0.8 mg Zn $kg^{-1}$; Mn is similar to other states that use soil extractable Mn concentration and pH | https://www.soiltest.vt.edu/content/dam/soiltest_vt_edu/PDF/recommendation-guidebook.pdf |

(*Continued*)

**Table 7.2** (Continued)

| State | Soil Extractant | Soil Micronutrient Concentration Below Which is Considered Deficient | Link (all verified January 26, 2019) |
|---|---|---|---|
| Washington | DTPA | For Corn silage and grain: < 0.6 mg Zn $kg^{-1}$ | http://www.cals.uidaho.edu/edComm/pdf/PNW/PNW0615.pdf |
| | DTPA | For Alfalfa: < 1.0 mg Zn $kg^{-1}$; < 0.2 mg Cu $kg^{-1}$ | (Koenig et al., 2009; Brown, Hart, Horneck, and Moore, 2010) http://cru.cahe.wsu.edu/CEPublications/PNW0611/PNW0611.pdf |
| | Hot Water | For Alfalfa: B < 0.4 to 0.8 mg $kg^{-1}$ depending on soil texture. | " |
| West Virginia | N/A | N/A | http://aesl.ces.uga.edu/sera6/PUB/MethodsManualFinalSERA6.pdf (general guidelines for the southeastern US) |
| Wisconsin | Hot Water | < 0.4 mg B $kg^{-1}$ | https://learningstore.uwex.edu/Assets/pdfs/A2809.pdf |
| | 0.1 N H3PO4 | < 10 mg Mn $kg^{-1}$ | " |
| | 0.1 N HCl | < 3.0 mg Zn $kg^{-1}$ | " |
| Wyoming | N/A | N/A | N/A |

The concept follows that F and Al in solution form complexes, reducing the likelihood of Al-P re-precipitation and thus increases the ability to quantify soil available P content. The Mehlich-3 also uses weak acids in conjunction with 0.001M EDTA to partially dissolve micronutrient soil phases and complex those micronutrients, essentially maintaining solution micronutrient concentrations and preventing their re-precipitation. As with P, this increases the ability to quantify soil available micronutrients. Other extractants currently utilized in acidic to neutral, noncalcareous soils (e.g., Indiana, Michigan, Ohio, Wisconsin; Table 7.2) are the 0.1M HCl extraction for determining available soil Mn and Zn concentrations, and 1.0M HCl for available Cu determination. The concept behind their use is identical to multi-element extractants: dissolve a portion of the soil phases that likely control nutrient availability.

The above extractants, used in acidic to noncalcareous soils, all utilize weak acids as extractants. They are typically not acceptable for use in calcareous soils as these acids would dissolve soil phases (e.g., Ca-P) that otherwise would not readily dissociate in calcareous soils. Or, they dissolve $CaCO_3$ and release occluded micronutrients (e.g., Zn; Reed and Martens, 1996) that otherwise would not be plant-available in calcareous soils. Thus, the use of the above extractants in calcareous soils may lead to erroneous results, and potentially overestimate nutrient availability, which would be a negative in terms of quantifying soil quality and health. It is obvious that other test methods are required.

### Neutral to Calcareous Soils

Soil macro- and micronutrient concentrations determined in neutral to calcareous soils (~ > pH 7.2) use limited element extractants, chelation, or multielement extractants developed specifically for these soils. For example, the Olsen extraction was designed to partially dissolve Ca–P mineral phases, representing a measure of plant-available P (Olsen et al., 1954). The extractant uses 0.5M $NaHCO_3$ at a pH of 8.5. This solution, at this pH, enhances the dissolution of Ca–P minerals by slightly dissolving them, allowing Ca and P to come into solution, and then precipitating solution Ca as $CaCO_3$. The P remains in solution and its concentration can be determined colorimetrically. The DTPA extraction is typically used to determine plant-available Cu, Fe, Mn, and Zn concentrations (Lindsay and Norvell, 1978) in neutral to calcareous soils west of the Mississippi River (e.g., Table 7.2). The use of DTPA under neutral to calcareous soil conditions leads to greater stability of micronutrients in solution versus other chelates (Lindsay and Norvell, 1978). This test was modified by Soltanpour and Schwab (1977) to include both 0.005M DTPA and 1.0M $NH_4HCO_3$, allowing for the simultaneous extraction of micronutrients and P with the use of $NH_4HCO_3$ acting similarly to $NaHCO_3$ in principle. This test is known as the AB-DTPA extraction, and others have also included K determination while using this extractant.

## Conclusions

Quantifying soil quality and health with respect to interpretation of soil macro- and micronutrient concentrations can be difficult. Paramount to the interpretation is having a strong understanding of the fine lines between deficiency and sufficiency, and between sufficiency and toxicity. Identifying macro- and micronutrient concentrations whereby deficiencies are overcome may be as simple as referring to state fertilizer recommendations. However, identifying the point whereby nutrients become toxic to plants, and thus reduce soil quality, is a challenge as the tipping point between sufficiency and toxicity can be a rather wide concentration range based on numerous factors. When working in most arable soils, more often than not a "more is better" approach should likely be used. However, in specific instances such as with soil P or micronutrient concentrations, scientists have enough information available to draw conclusions regarding the points between deficiency–sufficiency–toxicity to utilize a "somewhere in the middle" approach is best for interpreting soil quality and health.

## References


Adriano, A.C. (2001). *Trace elements in terrestrial environments: Biogeochemistry, bioavailability, and risk of metals*. 2nd ed. New York: Springer-Verlag. doi:10.1007/978-0-387-21510-5

Brown, B., Hart, J., Horneck, D., and Moore, A. (2010). *Nutrient management for field corn silage and grain. University of Idaho PNW* 615. Boise, ID: University of Idaho. http://www.cals.uidaho.edu/edComm/pdf/PNW/PNW0615.pdf

Davis, J.G., and Westfall, D.G. (2014). *Fertilizing corn*. Colorado State University Extension Bulletin 0.538. Fort Collins: Colorado State University. https://extension.colostate.edu/topic-areas/agriculture/fertilizing-corn-0-538/

Havlin, J.L., Tisdale, S.L., Nelson, W.L., and Beaton, J.D. (2014). *Soil fertility and fertilizers, an introduction to nutrient management* (8th ed). Boston, MA: Pearson Publishing.

Ippolito, J.A., Ducey, T., and Tarkalson, D. (2010). Copper impacts on corn, soil extractability, and the soil bacterial community. *Soil Sci.* 175, 586–592. doi:10.1097/SS.0b013e3181fe2960

Ippolito, J.A., Ducey, T., and Tarkalson, D. (2011). Interactive effects of copper on alfalfa growth, soil copper, and soil bacteria. *J. Agric. Sci.* 3, 138–148.

Kabata-Pendias, A. (2011). *Trace elements in soils and plants* (4th ed). Boca Raton, FL: CRC Press, Taylor and Francis Group.

Koenig, R.T., D. Horneck, T. Platt, P. Petersen, R. Stevens, S. Fransen, and B. Brown. (2009). *Nutrient management guide for dryland and irrigated alfalfa in the inland Northwest*. Washington State University PNW0611. http://cru.cahe.wsu.edu/CEPublications/PNW0611/PNW0611.pdf

Leikam, D.F., R.E. Lamond, and D.B. Mengel. (2003). *Soil test interpretations and fertilizer recommendations*. MF-2586. Manhattan, KS: Kansas State University. https://www.bookstore.ksre.ksu.edu/pubs/MF2586.pdf

Lindsay, W.L., and Norvell, W.A. (1978). Development of a DTPA test for zinc, iron, manganese, *and copper. Soil Sci. Soc. Am. J.* 42, 421–428. doi:10.2136/sssaj1978.03615995004200030009x

Moore, A., Satterwhite, M., and Ippolito, J. (2013). *Recommending soil copper thresholds for potato production in Idaho*. Nutrient digest– Nutrient Management Newsletter for the western US (5)1, 6–7. https://eprints.nwisrl.ars.usda.gov/1478/1/1443.pdf

Olsen, S.R., Cole, C.V., Watanabe, F.S., and Dean, L.A. (1954). *Estimation of available phosphorus in soils by extraction with sodium bicarbonate. USDA Circ.* 939. Washington, D.C.: USDA.

Reed, S.T., and Martens, D.C. (1996). Copper and zinc. In D.L. Sparks, editor, *Methods of soil analysis, part 3- Chemical methods* (p. 703–722). Madison, WI: SSSA.

Soltanpour, P.N., and Schwab, A.P. (1977). A new soil test for simultaneous extraction of macro- and micro-nutrient in alkaline soils. *Commun. Soil Sci. Plant Anal.* 8, 195–207. doi:10.1080/00103627709366714

# 8

# Assessment and Interpretation of Soil-Test Biological Activity

*Alan J. Franzluebbers*

**Core Idea**

Flush of $CO_2$ during 3 d following rewetting of dried soil (i.e. soil-test biological activity)

## Introduction

### Importance of Soil Biological Activity to Soil Health and Functioning

Soil biological activity is a broad term describing the heterotrophic activities of macro-fauna, micro-fauna, and microorganisms comprising the soil food web (Coleman et al., 2017). Soil microorganisms dominate soil biological activity and can be quantified efficiently by measuring soil respiration, a fundamental heterotrophic process that balances the autotrophic process of photosynthesis in the global carbon (C) cycle. Soil respiration is the process that transforms organic materials from plants and animals into carbon dioxide ($CO_2$), which is released back to the atmosphere. Key mediators in the process of respiration are soil microorganisms (i.e. bacteria, archaea, fungi, actinomycetes) using simple and complex organic compounds as energy to maintain biomass under limited-resource conditions and proliferate in soil when resources are abundant. Important soil health implications are derived from the process of soil respiration through exudates, sloughed root cells, dead microbial cells, and other organic compounds that can physically glue particles together into stable aggregates and eventually can become components of persistent organic matter.

Measurements of soil respiration can reflect the functional capability of soil to cycle nutrients, decompose organic amendments, and catalyze or stabilize ecosystem processes mediated by a diverse community of organisms.

*Soil Health Series: Volume 2 Laboratory Methods for Soil Health Analysis*, First Edition.
Edited by Douglas L. Karlen, Diane E. Stott, and Maysoon M. Mikha.

## Indicators of Soil Biological Activity

Soil respiration can be determined under field conditions by measuring the flux of $CO_2$ with various air-flow techniques, such as eddy covariance (Campioli et al., 2016; Runkle et al., 2017) after correcting for photosynthetic uptake or more directly by capturing the $CO_2$ emitted from within a chamber over a defined land area (Zibilske, 1994). The latter is most often accomplished using either static or dynamic chambers. Static chambers deploy either solid or liquid alkali as a trap to absorb $CO_2$ emitted from the soil during a defined period of time ranging from several hours to a full day. Alternatively, vented chambers can be deployed to accumulate headspace gas concentration emitted from soil during one to several points within a <2-h period with subsequent measurement of subsample gas concentrations on a gas chromatograph. Modification of the soil surface, especially temperature and air flow, are key limitations of chamber techniques. To overcome changes in microclimate due to chamber placement, dynamic chambers are often deployed for several minutes to an hour with a defined volume of air flowing through the chamber in order to displace $CO_2$, which is then detected using an infrared gas analyzer. Both static and dynamic chamber methods have been widely used and results reported for different ecosystems (Franzluebbers et al., 2002; Mosier et al., 2006; Venterea et al., 2006; Sun et al., 2015). Because chamber methods are deployed under field conditions, estimates are considered *in situ*. However, since temperature and moisture vary diurnally and seasonally, estimates must be made repeatedly over time to obtain reliable seasonal or yearly estimates.

Potential C mineralization is the process of soil respiration determined under standard temperature and moisture conditions in the laboratory. This measurement process begins by representatively sampling soil from a field of interest and sieving it to obtain a homogeneous sample. For estimates at field-moist condition, soil samples must be processed immediately, or stored briefly under refrigeration until they can be analyzed. Refrigeration for more than a couple of weeks before analysis is not recommended, because C and N mineralization can and does occur slowly even at relatively low temperature (Stott et al., 1986). When large numbers of samples are collected or processing time is limited, a viable approach is to dry soil and store until ready for analysis. Drying stabilizes soil to a uniform condition for all samples. Forced air at elevated temperature dries soil fast and this is desirable. Oven-drying soil at 60 °C was as non-intrusive on soil respiration as soil dried at 40 °C (Haney et al., 2004). This same study suggested that drying soil at 100 °C may cause undesired reduction in the flush of $CO_2$, at least for soils incubated for only 1 d. Similar conclusions were drawn from a diversity of soils in Oklahoma dried at 22, 45, 65, and 105 °C (McGowen et al., 2018). Air-drying soil at room temperature could be an acceptable approach, but using a fan blowing over samples is necessary to dry samples quickly and thoroughly. Even soil water content of 50 g water $kg^{-1}$ soil can reduce the flush of $CO_2$ by 15% or more (Franzluebbers, unpublished data). Drying soil was once considered an inappropriate step prior to estimation of soil microbial activity. However, despite C mineralization rate is significantly enhanced during the first few days following drying compared with field-moist soil, relative differences in C mineralization among different soils generally remains consistent (Franzluebbers 1999b; Haney et al., 2004).

## Laboratory-Based Soil Respiration Methods

Potential C mineralization is a relatively simple and robust indicator of soil respiration. By standardizing temperature, soil water content, and length of incubation, potential C mineralization can be a reliable indicator of readily available C substrates processed by the resident soil microbial community. There are several different ways of expressing potential C mineralization, that is, one is from the initial flush of $CO_2$ following rewetting of dried soil, one is from cumulative C mineralization over several weeks (typically 3 to 12 weeks or longer), and another is from basal soil respiration expressed as the near-linear rate of C mineralization sometime after the initial flush of $CO_2$ that can last for up to 10 d of initial incubation. Basal soil respiration is sometimes divided by soil microbial biomass C to obtain a metabolic quotient.

## Chosen Method – Flush of $CO_2$ During 3 d Following Rewetting of Dried Soil

The Soil Health Division of the USDA-Natural Resources Conservation Service selected a 4-d method of measuring the flush of $CO_2$ following rewetting of dried soil as a preferred approach (Stott, 2019). However, the chosen method presented here is based on the flush of $CO_2$ during 3 d following rewetting of dried soil, as defined while the author was in a PhD program at Texas A&M University. Some initial explanation is needed why the focus here is on the 3-d approach (Soil Ecology and Management Lab; SEM Method) rather than the 4-d approach (Cornell Assessment of Soil Health; CASH Method). A direct comparison of the two approaches was made in Franzluebbers and Veum (2020). One might presume that the only difference between methods is simply number of incubation days. However, with testing of soils from two long-term agronomic trials in Missouri and North Carolina, other unique differences were found between methods that have important implications on estimates. First, evaluation of the two methods was according to established protocols for each method at the time of experimentation. Second, C mineralization results were relatively well correlated between methods ($r^2 = 0.93$, n = 211 f or Missouri soils and $r^2 = 0.68$, n = 126 for North Carolina soils). However, based on regression slope, the CASH Method (4-d incubation) produced only 82% of the C mineralization of the SEM Method (3-d incubation) in both sets of soils, which was primarily due to using room temperature in the CASH Method (assumed ~20 °C) and 25 °C in the SEM Method. Greater variability in relationship for the sandier soils from North Carolina may have been due to use of capillary wetting in the CASH Method compared with 50% water-filled pore space in the SEM Method. Capillary wetting can lead to near saturated soil conditions (Franzluebbers and Haney, 2018; Wade et al., 2018) and cause greater variation and/or lower respiration in sandy loam soils. Finally, some fine-level soil management differences on C mineralization were not detected as significant with the CASH Method, but were found with the SEM Method, suggesting differences in sensitivity between methods. Therefore, the chosen protocol in this document continues to focus on the SEM Method that has been used consistently and routinely in a research laboratory for a couple of decades.

Variations of short-term C mineralization assays from that described here are being used commercially, most notably from 1-d $CO_2$ in the Haney Soil Health Tool using capillary wetting and infrared gas analysis (Haney et al., 2018a) and 1-d $CO_2$ in the Solvita $CO_2$ Burst Test using a pH-sensitive gel paddle (Haney et al., 2008). Reasonable correlation exists between these 1-d tests and the SEM Method described here (i.e. 3-d incubation), but saturation of soil with capillary wetting becomes a limiting factor for interpretation (Franzluebbers and Haney, 2018). Further comparison of methods is warranted to describe how variations in methodology might be overcome to relate to important soil functions through lab and field calibrations.

Soil respiration has been measured since at least the early 20$^{th}$ century to describe aspects of soil fertility, particularly for microbial activity as an agent of releasing N from organic matter (Waksman and Starkey, 1924). Drying soil was found to enhance the release of N for greater plant N uptake (Lebedjantzev, 1924). Birch and Friend (1956) found a repeatable pattern of high initial respiration following rewetting of dried soil followed by a slower rate of respiration. They proposed that the drying effect on respiration might be useful in soil fertility studies. Birch (1958) observed a repeatable flush of $CO_2$ during 43 consecutive drying (100 °C) and rewetting events, but with gradual decline in this flush thereafter in a neutral soil with 64 g C $kg^{-1}$ soil (releasing 38% of total C during these events). On an acidic soil with 42 g C $kg^{-1}$, the flush of $CO_2$ diminished to an indistinguishable level after 12 drying and rewetting events (releasing 9.5% of total C during these events). Carbon mineralized during 19 d following air-drying was proportional to the total organic C concentration of soils (Birch, 1960). Research of soil drying effects on organic matter diminished with emerging focus on the soil microbial biomass and how best to quantify it, particularly with chloroform fumigation (Jenkinson and Powlson, 1976). In addition, much more research since the 1950s was focused on inorganic N fertilization of crops, such that mineralization of organic matter as a source of soil fertility might have been viewed irrelevant.

A renewed focus on using the flush of $CO_2$ following rewetting of dried soil as an indicator of management-induced change in soil biological activity began during the course of two wheat (*Triticum aestivum* L.) growing seasons in Texas (Franzluebbers et al., 1995). The flush of $CO_2$ following rewetting of dried soil was investigated under both conventional and no tillage. With sampling every ~10 d, the flush of $CO_2$ peaked during maximum vegetative stage of wheat, suggesting that microbial proliferation in the rhizosphere was an important biological factor affecting the flush of $CO_2$. A wider set of soils from Texas was initially tested using the flush of $CO_2$ in 1 d compared with net N mineralization, soil microbial biomass C, and total organic C (Franzluebbers et al., 1996a). Although a strong association existed between the flush of $CO_2$ in 1 d with that in 3 d, the longer incubation time resulted in more reliable estimates of C mineralization (Franzluebbers et al., 2000).

The flush of $CO_2$ during 3 d was standardized following two important evaluations using a diverse set of soils from Georgia. Fifteen soils with varying texture were collected from an eroded landscape. The flush of $CO_2$ was optimized with water-filled pore space of $0.53 \pm 0.06$ $m^3$ $m^{-3}$ (Franzluebbers, 1999a). Optimum water-filled pore space for determination of net N mineralization (0-24 d) was $0.42 \pm 0.03$ $m^3$ $m^{-3}$. In a second study for soils with clay concentration of 3 to 35%, the flush of $CO_2$ was more similar between coarsely sieved and dried-intact soil cores than when sieved finely to <0.5 mm. This suggested that coarsely sieved soil would not adversely affect the flush of $CO_2$ estimates compared with intact cores, and therefore, might keep organic matter more protected from decomposition, similar to conditions in the field (Franzluebbers, 1999b). In fact, when soils from Alberta/British Columbia were crushed to <0.25 mm, enhanced emission of $CO_2$ occurred compared with intact soil aggregates of 0.25-1 mm and 1-5.6 mm (Franzluebbers and Arshad, 1997). Enhanced mineralization from crushed aggregates was greater for near-surface samples than samples collected deeper and for larger rather than smaller macro-aggregates, resulting in clear evidence of macro-aggregate protection. Therefore, management-induced changes in biologically active organic matter components are important and detectable with consistent C mineralization methodology. Although a 4-d incubation period (as proposed by Stott [2019]) might be equally valid as a 3-d incubation period, one has to question why longer than shorter when the SEM Method has been well documented, used extensively, and appears to be more sensitive to management and environmental characteristics (Franzluebbers and Veum, 2020). Rationale for a 1-d test to obtain an even faster result than the SEM Method appears enticing, but methodologies with shorter incubation time appear to have limitations with water saturation (Franzluebbers and Haney, 2018; Wade et al., 2018), routine use of limited soil mass (Franzluebbers, 2020b), and a general lack of calibration to important soil functions like net N mineralization that have been otherwise described for the SEM Method under laboratory conditions (Franzluebbers et al., 2018b), greenhouse growth conditions (Franzluebbers and Pershing, 2018), and field conditions (Franzluebbers, 2018b, 2020c; Franzluebbers et al., 2018a; Franzluebbers and Poore, 2020). The flush of $CO_2$ with the SEM Method was shown highly correlated with longer term net N mineralization of up to 150 d (Franzluebbers, 2020a). Longer term C mineralization assays have been used for routine adjustment of soil microbial biomass C with the chloroform fumigation-incubation method (Jenkinson and Powlson, 1976; Sainju et al., 2000) and separation of C pools defined by reaction rate (Collins et al., 2000). Short-term C mineralization assays have distinct time and resource efficiency advantages over longer term C mineralization assays, and the relatively low coefficient of variation and sensitivity of the flush of $CO_2$ during 3 d makes it an excellent choice.

The author has used the SEM Method proposed here consistently and routinely since ~2000. Therefore, the following section on how management might affect soil-test biological activity describes results using a well-documented and consistent approach in the laboratory.

## Management Factors Influencing Soil-Test Biological Activity

A standardized soil biological activity protocol is important for assessing soil health. Unlike other biological indicators, the flush of $CO_2$ does not exhibit enormous swings in values over days or weeks. This is likely due to its reliance on a broad range of C substrates from organic matter. To illustrate, the flush of $CO_2$ in the surface 20 cm of soil varied temporally by $\pm$20% within the growing season of wheat in a long-term tillage-rotation study in Texas (Franzluebbers et al., 1995, 1996b). The flush of $CO_2$ increased during the growing season until peak vegetative growth/initiation of flowering and then declined thereafter. These data illustrate the strong dependence of the flush of $CO_2$ on root growth and exudation of C substrates in the rhizosphere and surrounding soil.

Temporal analyses over yearly increments with the same management illustrate that the flush of $CO_2$ is strongly correlated with changes in several other soil C and N fractions, suggesting these substrates provide fuel for the flush of $CO_2$. For example, establishment and maturation of pasture on previously degraded cropland led to large increases in the flush of $CO_2$ and other biologically active C and N fractions during the first four years of transition (Franzluebbers and Stuedemann, 2003). Relative to a pre-pasture baseline, the flush of $CO_2$ from the surface 6 cm of soil increased by 47 and 62% $yr^{-1}$ under ungrazed and grazed conditions, respectively. These changes were in good agreement with annual changes in basal soil respiration (51 and 85% $yr^{-1}$), soil microbial biomass C (23 and 40% $yr^{-1}$), and particulate organic C (25 and 45% $yr^{-1}$) under ungrazed and grazed conditions, respectively. An evaluation following pasture termination in a different study in Georgia showed that the flush of $CO_2$ increased with time for all management systems (conventional and no tillage; ungrazed and grazed cover crops) and was sensitive to redistribution of organic matter with tillage treatment (Franzluebbers and Stuedemann, 2015).

The flush of $CO_2$ is also responsive to differences in soil type (Franzluebbers et al., 1996a, 2000, 2001), soil texture (Franzluebbers, 1999b), tillage type (Franzluebbers et al., 1995, 2007; Franzluebbers and Arshad, 1997; Franzluebbers and Studemann, 2008), crop rotation (Culman et al., 2013; Franzluebbers and Stuedemann, 2008), fertilization rate and source (Haney et al., 2001; Franzluebbers and Stuedemann, 2005), organic amendment type (Culman et al., 2013; Jangid et al., 2008), land use (Jangid et al., 2008, 2010, 2011), pasture management (Franzluebbers and Stuedemann, 2003, 2005, 2008, 2015), aggregate disruption (Franzluebbers and Arshad, 1997), and soil sampling depth (Franzluebbers and Stuedemann, 2003, 2008, 2015; Franzluebbers et al., 2007).

## Literature Review of Indicator and Method

### Consideration of Indicators for Determining Soil Biological Activity

Published approaches to measuring soil biological activity have varied by incubation time, amount of soil analyzed, soil water content and delivery method, temperature, soil sieving and drying approach, headspace volume, and $CO_2$ detection. Currently, four main approaches for short-term C mineralization have been used and reported in the scientific literature. All have the goal of assessing soil biological activity through potential C mineralization.

Potential C mineralization is non-linear, and therefore, defined time intervals are necessary to compare samples. Several-week-long incubations have been conducted and can give useful information on separation of temporally dynamic pools of C (Collins et al., 2000). Alternatively, 1-, 3-, 4-, and 10-d incubations have also been recently conducted (Franzluebbers et al., 2000; Haney et al., 2004; Sainju et al., 2013; Fine et al., 2017), all of which closely associate with management-induced differences in potential C mineralization among a variety of soils when analyzed with longer-term incubations.

Variations in estimates of potential C mineralization can occur due to inherent soil characteristics, management-induced factors controlling C substrates in soil, and laboratory conditions. Minimization of the latter should be possible. Soil and management factors should be quantifiable independently of laboratory conditions, but potential interactions between laboratory conditions and soil/ management factors have not been adequately characterized. Lab conditions of key importance are temperature, water content, time of incubation, and detection of $CO_2$. These factors are explored in more detail below.

Temperature is the key driver of soil respiration / potential C mineralization. Several studies have shown large increase in soil organic matter mineralization with increasing soil temperature (Ellert and Bettany, 1992; Zak et al., 1999). A $Q_{10}$ factor of ~2 is often assumed, i.e. doubling of respiratory activity with a 10 °C increase in temperature. This $Q_{10}$ factor may actually vary depending on the starting condition and various other factors (Ellert and Bettany, 1992). If soil temperature were set at 25 °C as a standard method and expected C mineralization were 150 mg $CO_2$-C $kg^{-1}$ soil 3 $d^{-1}$, but temperature reached only 22 °C (intentionally as part of protocol or unintentionally as malfunction), then C mineralization would be 106 mg $CO_2$-C $kg^{-1}$ soil 3 $d^{-1}$ with $Q_{10} = 2$. This 3 °C temperature difference would yield only 81% of the expected value, while incubation at 20 °C would yield 71% and at 24 °C would yield 93% of expected. Therefore, temperature should be standardized, or at least controlled or determined.

Soil water content is the second most important factor controlling soil respiration / potential C mineralization. Conditions of too dry or too wet limit C mineralization, while soil water content of 50 to 80% water-filled pore space is considered optimum (Linn and Doran, 1984; Franzluebbers, 1999a). Carbon mineralization appears to peak across a relatively broad range of soil water contents, but it is clear that optimum N mineralization occurs over a much narrower water-filled pore space range of 40 to 55% (Franzluebbers, 1999a).

Incubation time for potential C mineralization is another critical factor in obtaining an absolute value for a soil biological activity index. Values at 3 or 4 d of incubation are often 2 to 3 times greater than at 1 d of incubation. However, there is a strong and consistent relationship among all of these incubation times. In fact, the cumulative non-linear rate of C mineralization (Fig. 8.1) suggests that mathematically derived associations among different time periods could easily be developed (assuming similar methodology in temperature, water, and $CO_2$ detection methodologies). In a direct comparison of 1-d and 3-d incubations across a diversity of soils with a gradient in soil texture, random variation was slightly less with 3-d incubation than with 1-d incubation (Franzluebbers and Haney, 2018). This result is logical given that larger values would be more repeatable when using $CO_2$ detection methodology with a static detection limit.

Detection of $CO_2$ emission from a soil sample has been determined with a variety of approaches. These variations are summarized in Figure 8.2. Sensitivity of some detection methodologies has been compared. A more thorough comparison

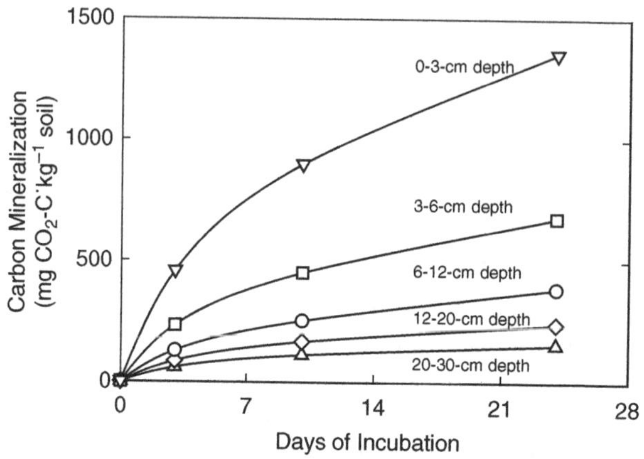


**Figure 8.1** Cumulative C mineralization as a function of incubation time for soils collected at five depths.

across all methodologies from a diversity of soil types and management conditions would be appropriate. In a diversity of soils from across North America, estimates of the flush of $CO_2$ were strongly related among different detection methodologies, including $r^2 = 0.95$ between acid-base titration and infrared gas analysis, $r^2 = 0.82$ between acid-base titration and gel paddle (Woods End Laboratories, Inc., Mt. Vernon, ME), and $r^2 = 0.79$ between infrared gas analysis and gel paddle (Haney et al., 2008). From a landscape-scale study in Colorado with different management treatments, estimates of the flush of $CO_2$ were also strongly related among different detection methodologies, including $r^2 = 0.93$ between acid-base titration and gas chromatograph, $r^2 = 0.93$ between acid-base titration and infrared gas analysis, and $r^2 = 0.94$ between infrared gas analysis and gas chromatograph (Sherrod et al., 2012). Evaluation of the electrical conductivity method with other more common detection methodologies remains to be verified but was recently tested against an alkali trap method (Franzluebbers and Veum, 2020). A key limitation of the gel paddle method (commercially available from Solvita) is that $CO_2$ saturation can occur at high levels of $CO_2$ emission (McGowen et al., 2018). Alkali traps also have an upper limit but can be adjusted with alkali concentration or quantity in individual studies depending on expected microbial activity. Maintaining low soil mass (e.g. 20-40 g) and short incubation time may be necessary with the gel paddle detection methodology, whereas soil mass and incubation time can be adjusted according to specific scientific needs with other detection methodologies. However, soil mass was considered optimum at 50 to 100 g to meet both central tendency and low random variation requirements (Franzluebbers, 2020b), as well as to maintain a practical laboratory protocol with moderate space requirement.

**Variations in protocol...**

| Component | Soil Ecology & Mgt NC State | Haney Soil Health Test | Solvita – Woods End | Cornell CASH |
|---|---|---|---|---|
| Soil processing | Dried 55 °C 3 days, sieved <4.75 mm | Dried 50 °C 1 day, sieved <2 mm | Shipped wet; dried, roller | Shipped wet; air dried, sieved <8 mm |
| Soil weight | Two 50-g subsamples | 40 g | 40 g | 20 g |
| Water | 50% WFPS in two 60 mL bottles | Capillary from bottom to saturation | From top to 50% WFPS; previously from bottom | Capillary from bottom to saturation |
| Incubation | 3 days at 25 °C in 1-L jar | 1 day at 25 °C in 0.25-L jar | 1 day at room temperature in 0.25-L jar | 4 days at room temperature in 0.5-L jar |
| $CO_2$ detection | Acid titration of 1 M NaOH trap to phenolphthalein endpoint | Infrared gas analysis of headspace | Gel paddle with digital color reader | Electrical conductivity of 0.5 M KOH trap |

**Figure 8.2** Variations in lab protocol for short-term C mineralization.
Note: Gas chromatography has also been used for $CO_2$ detection and is often considered the most precise, but requires specialized equipment and handling.

Repeatability of C mineralization estimates from the same soil sample using acid-base titration from coarsely sieved soil (<4.75 mm) was recently evaluated in two different experiments. First, coefficient of variation (CV) in the flush of $CO_2$ during 3 d among 10 subsamples of five different soils was 6.4 ± 5.8% when soil mass was 50 to 100 g (Franzluebbers, 2020b). When soil mass was ≤20 g, CV was 11.7 ± 8.9%. Second, a total of 42 experimental units from a long-term field experiment were sampled from 0-6 cm soil depth after removing surface residue (Franzluebbers, unpublished data). Field-moist soil was sieved to <4.75 mm and 60-g portions of soil (~51 g dry weight equivalent) were weighed into a total of 6 replicate bottles. Two bottles were placed into each of three incubation vessels for determination of C mineralization during 0-3, 3-10, 10-24, and 24-41 d. Triplicate observations were made for all incubation periods, except duplicate observations were made for 24-41 d. Carbon mineralization estimates ranged from 34 to 463 mg $CO_2$-C $kg^{-1}$ soil. Coefficient of variation had a median value of 4.4% with the middle 50% of CVs ranging from 2.5 to 7.0%. Both of these examples gave CVs that represent true repeatability of the method. Estimates of CV among field replicates introduces another level of geospatial variation; CVs should be larger. Coefficient of variation among four field replicates of the flush of $CO_2$ during 3 d was reported as 14.4% (middle 50% of CVs from 147 means ranging from 8.9 to 22.4%) (Franzluebbers et al., 2018b). When using four soils from different geographic locations as multiple observations within each of three broad soil textural classes, CV of the flush of $CO_2$ during 3 d was 36% (Franzluebbers and Haney, 2018). Increasing geospatial distance among field replicates should be expected to increase CV. Repeatability of C mineralization was improved with longer incubation time from five sites in the Pacific Northwest region, in which CV among triplicate samples of 16 treatments was 17% at 0-1 and 0-3 d periods, 15% at 0-10 d period, and 13% at 0-24 d period (Morrow et al., 2016). The procedure used only 5 g samples incubated at 20 °C and $CO_2$ was detected by gas chromatograph during a 2-h period at each time interval. The larger CV in this study with low soil mass was consistent with greater CV from low soil mass when specifically assessing soil mass influence on the flush of $CO_2$ (Franzluebbers, 2020b).

## Relationship of Soil Biological Activity to Important Soil Functions / Ecosystem Services

Soil-test biological activity relates very closely with basal soil respiration, microbial biomass C, particulate organic C and N, and potential C and N mineralization, and also has reasonable association with soil aggregation, soil biological diversity, and soil organic C sequestration (Franzluebbers, 2018a). Two functions of utmost importance in agriculture will be highlighted here – basal soil respiration as a general process of soil biological activity and net N mineralization as a specific process of supplying N to crops.

Association of the flush of $CO_2$ with basal soil respiration has been made in several long-term field studies, as well as surveys of farm fields in the eastern United States (Table 8.1). In each individual evaluation, coefficient of determination was at least 80% and in many cases >90%. These strong individual sets of association and the equally strong overall association suggest that the flush of $CO_2$ is a very good indicator of basal soil respiration across a range of conditions.

As an indicator of soil N availability via net N mineralization with aerobic incubation, the flush of $CO_2$ has exhibited reasonably strong associations in a number of different soil types and agricultural management conditions (Table 8.2). Of course, there are other studies that show poor correlation between soil C and N mineralization, but this can often be attributed to the type of method deployed (Chu et al., 2019). Slope estimates of the regression of net N mineralization on the flush of $CO_2$ were $0.17 \pm 0.09$ mg N $mg^{-1}$ $CO_2$-C with 3-d incubation and $0.77 \pm 0.43$ mg N $mg^{-1}$ $CO_2$-C with 1-d incubation. In a study directly comparing length of incubation from a dozen soils varying in soil texture from North Carolina, Pennsylvania, and Virginia, slope estimate of the

**Table 8.1** Compilation of studies reporting associations between the flush of $CO_2$ following rewetting of dried soil (FCO2; mg $CO_2$-C $kg^{-1}$ soil 3 $d^{-1}$) and basal soil respiration (mg $CO_2$-C $kg^{-1}$ soil $d^{-1}$). Note: Linear regression followed the form: BSR = $b_0 + b_1 \times$ FCO2. SE = standard error.

| Description | n | Range of FCO2 | Intercept (SE) | Slope (SE) | $r^2$ | SE of estimate |
|---|---|---|---|---|---|---|
| Long-term crop study in Georgia with tillage, grazing, and rotation as management variables; Sampled at 0-3, 3-6, 6-12, 12-20, and 20-30 cm depths at 0, 1, 2, 3, 5, and 7 years; Means of 4 replications; Franzluebbers and Stuedemann (2008, 2015) | 266 | 28 – 864 | -0.4 (0.4) | 0.072 (0.001) | 0.90 | 4.2 |
| Long-term crop study in Georgia with tillage as management variable; Sampled at 0-3, 3-6, 6-12, and 12-20 cm depths during 5 years; Means of 5 pseudo-replications; Franzluebbers et al. (2007) | 80 | 57 – 599 | -2.3 (0.5) | 0.071 (0.002) | 0.96 | 2.3 |
| Long-term pasture study in Georgia with fertilization source and endophyte as management variables; Sampled at 0-3, 3-6, 6-12, and 12-20 cm depths; Means of 2 replications and 3 pseudo-replications; Franzluebbers et al. (2012); Franzluebbers (unpublished data) | 208 | 27 – 1079 | -0.6 (0.6) | 0.082 (0.002) | 0.93 | 6.1 |
| Long-term crop study in North Carolina with silage cropping intensity as management variable; Sampled at 0-3, 3-6, 6-12, and 12-20 cm depths during 5 years; Means of 2 replications; Franzluebbers and Brock (2007) | 60 | 63 – 893 | -2.0 (0.4) | 0.073 (0.003) | 0.92 | 4.2 |
| On-farm survey of crop fields in North Carolina and Virginia; Sampled at 0-10, 10-20, and 20-30 cm depths; Franzluebbers et al. (2018b) | 539 | 13 – 865 | -0.5 (0.2) | 0.063 (0.001) | 0.88 | 3.2 |
| On-farm survey of crop and pasture fields in Alabama, Georgia, South Carolina, North Carolina, and Virginia; Sampled at 0-5, 5-12.5, and 12.5-20 cm depths; Causarano et al. (2008) | 261 | 7 – 566 | 0.4 (0.2) | 0.069 (0.001) | 0.91 | 2.1 |
| Long-term crop study in Illinois with municipal biosolid application as management variable; Sampled at 0-15 and 15-30 cm depths; Means of 2 replications; Tian et al. (2013) | 88 | 32 – 432 | 3.8 (0.7) | 0.059 (0.003) | 0.84 | 2.3 |
| All data | 1502 | 7 – 1079 | -1.0 (0.2) | 0.074 (0.001) | 0.91 | 4.2 |

**Table 8.2** Compilation of studies reporting associations between the flush of $CO_2$ following rewetting of dried soil (FCO2) and net N mineralization. Note: Linear regression followed the form: NMIN = $b_0 + b_1 \times$ FCO2. Polynomial regression followed the form: NMIN = $b_0 + b_1 \times$ FCO2 + $b_2 \times$ FCO2$^2$.

| Description | Condition | $b_0$ | $b_1$ | $r^2$ |
|---|---|---|---|---|
| Texas; 5 soil samples with range of soil organic matter content; Franzluebbers et al. (1996); 1-d FCO2 used | 0-15 d | 3 | 0.35 | 0.87 |
| Texas; 8 locations; total of 57 soil samples; range of soil organic C conditions and sampling depths to 20 cm; Franzluebbers et al. (1996); 1-d FCO2 used | 0-21 d | -4 | 0.89 | 0.85 |
| Georgia; 10 soil samples with 3-35% clay; depths of 0-4 and 4-8 cm; pasture for 5 years; Franzluebbers (1999b) | 0.5 mm | 14 | 0.09 | 0.81 |
| | 2.0 mm | 17 | 0.12 | 0.42 |
| | 4.7 mm | 12 | 0.15 | 0.90 |
| | 7.9 mm | 15 | -0.01 | 0.00 |
| | intact | 9 | 0.18 | 0.43 |
| | moist | 0 | 0.37 | 0.88 |
| Multiple states; total of 2131 soil samples; various conditions; 30 individual relationships reported; associations for states were from 0-500 mg $CO_2$-C $kg^{-1}$ 3 $d^{-1}$ only; Franzluebbers et al. (2000) | Alberta | 5 | 0.11 | 0.50 |
| | Georgia | 15 | 0.11 | 0.48 |
| | Maine | -15 | 0.31 | 0.87 |
| | Texas | 4 | 0.26 | 0.47 |
| Texas; 8 sample means; bermudagrass pasture; dairy manure application rates; 0-7.5 cm depth; Haney et al. (2001); 1-d FCO2 used | 1995 | 8 | 1.31 | 0.78 |
| | 1996 | 18 | 0.51 | 0.93 |
| Georgia; 60 soil samples; tillage type and frequency; 2.5-7.5 and 7.5-15 cm depths; Picone et al. (2002) | | -2 | 0.19 | 0.64 |
| Wyoming; 22 sample means; native grassland and reclaimed mine spoils; 0-2.5 and 2.5-15 cm depths; Ingram et al. (2005) | 0-21 d | -1 | 0.13 | 0.72 |
| North Carolina; 60 soil samples; 3 silage cropping intensities; 0-3, 3-6, 6-12, and 12-20 cm depths; 7 years; Franzluebbers and Brock (2007) | | 16 | 0.17 | 0.72 |
| Georgia; 80 sample means; 4 watersheds; paraplow and no-tillage; 5 years; Franzluebbers et al. (2007) | | -3 | 0.19 | 0.85 |
| Brazil; 27 samples; tillage management as variable; 0-5, 5-20, and 20-30 cm depths; Green et al. (2007) | | NA | NA | 0.71 |
| Georgia; 646 soil samples; conventional and no-tillage; summer and winter cover cropping; with and without grazing; 0-3, 3-6, 6-12, 12-20, and 20-30 cm depths; 4 years; Franzluebbers and Stuedemann (2008) | | 5 | 0.19 | 0.82 |

NA, not available.

regression of net N mineralization on the flush of $CO_2$ was 0.29 mg N $mg^{-1}$ $CO_2$-C with 3-d incubation and 0.58 mg N $mg^{-1}$ $CO_2$-C with 1-d incubation (Franzluebbers and Haney, 2018). All of these data clearly suggest a unique calibration would be needed for different short-term C mineralization methods varying by incubation period, but they also clearly suggest that the flush of $CO_2$ is a very strong indicator of net N mineralization.

Across 47 farms in Coastal Plain, Piedmont, Great Valley, and Blue Ridge physiographic regions in North Carolina and Virginia, strong association of the flush of $CO_2$ with net N mineralization in 24 d was found independent of soil texture (Franzluebbers et al., 2018b). The regression of net N mineralization on the flush of $CO_2$ was 0.24 mg N $mg^{-1}$ $CO_2$-C with 3-d incubation – very much aligned with previous investigations. A companion study showed a very strong relationship between the flush of $CO_2$ during 3 d and plant N uptake by an unfertilized annual grass grown in the greenhouse (Franzluebbers and Pershing, 2018). In fact, the relationship across 537 observations (47 fields with 3 depths) ($r^2$ = 0.76) was nearly equal to that of plant N uptake with net N mineralization ($r^2$ = 0.81) and plant available N (summation of residual inorganic N and net N mineralization) ($r^2$ = 0.85).

In the survey of 47 corn fields in North Carolina and Virginia, relative yield without sidedress N application was indicated reasonably well with plant available N measured at 0-10 cm depth ($r^2$ = 0.64) (Franzluebbers, 2018b). As well, the flush of $CO_2$ was a reasonably good indicator of relative yield ($r^2$ = 0.60). The flush of $CO_2$ was also predictive of grain yield response to initial dose of N fertilizer ($r^2$ = 0.63) and to economically optimum N fertilizer rate ($r^2$ = 0.45). Based on these results, soil sampled at a depth of 0-10 cm, or as deep as 30 cm (and any depth increment between these limits with reworking of existing data), could be used to estimate the economically optimum N fertilizer rate at a given cost-to-value threshold (i.e. cost of fertilizer to value of grain). For example, soil with low soil-test biological activity of 100 mg $CO_2$-C $kg^{-1}$ soil 3 $d^{-1}$ would be recommended to have N application of 17.8 and 9.6 kg N $Mg^{-1}$ grain (1.09 and 0.54 lb N $bu^{-1}$ grain) to achieve low cost-to-value threshold of 5 kg grain $kg^{-1}$ N and high cost-to-value threshold of 20 kg grain $kg^{-1}$ N, respectively. Soil with high soil-test biological activity of 500 mg $CO_2$-C $kg^{-1}$ soil 3 $d^{-1}$ would be recommended to have N application of only 3.0 and 1.8 kg N $Mg^{-1}$ grain (0.17 and 0.10 lb N $bu^{-1}$ grain) to achieve low and high cost-to-value threshold, respectively. These interpretations are based on biologically active N in surface soil being available for N uptake by crops during the growing season. Current recommendations that do not take this biologically active N into account are effectively allowing N to be more susceptible to environmental loss through denitrification, leaching, runoff, and/or volatilization.

## Issues of Concern With Various C Mineralization Indicators

The following derivations of short-term C mineralization assays have been used in various research and service labs:

- "One-day $CO_2$" at 25 °C (or room temperature in some cases) following rewetting of dried soil with water delivery from bottom via capillarity and $CO_2$ detected with Solvita gel paddle (Haney et al., 2008; Haney and Haney, 2010) or with infrared gas analyzer (Culman et al., 2013; Haney et al., 2018b)
- "Soil-test biological activity with the flush of $CO_2$" during 3 d at 25 °C following rewetting of dried soil with water delivery from top to 50% water-filled pore space and $CO_2$ detected with acid-base titration (Franzluebbers et al., 2000; Franzluebbers, 2016)
- "Respiration" during 4 d at room temperature following rewetting of dried soil with water delivery from bottom via capillarity and $CO_2$ detected with electrical conductivity of KOH solution (Fine et al., 2017; Moebius-Clune et al., 2017)
- "Potential C mineralization" during 10 d at 21 °C following rewetting of dried soil with water delivery from top to 50 or 70% water-holding capacity and $CO_2$ detected with acid-base titration (Sainju et al., 2000, 2013)
- "Mineralizable C" during 10 d at 25 °C following adjustment of field-moist soil to −0.03 MPa water retention and $CO_2$ detected with acid-base titration (Franzluebbers et al., 1994)
- "Respired $CO_2$-C" during 1-21 d at 30 °C following rewetting of dried soil with water delivery from top to 50% water-filled pore space and $CO_2$ detected with infrared gas analyzer or gas chromatograph (Sherrod et al., 2012)

Effective soil biological activity indicators should (i) be easy to measure, (ii) detect changes in soil function, (iii) integrate soil physical, chemical, and biological properties and processes, (iv) be accessible to many users and applicable to field conditions, (v) be sensitive to variations in management and climate, (vi) encompass ecosystem processes and relate to process-oriented modeling, and (vii) where possible, be components of existing soil data bases (Franzluebbers, 2016). All of the above-listed methods are considered easy to measure and most likely able to detect changes in soil function.

Some comparisons of methods reveal limitations of various methodological details and somewhat greater sensitivity with other aspects. Although capillary rewetting makes the process of water delivery easy and without need for individual soil sample adjustment (Haney and Haney, 2010), it has been shown to result in lower $CO_2$ flux (Wade et al., 2018), but more so in coarse-textured soil than in other soil textures (Franzluebbers and Haney, 2018).

Incubation temperature must be stated in any method, as it is a major controlling factor. Temperature should be a scalable factor that does not necessarily interact with other components of the method, so once set in the method should not be allowed to be variable. Incubation in a controlled-temperature environment is important, so room temperature with diurnal and seasonal variations will not be acceptable. As shown earlier, even a 1 °C difference in temperature will lead to a 7% difference in $CO_2$ flux.

Mass or volume of soil should be standardized if at all possible to avoid unnecessary biases in estimation and to obtain the most repeatable estimate possible (Franzluebbers, 2020b). Greater soil mass leads to greater $CO_2$ flux, but some methods of $CO_2$ detection may become saturated and unable to distinguish flux rate. For example, an acid-base trap has a limit of how much $CO_2$ can be captured and the gel paddle is likely to become saturated when soil mass exceeds the recommended 40-g aliquot, particularly in soils with high biological activity. Low soil mass leads to high variability in repeated estimations from the same sample.

The $CO_2$ detection methodology has been compared and there are indications that most methods are acceptable under normal circumstances. However, there may be slight differences, such as inefficient trapping of $CO_2$ with the acid-base trap when incubation is shorter than a few days (Franzluebbers and Haney, 2018). Sherrod et al. (2012) found that gas chromatograph, infrared gas analyzer, and acid-base titration were all effective, but reported that analysis time was 150 $h^{-1}$ for infrared gas analyzer, 17 $h^{-1}$ for gas chromatograph, and 10 $h^{-1}$ for titration. Practicality is important, but it should not override the need to obtain quality data.

## Method

The flush of $CO_2$ during 3 d following rewetting of dried soil follows the method described in Franzluebbers (2016), which is essentially the same as that described in detail in Franzluebbers and Stuedemann (2008). The method was developed based on features described in Zibilske (1994) and Anderson (1982).

Alternatives to the flush of $CO_2$ during 3 d could include incubation for 1 or 4 d, but these methods should ensure that water is added to 50% water-filled pore space. Water-filled pore space is a large controlling factor on the flush of $CO_2$ (Franzluebbers, 2016) and soil microbial activity in general (Linn and Doran, 1984; Franzluebbers, 1999a). Calibration of different incubation time periods to the 3-day standard protocol should be made using all of the variations in other methodological approaches specific to the alternative method. Several calibrations of the 1-d and 3-d methods have been made in the past, but these were with the 1-d method using acid-base titration (Franzluebbers et al., 2000; Franzluebbers and Haney, 2018). $CO_2$ detection with the 1-d method is more often with gel paddle (Solvita), infrared gas analyzer, or gas chromatograph (Haney et al., 2008; Haney and Haney, 2010). Differences in amount of soil, soil water content, and how water is delivered are other important methodological differences to consider.

## Materials

- 60-mL graduated wide-mouth (3-cm diam) glass containers with lab identification number – to contain soil
- 30-mL Nalgene screw-top high-density polyethylene (HDPE) vials plus screw-cap – to contain alkali
- Auto-dispenser – to accurately dispense 10 mL alkali
- 7-dram plastic vial – to contain water for humidity
- 1-L mason jars with rubber-seal lids – as incubation vessel
- Wooden dowel – to press soil to desired density
- 10-mL adjustable hand pipette – to dispense water to soil
- Incubation cabinet – to maintain temperature at 25 °C
- Digital dispensing burette – to dispense acid for neutralizing base
- Stir plate and stir bar – to mix alkali solution and keep $BaCO_3$ away from acid

## Reagents

- 1 mol $L^{-1}$ sodium hydroxide (NaOH) – dissolve 40 g NaOH in deionized $H_2O$, dilute to 1 L – to trap $CO_2$
- 1.5 mol $L^{-1}$ barium chloride ($BaCl_2$) – dissolve 366 g $BaCl_2$ in deionized $H_2O$ and dilute to 1 L – to precipitate absorbed carbonate in alkali solution as $BaCO_3$
- Phenolphthalein solution – dissolve 1.0 g phenolphthalein in 100 mL 95% ethanol – to serve as acid/base color indicator
- 1 mol $L^{-1}$ hydrochloric acid (HCl) – standardize against ~0.5 g (to nearest 0.0001) oven-dried TRIS [$(HOCH_2)_3CNH_2$] using THAM indicator solution (0.1 g BromoCresol Green + 0.1 g Alizarin Red S in 100 mL water) – to neutralize alkali

## Method

1) Obtain representative soil sample from field of interest at defined depth (recommended depth of 0-10 cm, but other depths can be used if desired for specific purpose).
2) Sieve oven-dried (55 °C, 48-72 h) soil to pass a 4.75-mm screen (all material should pass the screen, except stones and large pieces of residues).
3) Weigh a 50 g portion of soil into a 60-mL graduated glass container. The SEM lab typically uses two 50-g portions to obtain the flush of $CO_2$ from 100 g soil.

(This protocol describes an approach with standardized soil weight and variable volume, but an alternative approach to standardize volume and vary soil weight would be to use a 50 to 70 mL vessel to which soil could be lightly packed and subsequently weighed to determine total porosity; both approaches require variable water delivery).

4) After all soil samples have been placed into glass containers, determine volume occupied by soil and lightly pack similar soils to the same volume, that is, within nearest 2.5 mL. Add defined quantity of deionized water to surface of soil and place glass container with soil into incubation jar. Amount of water should be 50% water-filled pore space (WFPS):

$$\text{mL } H_2O = 0.50 \times \text{porosity} \times \text{volume}$$

where porosity is determined from weight and volume occupied by soil, e.g., 50 g soil occupying 40 mL volume has bulk density of 1.25 Mg $m^{-3}$; resulting in 0.53 $m^3$ $m^{-3}$ porosity, therefore 40 mL volume requires 10.6 mL water to achieve 50% WFPS:

$$\text{Total porosity} = \left(1 - \text{BD/PD}\right)$$

where BD = bulk density and PD = particle density (assume 2.65 Mg $m^{-3}$)

5) Add 10 mL of water to plastic vial and place in incubation jar (this is only to maintain humidity in jar)
6) Dispense 10 mL of 1 mol $L^{-1}$ NaOH into a labelled 30-mL Nalgene vial and place it along with glass container of soil into incubation jar (At this point the incubation jar contains a vial of water to maintain humidity, 1 or 2 glass containers of wetted soil to assess microbial activity, and a vial of alkali [NaOH solution] to capture evolved $CO_2$)
7) Place completed box of 12 jars into a constant temperature room at 25 $\pm$ 0.5 °C. Note: include one jar of an alkali blank without soil for every ~25 jars with soil, e.g., 5 blanks for every 100-sample set
8) At the end of 3 d of incubation, remove vial of alkali and immediately cap with air-tight screw-top cap
9) Once alkali vials from all incubation jars have been removed, open an alkali vial to titrate with HCl to a phenolphthalein endpoint (pink to colorless). First, add $BaCl_2$ solution in excess (e.g., 1 to 3.3 mL) to alkali solution to form $BaCO_3$ precipitate. Place a small magnetic stir bar into container and place vial on magnetic stir plate (stirring of solution is necessary to avoid reaction of HCl with $BaCO_3$). Add 3 drops of phenolphthalein indicator solution. Titrate with HCl until color changes from pink to clear/colorless. Record volume to nearest 0.01 mL

10) Calculate the quantity of C mineralized between 0 and 3 d from the equation:

$$\mathrm{CMIN}\left(\mathrm{mg\,kg^{-1}\,soil}\right)=\left(\mathrm{mL_{BLANK}}-\mathrm{mL_{SAMPLE}}\right)\times \mathrm{mol\,L^{-1}\,HCl}\times 6\times 1000/\left(\mathrm{g\,soil}\right)$$

where 6 = equivalent weight of C and 1000 is a conversion factor (g to kg)

Note: Each drop of acid (0.03 mL) is equivalent to 3.6 mg $CO_2$-C $kg^{-1}$ soil (50 g sample) or 1.8 mg $CO_2$-C $kg^{-1}$ soil (100 g sample), and therefore, this can be considered the minimum detection limit.

## Calibration of the Flush of $CO_2$ During 3 d With Other Methods

Limited data comparisons have yet to be made for various soil biological activity methods across a wide range of soil conditions. More direct comparisons of methods are needed to understand meaningful differences and practical similarities among methods. Nonetheless, a few comparisons are available as described in the following. Across 8 soils varying in texture in Texas, the flush of $CO_2$ during 3 d was highly related to that during 1 d (both detected by acid-base titration) [$FCO2_{(1d)} = 10.8 + 0.39 \times FCO2_{(3d)}$, $r^2 = 0.93$] (Franzluebbers et al., 2000). From 12 soils varying in texture in North Carolina, Pennsylvania, and Virginia, the flush of $CO_2$ during 3 d was also highly related to that during 1 d (both detected by acid-base titration) when water was added to the top to 50% water-filled pore space [$FCO2_{(1d)} = -3.8 + 0.45 \times FCO2_{(3d)}$, $r^2 = 0.95$] and when water was added from the bottom via capillary action to a generally more saturated condition (0-1 d incubation only) [$FCO2_{(1d)} = -12.6 + 0.44 \times FCO2_{(3d)}$, $r^2 = 0.73$] (Franzluebbers and Haney, 2018).

Association of the flush of $CO_2$ during 3 d to basal soil respiration, cumulative C mineralization during 24 d, soil microbial biomass C, and net N mineralization has been assembled in a review of the method (Franzluebbers, 2018a). Across 11 states in the United States and from two tropical countries (Alabama, California, Georgia, Guam, Illinois, Iowa, Minnesota, Oklahoma, North Carolina, Pennsylvania, Sao Paulo Brazil, South Carolina, and Virginia), basal soil respiration (BSR; mg $CO_2$-C $kg^{-1}$ soil $d^{-1}$) was closely associated with the flush of $CO_2$ (FCO2; mg $CO_2$-C $kg^{-1}$ 3 $d^{-1}$) according to the following:

$$\mathrm{BSR}=1.2+0.060\times \mathrm{FCO2},\ r^2=0.89,\ n=130$$

Similarly, cumulative C mineralization (CMIN; mg $CO_2$-C $kg^{-1}$ soil 24 $d^{-1}$) and net N mineralization (NMIN; mg N $kg^{-1}$ soil 24 $d^{-1}$) were highly associated with the flush of $CO_2$ according to the following:

$$\mathrm{CMIN}_{(0-24\,d)}=22+2.66\times \mathrm{FCO2},r^2=0.98,n=130$$

$$NMIN_{(0-24\,d)} = 16 + 0.196 \times FCO2, r^2 = 0.62, n = 130$$

From 47 field sites in North Carolina and Virginia, the flush of $CO_2$ was highly associated with net N mineralization under laboratory conditions, and the relationship unaffected by soil texture (Franzluebbers et al., 2018b). From these same sites with soil incubated in the greenhouse without amendment using sorghum-sudangrass (*Sorghum bicolor* ssp. Drummondii) as test crop, plant N uptake during 6-8 weeks was strongly associated with the flush of $CO_2$ ($r^2$ = 0.76, n = 537) (Franzluebbers and Pershing, 2018). Finally, from evaluation of corn yield responses to sidedress N application in these same 47 trials, the flush of $CO_2$ was highly negatively associated with yield response to initial dose of N fertilizer and economically optimum N fertilizer rate (Franzluebbers, 2018b). Further field data supported this initial field evaluation, but considerable variation existed among the diversity of sites tested (Franzluebbers, 2020c). Large discrepancy between lab data and field observations could be expected, as temperature and moisture are variable in the field, and especially among fields within a large geographic region across several states. Of course, many other reasons can be expected as to why a lab-based prediction might not correspond directly to field-based observation of yield response to soil N availability, including excessively wet conditions that could cause denitrification of previously mineralized N, spatial contact between mineralized N of surface soil and roots penetrating deeper in soil, and synchrony between time when N is actually mineralized in the field and peak N uptake demand of crops. The observation that some association occurs is a major first step toward better in-field management. Additional field studies in the midwestern United States suggest that the biologically active component of soil was also a key component of soil N availability and may help predict optimum N fertilizer inputs (Yost et al., 2018; Bean et al., 2020). All of these results suggest that soil-test biological activity via the flush of $CO_2$ is a robust and appropriate indicator for assessing soil N availability. Therefore, the flush of $CO_2$ during 3 d is strongly recommended for routine analysis of soil-test biological activity.

## Interpretations

Soil-test biological activity reflects the functional capability of soil to cycle nutrients, decompose organic amendments, and catalyze and stabilize ecosystem processes through the interactions among a diversity of organisms (Franzluebbers, 2016). Combined with total organic matter, water-stable aggregation, routine inorganic nutrient extraction, and some measure of soil microbial diversity, soil-test biological activity will yield a good assessment of the nutrient supplying capacity of many soils. In describing total plant N uptake in the greenhouse without soil amendment, the most important variables (with additional criteria of simple, rapid, and inexpensive) were the flush of $CO_2$ (72% of variation explained), residual inorganic N (7% of variation explained), and total soil N (3% of variation explained) (Franzluebbers and Pershing, 2018). Soil-test biological activity is a reflection of total soil organic matter, but focuses on the most dynamic portion of that organic matter. Some soils have a greater proportion of older, unreactive soil organic matter, so evaluating the active fraction will be more discerning of nutrient cycling. Soil-test biological activity contributes to water-stable aggregation to varying degrees depending on soil type. Soil-test biological activity is an important indicator of biologically active N supply, which has until now been inadequately characterized for making N fertility recommendations.

## References

Anderson, J.P.E. (1982). Soil respiration. In A.L. Page, R.H. Miller, and D.R. Keeney, (Eds.), *Methods of soil analysis, Part 2* (p. 837–871). 2nd Ed. Madison, WI: SSSA.

Bean, M., Kitchen, N.R., Veum, K.S., Camberato, J.J., Ferguson, R.B., Fernandez, F.G., Franzen, D.W., Laboski, C.A.M., Nafziger, E.D., Sawyer, J.E., and Yost, M. (2020). Relating four-day soil respiration to corn nitrogen fertilizer needs across 49 U.S. Midwest fields. *Soil Sci. Soc. Am. J.* https://doi.org/10.1002/saj2.20091

Birch, H.F. (1958). The effect of soil drying on humus decomposition and nitrogen availability. *Plant Soil* 10, 9–31. doi:10.1007/BF01343734

Birch, H.F. (1960). Nitrification in soils after different periods of dryness. *Plant Soil* 12, 81–96. doi:10.1007/BF01377763

Birch, H.F., and Friend, M.T. (1956). Humus decomposition in East African soils. *Nature* 178, 500–501. doi:10.1038/178500a0

Campioli, M., Malhi, Y., Vicca, S., Luyssaert, S., Papale, D., Penuelas, J., Reichstein, M., Migliavacca, M., Arain, M.A., and Janssens, I.A. (2016). Evaluating the convergence between eddy-covariance and biometric methods for assessing carbon budgets of forests. *Nature Commun.* 7, 13717.

Causarano, H.J., Franzluebbers, A.J., Shaw, J.N., Reeves, D.W., Raper, R.L., and Wood, C.W. (2008). Soil organic carbon fractions and aggregation in the Southern Piedmont and Coastal Plain. *Soil Sci. Soc. Am. J.* 72, 221–230. doi:10.2136/sssaj2006.0274

Chu, M., Singh, S., Walker, F.R., Eash, N.S., Buschermohle, M.J., Duncan, L.A., and Jagadamma, S. (2019). Soil health and soil fertility assessment by the Haney soil health test in an agricultural soil in West Tennessee. *Commun. Soil Sci. Plant Anal.* 50, 1123–1131.

Coleman, D.C., Callaham, Jr., M.A, and Crossley, Jr., D.A. (2017). *Fundamentals of soil ecology*, 3rd Ed. San Diego, CA: Academic Press.

Collins, H.P., Elliott, E.T., Paustian, K., Bundy, L.G., Dick, W.A., Huggins, D.R., Smucker, A.J.M., and Paul, E.A. (2000). Soil carbon pools and fluxes in long-term corn belt agroecosystems. *Soil Biol. Biochem.* 32, 157–168. doi:10.1016/S0038-0717(99)00136-4

Culman, S.W., Snapp, S.S., Green, J.M., and Gentry, L.E. (2013). Short- and long-term labile soil carbon and nitrogen dynamics reflect management and predict corn agronomic performance. *Agron. J.* 10, 493–502.

Ellert, B.H., and Bettany, J.R. (1992). Temperature dependence of net nitrogen and sulfur mineralization. *Soil Sci. Soc. Am. J.* 56, 1133–1141. doi:10.2136/sssaj1992.03615995005600040021x

Fine, A.K., van Es, H., and Schindelbeck, R.R. (2017). Statistics, scoring functions, and regional analysis of a comprehensive soil health database. *Soil Sci. Soc. Am. J.* 81, 589–601. doi:10.2136/sssaj2016.09.0286

Franzluebbers, A.J. (1999a). Microbial activity in response to water-filled pore space of variably eroded southern Piedmont soils. *Appl. Soil Ecol.* 11, 91–101. doi:10.1016/S0929-1393(98)00128-0

Franzluebbers, A.J. (1999b). Potential C and N mineralization and microbial biomass from intact and increasingly disturbed soils of varying texture. *Soil Biol. Biochem.* 31, 1083–1090. doi:10.1016/S0038-0717(99)00022-X

Franzluebbers, A.J. (2016). Should soil testing services measure soil biological activity? *Agric. Environ. Lett.* 1, 1–5. doi:10.2134/ael2015.11.0009

Franzluebbers, A.J. (2018a). Short-term C mineralization (aka the flush of $CO_2$) as an indicator of soil biological health. *CAB Reviews* 13, 1–14. doi:10.1079/PAVSNNR201813017

Franzluebbers, A.J. (2018b). Soil-test biological activity with the flush of $CO_2$: III. Corn yield responses to applied nitrogen. *Soil Sci. Soc. Am. J.* 82, 708–721. doi:10.2136/sssaj2018.01.0029

Franzluebbers, A.J. (2020a). Soil carbon and nitrogen mineralization after the initial flush of $CO_2$. *Agric. Environ. Lett.* 5, e20006.

Franzluebbers, A.J. (2020b). Soil mass and volume affect soil-test biological activity estimates. *Soil Sci. Soc. Am. J.* 84, 502–511.

Franzluebbers, A.J. (2020c). Soil-test biological activity with the flush of $CO_2$: V. Validation of nitrogen prediction for corn production. *Agron. J.* 112, 2188–2204.

Franzluebbers, A.J., and Arshad, M.A. (1997). Soil microbial biomass and mineralizable carbon of water-stable aggregates. *Soil Sci. Soc. Am. J.* 61, 1090–1097.

Franzluebbers, A.J., and Brock, B.G. (2007). Surface soil responses to silage cropping intensity on a Typic Kanhapludult in the piedmont of North Carolina. *Soil Tillage Res.* 93, 126–137.

Franzluebbers, A.J., Endale, D.M., Buyer, J.S., and Stuedemann, J.A. (2012). Tall fescue management in the Piedmont: Sequestration of soil organic carbon and total nitrogen. *Soil Sci. Soc. Am. J.* 76, 1016–1026. doi:10.2136/sssaj2011.0347

Franzluebbers, A.J., and Haney, R.L. (2018). Evaluation of soil processing conditions on mineralizable C and N across a textural gradient. *Soil Sci. Soc. Am. J.* 82, 354–361. doi:10.2136/sssaj2017.08.0275

Franzluebbers, A.J., Haney, R.L., Honeycutt, C.W., Arshad, M.A., Schomberg, H.H., and Hons, F.M. (2001). Climatic influences on active fractions of soil organic matter. *Soil Biol. Biochem.* 33, 1103–1111. doi:10.1016/S0038-0717(01)00016-5

Franzluebbers,

A.J., Haney, R.L., Honeycutt, C.W., Schomberg, H.H., and Hons, F.M. (2000). Flush of carbon dioxide following rewetting of dried soil relates to active organic pools. *Soil Sci. Soc. Am. J.* 64, 613–623. doi:10.2136/sssaj2000.642613x

Franzluebbers, A.J., Haney, R.L., Hons, F.M., and Zuberer, D.A. (1996a). Determination of microbial biomass and nitrogen mineralization following rewetting of dried soil. *Soil Sci. Soc. Am. J.* 60, 1133–1139. doi:10.2136/sssaj1996.03615995006000040025x

Franzluebbers, A.J., Hons, F.M., and Zuberer, D.A. (1994). Season changes in soil microbial biomass and mineralizable C and N in wheat management systems. *Soil Biol. Biochem.* 26, 1469–1475. doi:10.1016/0038-0717(94)90086-8

Franzluebbers, A.J., Hons, F.M., and Zuberer, D.A. (1995). Tillage and crop effects on seasonal soil carbon and nitrogen dynamics. *Soil Sci. Soc. Am. J.* 59, 1618–1624. doi:10.2136/sssaj1995.03615995005900060016x

Franzluebbers, A.J., Hons, F.M., and Zuberer, D.A. (1996b). Seasonal dynamics of active soil carbon and nitrogen pools in conventional and no tillage under intensive cropping. *J. Plant Nutr. Soil Sci.* 159, 343–349.

Franzluebbers, A.J., and Pershing, M.R. (2018). Soil-test biological activity with the flush of $CO_2$: II. Greenhouse growth bioassay from soils in corn production. *Soil Sci. Soc. Am. J.* 82, 696–707. doi:10.2136/sssaj2018.01.0024

Franzluebbers, A.J., Pehim-Limbu, S., and Poore, M.H. (2018a). Soil-test biological activity with the flush of CO2: IV. Fall-stockpiled tall fescue yield response to applied nitrogen. *Agron. J.* 110, 2033–2049.

Franzluebbers, A.J., Pershing, M.R., Crozier, C., Osmond, D., and Schroeder-Moreno, M. (2018b). Soil-test biological activity with the flush of $CO_2$: I. C and N characteristics of soils in corn production. *Soil Sci. Soc. Am. J.* 82, 685–695. doi:10.2136/sssaj2017.12.0433

Franzluebbers, A.J., and Poore, M.H. (2020). Soil-test biological activity with the flush of $CO_2$: VII. Validating nitrogen needs for fall-stockpiled forage. *Agron. J.* 112, 2240–2255.

Franzluebbers, A.J., Schomberg, H.H., and Endale, D.M. (2007). Surface-soil responses to paraplowing of long-term no-tillage cropland in the Southern Piedmont USA. *Soil Tillage Res.* 96, 303–315.

Franzluebbers, A.J., and Stuedemann, J.A. (2003). Bermudagrass management in the Southern Piedmont USA. III. Particulate and biologically active soil carbon. *Soil Sci. Soc. Am. J.* 67, 132–138. doi:10.2136/sssaj2003.1320

Franzluebbers, A.J., and Stuedemann, J.A. (2005). Soil carbon and nitrogen pools in response to tall fescue endophyte infection, fertilization, and cultivar. *Soil Sci. Soc. Am. J.* 69, 396–403.

Franzluebbers, A.J., and Stuedemann, J.A. (2008). Early response of soil organic fractions to tillage and integrated crop–livestock production. *Soil Sci. Soc. Am. J.* 72, 613–625.

Franzluebbers, A.J., and Stuedemann, J.A. (2015). Does grazing of cover crops impact biologically active soil C and N fractions under inversion and no tillage management? *J. Soil Water Conserv.* 70, 365–373. doi:10.2489/jswc.70.6.365

Franzluebbers, A.J., and Veum, K.S. (2020). Comparison of two alkali trap methods for measuring the flush of $CO_2$. *Agron. J.* 112, 1279–1286.

Franzluebbers, K., Franzluebbers, A.J., and Jawson, M.D. (2002). Environmental controls on soil and whole-ecosystem respiration from a tallgrass prairie. *Soil Sci. Soc. Am. J.* 66, 254–262. doi:10.2136/sssaj2002.2540

Green, V.S., Stott, D.E., Cruz, J.C., and Curi, N. (2007). Tillage impacts on soil biological activity and aggregation in a Brazilian Cerrado Oxisol. *Soil Tillage Res.* 92, 114–121. doi:10.1016/j.still.2006.01.004

Haney, R.L., Brinton, W.H., and Evans, E. (2008). Soil CO2 respiration: Comparison of chemical titration, CO2 IRGA analysis and the Solvita gel system. *Renew. Agric. Food Syst.* 23, 171–176. doi:10.1017/S174217050800224X

Haney, R.L., Franzluebbers, A.J., Porter, E.B., Hons, F.M., and Zuberer, D.A. (2004). Soil carbon and nitrogen mineralization: Influence of drying temperature. *Soil Sci. Soc. Am. J.* 68, 489–492.

Haney, R.L., and Haney, E.B. (2010). Simple and rapid laboratory method for rewetting dry soil for incubations. *Commun. Soil Sci. Plant Anal.* 41, 1493–1501.

Haney, R.L., Haney, E.B., Smith, D.R., Harmel, R.D., and White, M.J. (2018a). The soil health tool – Theory and initial broad-scale application. *Appl. Soil Ecol.* 125, 162–168.

Haney, R.L., Haney, E.B., White, M.J., and Smith, D.R. (2018b). Soil $CO_2$ response to organic and amino acids. *Appl. Soil Ecol.* 125, 297–300.

Haney, R.L., Hons, F.M., Sanderson, M.A., and Franzluebbers, A.J. (2001). A rapid procedure for estimating nitrogen mineralization in manured soil. *Biol. Fertil. Soils* 33, 100–104.

Ingram, L.J., Schuman, G.E., Stahl, P.D., and Spackman, L.K. (2005). Microbial respiration and organic carbon indicate nutrient cycling recovery in reclaimed soils. *Soil Sci. Soc. Am. J.* 69, 1737–1745.

Jangid, K., Williams, M.A., Franzluebbers, A.J., Blair, J.M., Coleman, D.C., Whitman, W.B. (2010). Development of soil microbial communities during tallgrass prairie restoration. *Soil Biol. Biochem.* 42, 302–312.

Jangid, K., Williams, M.A., Franzluebbers, A.J., Sanderlin, J.S., Reeves, J.H., Jenkins, M.B., Endale, D.M., Coleman, D.C., and Whitman, W.B. (2008). Relative impacts of land-use, management intensity and fertilization upon soil microbial community structure in agricultural systems. *Soil Biol. Biochem.* 40, 2843–2853.

Jangid, K., Williams, M.A., Franzluebbers, A.J., Schmidt, T.M., Coleman, D.C., and Whitman, W.B. (2011). Land-use history has a stronger impact on soil microbial community composition than aboveground vegetation and soil properties. *Soil Biol. Biochem.* 43, 2184–2193.

Jenkinson, D.S., and Powlson, D.S. (1976). The effects of biocidal treatments on metabolism in soil– I. Fumigation with chloroform. *Soil Biol. Biochem.* 8, 167–177. doi:10.1016/0038-0717(76)90001-8

Lebedjantzev, A.N. (1924). Drying of soil as one of the natural factors in maintaining soil fertility. *Soil Sci.* 18, 419–448. doi:10.1097/00010694-192412000-00001

Linn, D.M., and Doran, J.W. (1984). Effect of water-filled pore space on carbon dioxide and nitrous oxide production in tilled and nontilled soils. *Soil Sci. Soc. Am. J.* 48, 1267–1272. doi:10.2136/sssaj1984.03615995004800060013x

McGowen, E.B., Sharma, S., Deng, S., Zhang, H., and Warren, J.G. (2018). An automated laboratory method for measuring $CO_2$ emissions from soils. *Agric. Environ. Lett.* 3, 1–5. doi:10.2134/ael2018.02.0008

Moebius-Clune, B.N., Moebius-Clune, D.J., Gugino, B.K., Idowu, O.J., Schindelbeck, R.R., Ristow, A.J., van Es, H.M., Thies, J.E., Shayler, H.A., McBride, M.B., Kurtz, K.S.M., Wolfe, D.W., and Abawi, G.S. (2017). *Comprehensive assessment of soil health – the Cornell framework.* Ed. 3.2. Geneva, NY: Cornell University.

Morrow, J.G., Huggins, D.R., Carpenter-Boggs, L.A., and Reganold, J.P. (2016). Evaluating measures to assess soil health in long-term agroecosystem trials. *Soil Sci. Soc. Am. J.* 80, 450–462. doi:10.2136/sssaj2015.08.0308

Mosier, A.R., Halvorson, A.D., Reule, C.A., and Liu, X.J. (2006). Net global warming potential and greenhouse gas intensity in irrigated cropping systems in northeastern Colorado. *J. Environ. Qual.* 35, 1584–1598. doi:10.2134/jeq2005.0232

Picone, L.I., Cabrera, M.L., and Franzluebbers, A.J. (2002). A rapid method to estimate potentially mineralizable nitrogen in soil. *Soil Sci. Soc. Am. J.* 66, 1843–1847. doi:10.2136/sssaj2002.1843

Runkle, B.R.K., Rigby, J.R., Reba, M.L., Anapalli, S.S., Bhattacharjee, J., Krauss, K.W., Liang, L., Locke, M.A., Novick, K.A., Sui, R., Suvocarev, K., and White, P.M., Jr. (2017). Delta-flux: An eddy covariance network for a climate-smart lower Mississippi Basin. *Agric. Environ. Lett.* 2, 170003.

Sainju, U.M., Singh, B.P., and Whitehead, W.F. (2000). Cover crops and nitrogen fertilization effects on soil carbon and nitrogen and tomato yield. *Can. J. Soil Sci.* 80, 523–532. doi:10.4141/S99-107

Sainju, U.M., Stevens, W.B., Evans, R.G., and Iversen, W.M. (2013). Irrigation system and tillage effects on soil carbon and nitrogen fractions. *Soil Sci. Soc. Am. J.* 77, 1225–1234. doi:10.2136/sssaj2012.0412

Sherrod, L.A., Reeder, J.D., Hunter, W., and Ahuja, L.R. (2012). Rapid and cost-effective method for soil carbon mineralization in static laboratory incubations. *Commun. Soil Sci. Plant Anal.* 43, 958–972. doi:10.1080/00103624.2012.653031

Stott, D.E. (2019). *Recommended soil health indicators and associated laboratory procedures.* Soil Health Technical Note No. 450-03. Washington, D.C.: United States Department of Agriculture, Natural Resources Conservation Service.

Stott, D.E., Elliott, L.F., Papendick, R.I., and Campbell, G.S. (1986). Low temperature or low water potential effects on the microbial decomposition of wheat residue. *Soil Biol. Biochem.* 18, 577–582. doi:10.1016/0038-0717(86)90078-7

Sun, S.-Q., Bhatti, J.S., Jassal, R.S., Chang, S.X., Arevalo, C., Black, T.A., and Sidders, D. (2015). Stand age and productivity control soil carbon dioxide efflux and organic carbon dynamics in poplar plantations. *Soil Sci. Soc. Am. J.* 79, 1638–1649. doi:10.2136/sssaj2015.06.0233

Tian, G., Franzluebbers, A.J., Granato, T.C., Cox, A.E., and O'Connor, C. (2013). Stability of soil organic matter under long-term biosolids application. *Appl. Soil Ecol.* 64, 223–227. doi:10.1016/j.apsoil.2012.12.001

Venterea, R.T., Baker, J.M., Dolan, M.S., and Spokas, K.A. (2006). Carbon and nitrogen storage are greater under biennial tillage in a Minnesota corn-soybean rotation. *Soil Sci. Soc. Am. J.* 70, 1752–1762. doi:10.2136/sssaj2006.0010

Wade, J., Culman, S.W., Hurisso, T.T., Miller, R.O., Baker, L., and Horwath, W.R. (2018). Sources of variability that compromise mineralizable carbon as a soil health indicator. *Soil Sci. Soc. Am. J.* 82, 243–252. doi:10.2136/sssaj2017.03.0105

Waksman, S.A., and Starkey, R.L. (1924). Microbiological analysis of soil as an index of soil fertility: VII. *Carbon dioxide evolution. Soil Sci.* 17, 141–162. doi:10.1097/00010694-192402000-00004

Yost, M.A., K.S. Veum, N.R. Kitchen, J.E. Sawyer, J.J. Camberato, P.R. Carter, R.B. Ferguson, F.G. Fernandez, D.W. Franzen, C.A. Laboski., and E.D. Nafziger. (2018). Evaluation of the Haney soil health tool for corn nitrogen recommendations across eight Midwest states. *J. Soil Water Conserv.* 73, 587–592.

Zak, D.R., Holmes, W.E., MacDonald, N.W., and Pregitzer, K.S. (1999). Soil temperature, matric potential, and the kinetics of microbial respiration and nitrogen mineralization. *Soil Sci. Soc. Am. J.* 63, 575–584. doi:10.2136/sssaj1999.03615995006300030021x

Zibilske, L.M. (1994). Carbon mineralization. In Weaver, R.W., Angle, J.S., and Bottomley, P.J. editors, *Methods of soil analysis, Part 2. Microbiological and biochemical properties* (p. 835–863). Book series 5.2. Madison, WI: SSSA.

# 9

# Permanganate Oxidizable Carbon: An Indicator of Biologically Active Soil Carbon

*Steve W. Culman, Tunsisa T. Hurisso, and Jordon Wade*

Soil organic matter (SOM) is a pivotal component of healthy and functioning soils, as it links soil physical, chemical, and biological properties and processes into a coherent and living matrix. Soil C is the primary constituent of organic matter (~50%; Pribyl, 2010), and therefore soil organic C (SOC) has become the primary or even universal currency of soil health. The SOC quantity in every soil reflects the net effect of inputs (i.e., the supply of plant-derived C) minus outputs (i.e., respiration of $CO_2$), mediated through microbial and mineral complex interactions (Cotrufo, Wallenstein, Boot, Denef, & Paul, 2013).

Commercial soil testing laboratories typically measure the total SOM pool, making it one of the most important indicators of overall soil health (Blanco-Canqui & Benjamin, 2015; Byrnes, Eastburn, Tate, & Roche, 2018; Gregorich, Monreal, Carter, Angers, & Ellert, 1994; Oldfield, Bradford, & Wood, 2019; Wander & Drinkwater, 2000). But SOM is often slow to respond or even insensitive to new soil and crop management practices, and thus it may require several years for changes to be detected (Culman et al., 2013; Drinkwater, Wagoner, Sarrantonio, 1998; Haynes, 2005; Robertson, Paul, & Harwood, 2000). Furthermore, total SOM is strongly influenced by clay content (Schimel et al., 1994; Six, Conant, Paul, & Paustian, 2002) and soil mineral composition (Rasmussen et al., 2018), which means that soil type can have greater effects on total SOM than management.

The active or biologically active pool of SOM is small but extremely important because it reflects relatively recent C additions as well as the size and composition of the microbial community (McDaniel, Tiemann, & Grandy, 2014; Wander, 2004).

*Soil Health Series: Volume 2 Laboratory Methods for Soil Health Analysis*, First Edition.
Edited by Douglas L. Karlen, Diane E. Stott, and Maysoon M. Mikha.

Biologically active C is thus more sensitive to recent management practices than total SOM and has been identified as a measurement that integrates soil C dynamics. This short-term responsiveness has made biologically active C an important early indicator of potential SOC accrual and capable of assessing overall soil biological health.

## Soil Organic Carbon Pools

Soil organic matter is primarily derived from plant and microbial products and is incredibly complex in both form and function (Lavallee, Soong, & Cotrufo, 2020; Lehmann & Kleber, 2015). To better quantify, understand, and model C dynamics, scientists have historically separated total SOM into multiple pools, each with different properties. A common three-pool conceptualization is: (a) stable, (b) slow, and (c) active, for which mean residence times (i.e., turnover rates) increase as pool size (% of total) decreases (Figure 9.1; Wander, 2004).

Recent advances in the understanding of SOM have revised and even overturned many long-standing paradigms in soil science. Most notably are the importance of microbial processing and subsequent physical protection of SOM rather than chemical recalcitrance as primary mechanisms leading to soil C storage, accrual, and sequestration (Cotrufo et al., 2013; Lavallee et al., 2020; Lehmann & Kleber, 2015; Schmidt et al., 2011). While these advances have necessitated revisiting and refining some conceptual frameworks, fractionating organic matter remains an important tool for advancing our understanding of SOM dynamics.

The labile or biologically active pool of SOM is of particular interest for soil health testing and assessing agroecosystem performance because it strongly influences critical soil processes such as nutrient cycling and availability, soil aggregation, and soil C accrual or loss (Jensen et al., 2019; Six, Elliott, Paustian, & Doran, 1998; von Lützow et al., 2007; Wander, 2004; Wardle, 1992). Decades of research have confirmed that biologically active soil C is sensitive to tillage, crop rotation, and a variety of soil amendments (Awale, Emeson, & Machado, 2017; Haynes, 2005). Labile soil C provides an important soil nutrient pool for plants and the soil food web. The size of this pool has important implications not only for plant nutrition, but also for nutrient storage and the synchrony of nutrient release relative to plant demand (von Lützow et al., 2007).

## Indicators of Biologically Active soil Carbon

There are numerous ways to measure labile or active SOC pools, as several previous reports have reviewed (Haynes, 2005; von Lützow et al., 2007; Wander, 2004). The methods vary conceptually based on their target, i.e., physical, chemical, or biological activity (Figure 9.1). Nearly all biologically active soil C measurements are operationally defined based on the C fraction being quantified. This chapter focuses on permanganate-oxidizable C (POXC), which originally described by Weil, Islam, Stine, Gruver, and Sampson-Liebig (2003) and is recommended for rapid soil health assessment of biologically active C. Permanganate-oxidizable C is a rapid and inexpensive measurement, easily performed in the laboratory, and also suitable for on-site soil testing (Stiles et al., 2011). Here, we compare POXC with three other widely deployed methods including particulate organic matter (POM), microbial biomass, and extractable organic C. These methods pre-date POXC as reliable indicators of biologically active or labile organic C, but they are labor intensive and expensive to perform. For example, Bongiorno et al. (2019) estimated that various extractable organic C and POM fractions took 3 to 32 times longer to run and cost 1.4 to 2.7 times more in analytical expenses than POXC. As a result, few commercial soil testing laboratories offer these analyses for routine testing of soils. Two other important measurements that directly reflect biologically active C pools (microbial enzymes and mineralizable C) are discussed in chapters 11 and 8, respectively.

## Particulate Organic Matter

Fractionating soil C pools significantly improved our understanding of how management influences soil C dynamics. One of the first products identified was POM, which is commonly fractionated by size, density, or even chemistry. A universally accepted definition of POM is lacking, but it is typically defined as being >53 μm in size, lighter than 1.6–1.85 g $cm^{-3}$, and a non-water-extractable fraction (Lavallee et al., 2020). As SOM science has evolved, numerous other methods to fractionate, quantify, and define SOM pools have proliferated. For example,

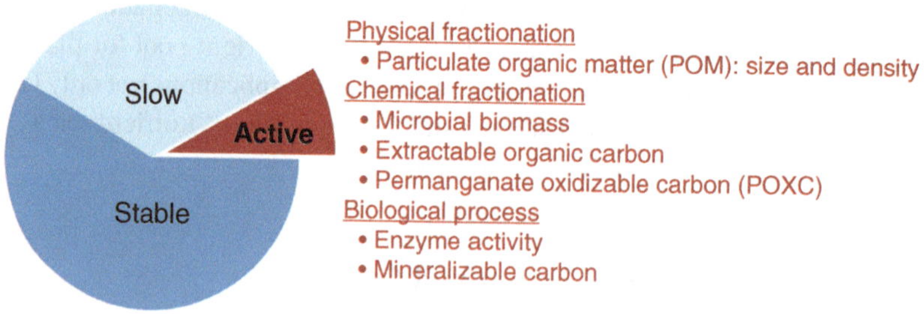


**Figure 9.1** Conceptual pools and potential methods to measure active soil organic matter.

Poeplau et al. (2018) systematically evaluated 20 distinct SOC fractionation methods based on physical (aggregate, size, density) and chemical properties. See https://www.somfractionation.org/ for an extensive description of available SOM fractionation methodologies.

As an established indicator of biologically active C, POM has successfully reflected changes in management practices across a wide range of soils and ecosystems (Haynes, 2005; Poeplau et al., 2018; Six et al., 1998; von Haden, Kucharik, Jackson, & Marín-Spiotta, 2019; Wander, 2004). Recently, Lavallee et al. (2020), noting the diversity of POM methodologies and lack of standardization, proposed to simplify our conceptualization of SOM into two pools: POM and mineral-associated organic matter. They argued that simplification would help move SOC assessment toward more standardized methods, thus enabling more robust, multi-study comparisons.

## Microbial Biomass Carbon

Microbial biomass C (MBC) reflects the size of the soil microbial pool and has served as an important indicator, advancing our understanding of the role microbial communities play in stabilizing SOC. Microbial biomass C can be measured via numerous established methods including chloroform fumigation–incubation (Jenkinson & Powlson, 1976), chloroform fumigation–extraction (Vance, Brookes, & Jenkinson, 1987), direct chloroform extraction (Fierer, Schimel, & Holden, 2003), substrate induced respiration (Anderson & Domsch, 1978), microwave digestion (Islam & Weil, 1998), and by quantifying total phospholipid fatty acid content within the soil profile (Buyer & Sasser, 2012). Measurements of MBC can be adjusted with varying $K_{ec}$ factors to adjust for different extraction efficiencies (Joergensen, 1996; Wardle, 1992), which further complicates robust, multi-study comparisons. Numerous reviews and meta-analyses have demonstrated that MBC is sensitive to changes in management practices including, tillage (Zuber & Villamil, 2016), N fertilization (Geisseler & Scow, 2014; Treseder, 2008), cover crops (Kim, Zabaloy, Guan, & Villamil, 2020), and crop rotational diversity (McDaniel et al., 2014).

## Extractable Organic Carbon

Extractable organic C (EOC) is probably one of the most labile SOM fractions (Marschner & Kalbitz, 2003). Several extractants and methods can be used to measure EOC; the most common are $K_2SO_4$—which is often referred to as dissolved organic C—and deionized water, which is referred to as water-extractable organic C (WEOC). Water extractions are further divided into hot-water (~80 °C) and cold-water extractions (~22 °C). Often, the suspended organic matter is further clarified by high-speed centrifugation or filtering through 0.45-mm filters (Marschner & Kalbitz, 2003). Generally, the amount of organic C from these extracts is: hot WEOC > cold WEOC > dissolved organic C (Bongiorno et al., 2019; Haynes, 2005).

Recent proposals to include WEOC and water-extractable organic N as components of a soil health assessment framework have demonstrated mixed results with regard to agronomic outcomes (Haney, Haney, Smith, Harmel, & White, 2018; Yost et al., 2018). Water extractions are highly susceptible to variations in shaking time, temperature, or ion species and activity in solution (Li & Hur, 2017; Marschner & Kalbitz, 2003; Rousk & Jones, 2010), which could account for inconsistent results across studies (Haney et al., 2018; Morrow, Huggins, Carpenter-Boggs, & Reganold, 2016). Coupling extractable organic C measurements with absorbance or fluorescence spectroscopy can yield insight into SOM quality and soil fertility outcomes (Jaffrain, Gérard, Meyer, & Ranger, 2007; Rinot et al., 2018), although analyses of those data are often quantitatively complex. Therefore, while EOC is theoretically a viable measurement of organic matter quality, it has not yet been widely implemented.

## Permanganate-Oxidizable Carbon

### History of POXC

Metabolism or decomposition of SOM generally proceeds from the most (oxidized) to least (reduced) thermodynamically stable form (Barré et al., 2016). As a general oxidant of reduced soil C, potassium permanganate ($KMnO_4$) is a useful approximation of active organic matter. Potassium permanganate has been used to fractionate and characterize SOM (Blair, Lefroy, & Lisle, 1995; Matsuda & Schnitzer, 1972) for both C (Lefroy, Blair, & Strong, 1993; Loginow, Wisniewski, Gonet, & Ciescinska, 1987) and N (Bundy & Bremner, 1973; Carski & Sparks, 1987). Earlier studies attempted to use $KMnO_4$ to oxidize all soil C, but subsequent studies documented that weaker concentrations (0.02–0.033 mol $L^{-1}$ $KMnO_4$) were consistently more sensitive to changes in soil and crop management or environmental conditions (Bell, Moody, Connolly, & Bridge, 1998; Lefroy et al., 1993; Vieira et al., 2007; Weil et al., 2003).

Weil et al. (2003) reported modifications using $KMnO_4$ intended to streamline the approach, namely reducing reactant concentration to 0.02 mol $L^{-1}$ $KMnO_4$, using 1.0 mol $L^{-1}$ $CaCl_2$ to flocculate the soil, eliminating a centrifugation step, and reducing the overall shaking time. They defined this new method as a measure of "active C". However, as Culman et al., (2012) noted "active C" is vague and lacks precision for a scientific audience. Furthermore, evidence suggests that "active C" may be a misleading term for what this measurement functionally reflects (see below). Since their original publication, hundreds of studies have used POXC to evaluate soil health and C dynamics (i.e., as of early 2020, Scopus = 413 citations; Google Scholar = 740).

## Functional Role of POXC

Like most soil C fractions, POXC is an operationally defined pool of easily oxidized C. Early work using dilute $KMnO_4$ considered POXC to be "active" or labile C (Weil et al., 2003). Subsequent studies using the same $KMnO_4$ concentration found that POXC did not oxidize the most readily available carbohydrates (e.g., glucose) but preferentially oxidized slightly less readily available compounds, such as glycol-containing compounds (Tirol-Padre & Ladha, 2004). With current $KMnO_4$ concentrations (0.02 mol $L^{-1}$), a routine POXC analysis oxidizes much less SOC than other common soil oxidants (i.e., NaOCl or $H_2O_2$) (Margenot, Calderón, Magrini, & Evans, 2017) and therefore represents 1–4% of the total organic C pool (Culman et al., 2012; Hurisso et al., 2016). Furthermore, studies using a variety of spectroscopic techniques found that POXC oxidizes both aliphatic and aromatic compounds (Romero et al., 2018), although aliphatic compounds (e.g., lipids and proteins) tend to be preferentially oxidized (Arachchige, 2016; Calderón et al., 2017; Ramírez, Calderón, Fonte, Santibáñez, & Bonilla, 2020; Romero et al., 2018). Therefore, while not a compositionally distinct pool, POXC seems to be functionally representative of SOC that is slightly processed rather than the most labile C forms (Culman et al., 2012).

The emergent view of POXC functionality is that it represents a moderately stable, potentially mineral-associated SOC pool that is slightly processed, presumably representing microbial metabolites (Culman et al., 2012). These microbial byproducts—which readily form associations with the soil mineral phase—comprise a considerable amount of the more-stable SOC and are relatively enriched with lipids and proteins (Kallenbach, Frey, & Grandy, 2016; Kleber, Sollins, & Sutton, 2007; Kopittke et al., 2018; Liang, Schimel, & Jastrow, 2017). As mentioned above, POXC has a slight preference for oxidizing these aliphatic-rich compounds, suggesting that POXC should also be associated with more-stable C. A wide-ranging POXC study found that POXC was associated with a suite of management practices that tend to stabilize SOC (Hurisso et al., 2016). Although that study was associative, recent studies have supported this view. For example, POXC has been found to strongly correlate with mineral-associated organic C ($R^2 = .92$; $p < .001$; Sherrod, Virgil, & Stewart, 2019) as well as reactive soil minerals (i.e., Al and Fe) that bind soil C (Arachchige, 2016).

In addition to the chemical evidence that POXC is a more stable form of SOC, recent studies have shown POXC to be associated with greater aggregate stability and non-dispersible clays (Fine, van Es, & Schindelbeck, 2017; Jensen et al., 2019, 2020; Wade et al., 2019).

From a practical soil health testing perspective, Hurisso et al. (2016) proposed a simple and intuitive reporting framework using two common indicators of biologically active C—POXC and soil mineralizable C (Chapter 8)—that can collectively provide information about SOC mineralization vs. stabilization processes (Figure 9.2). This framework proposes that POXC is more representative of soil C stabilization processes, while mineralizable C is more reflective of mineralization processes. Typically, these two indicators are related, and as SOC increases, both indicators increase (Hurisso et al., 2016; Morrow et al., 2016). However, Hurisso et al. (2016) showed that the relative rate of increase between the two indicators is reflective of specific management practices. For example, across 13 diverse studies, conservation tillage and compost additions, two practices known to increase soil organic matter stabilization, were both associated with positive residual values, indicating greater increases in POXC relative to mineralizable C. Conversely, conventional tillage and manure additions were both associated with negative residual values, indicating greater influence of mineralizable C and mineralization processes (Hurisso et al., 2016). These relationships have since

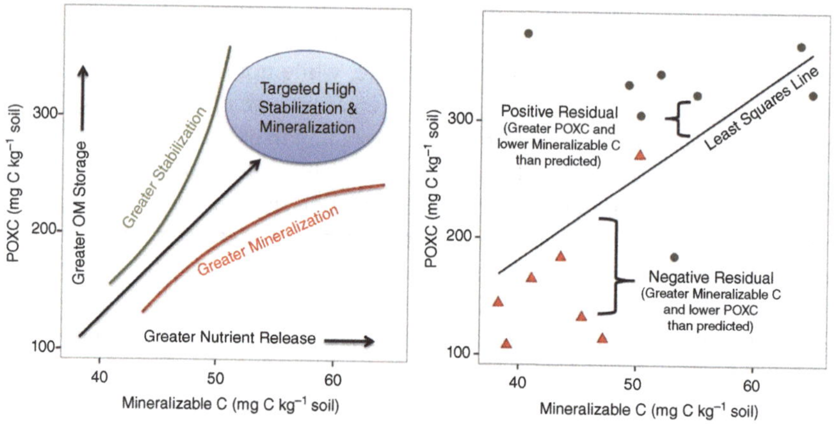


**Figure 9.2** A proposed soil health reporting framework using two recommended soil health indicators, permanganate-oxidizable C (POXC) and mineralizable C. Soil management practices that promote greater soil organic matter (OM) stabilization will be more strongly influenced by POXC values, while practices that result in more soil OM mineralization will be more influenced by mineralizable C (left panel). These relationships can be evaluated by extracting residuals from a regression analysis and linking management practices to positive or negative residuals (right panel) (adapted from Hurisso et al., 2016).

been corroborated with long-term maize (*Zea mays* L.) trials in Kenya (Sprunger, Culman, Palm, Thuita, & Vanlauwe, 2019), long-term biofuel trials in the upper Midwest (Sprunger, Martin, & Mann, 2020), and for corn (*Zea mays* L.) production systems in Ohio (Wade et al., 2019). The soil health reporting framework of Hurisso et al. (2016) seems to be robust across a range of agroecosystems, but additional studies are needed for site-specific or regional calibration. To facilitate those efforts, R code for this analysis can be found at https://github.com/jordon-wade/POXC-chapter.

**Methodological Standardization of POXC**

A significant advantage of POXC is that it is a relatively simple method to use. There are essentially two steps: sample reaction and sample dilution. Few studies have deviated from the original Weil et al. (2003) method, although one significant change was reducing the amount of soil from 5.0 to 2.5 g. This recommendation came from Dr. Weil and was personally communicated to several scientists (e.g., Culman et al., 2012; Moebius-Clune et al., 2017). Another change has been a modification of the settling time. The original method used a 2-min shaking time and 10-min settling time (Weil et al., 2003), which were further articulated and standardized by Culman et al. (2012). However, the Cornell Assessment of Soil Health (CASH) protocol, which the NRCS through Technical Note 450-03 (Stott, 2019) has adopted as the recommended procedure, uses an 8-min settling time (Moebius-Clune et al., 2017). Currently, it is unclear how the 8 versus 10 min of settling time impacts the final POXC measurements, but given the time-sensitive nature of the reaction, it is probable that an 8-min setting time will result in lower total POXC values. Despite these procedural differences, the relatively consistent methodology of POXC is a major advantage that facilitates comparisons across many studies. In contrast, the complexity of other soil C fractionation methods are not as amenable, as discussed above.

**Sensitivity of POXC to Management Practices**

Weil et al. (2003) reported that POXC was sensitive to tillage differences. Since that time, numerous studies have confirmed that POXC can detect differences associated with soil and crop management as well as environmental conditions across a variety of soils (Bongiorno et al., 2019; Culman et al., 2012; Hurisso et al., 2016; Morrow et al., 2016; Nunes, Karlen, Veum, Moorman, & Cambardella, 2020; Ramírez et al., 2020). Relative to other SOC indicators, POXC has consistently ranked in the top set of indicators for detecting management differences. For example, in 12 studies across 53 sites, Culman et al. (2012) found that POXC was a more sensitive indicator than POM, MBC, or total SOC. Morrow et al. (2016) also measured numerous labile C and N indicators across five field experiments and concluded that POXC was the most reliable and sensitive indicator for assessing soil health. Likewise, Awale et al. (2017) measured a suite of C and N indicators and also concluded that POXC and WEOC were the most sensitive indicators for detecting tillage-induced SOM changes. Utilizing the large CASH dataset, Fine et al. (2017) used best subset regression to determine which indicators best reflected the constructed, overall soil health score. Their results indicated that POXC accounted for 45% of the variation and was the overall best indicator of soil health.

In addition to sensitivity, analytical variability is an important measure of an indicator's ability to detect significant differences among management practices. Hurisso, Culman, and Zhao (2018) evaluated the analytical, temporal, and spatial variability of POXC relative to routine soil nutrient measurements (pH, extractable P and K, organic matter via loss-on-ignition). They concluded that (a) POXC had similar analytical variability to loss-on-ignition estimates of SOM, (b) POXC exhibited temporal variability similar to those associated with routine soil nutrient tests, and (c) POXC did not require greater soil sampling densities within a field than used for routine nutrient testing (Hurisso et al., 2018). Another interlaboratory comparison among 12 laboratories with diverse soils (Wade, Maltais-Landry, et al., 2020) found median within-laboratory coefficients of variation for POXC to be 6.5%, while interlaboratory variability averaged 13.4%. Those studies confirm that as a milligram per kilogram (ppm) C measurement, POXC does exhibit intra- and interlaboratory variability, but this variability is similar to that for other routine C-based measurements (see procedural notes below for details of POXC variability over 5 yr).

### Relationship of POXC to Crop Yield and Productivity

Soil health indicators that are related to crop productivity are of great interest to farmers. There are several reports of positive relationships between POXC and crop productivity. Weil et al. (2003) reported a positive relationship with POXC and corn dry matter yield ($R^2 = 0.4$). This has been supported by several other studies (e.g., Culman et al., 2013; de Moraes Sá et al., 2014; Majumder, Mandal, Bandyopadhyay, & Chaudhury, 2007; Stine & Weil, 2002), although sometimes those relationships were no stronger than the crop yield response to total SOC (Lucas & Weil, 2012). When regression analysis has been used with multiple soil health indicators, POXC has sometimes, but not always, been selected as being related to yield (Hurisso et al., 2016; Sprunger et al., 2019). However, one study of 29 replicated field trials across the midwestern United States showed that—independent of the relationship with SOM content—POXC was an effective soil health metric for predicting crop productivity across the Corn Belt (Wade, Culman, et al., 2020). Presumably there is a suite of direct and indirect mechanisms responsible for the overall positive relationship between POXC and crop productivity, but most are probably site specific and will require independent validation to advance our understanding.

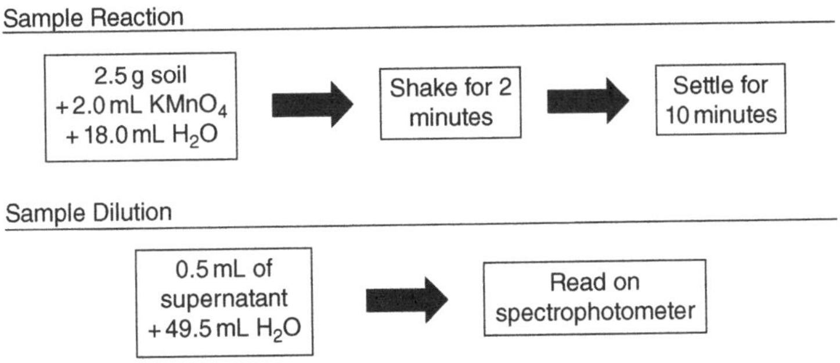


**Figure 9.3** Overview of permanganate-oxidizable C method.

## Procedure to Quantify Poxc

The POXC procedure involves two steps: (a) sample reaction, and (b) sample dilution (Figure 9.3).

### Materials and Reagents

$KMnO_4$ stock solution preparation:

- Reagent-grade $KMnO_4$ (formula weight = 158.03 g $mol^{-1}$)
- Reagent-grade $CaCl_2{\cdot}2H_2O$ (formula weight = 147.01 g $mol^{-1}$)
- Magnetic stir plate and stir bars
- Laboratory glassware for reagent preparation and waste collection
- Brown laboratory glassware for reagent storage

Standard preparation:

- 50-mL disposable polypropylene centrifuge tubes with caps
- Adjustable bottle-top dispensers fitted to a bottle of deionized water and calibrated to deliver 18.0 and 49.5 mL
- Adjustable 10-mL and 100–1,000-µL pipettor and tips

Sample reaction and dilution:

- 50 mL-disposable polypropylene centrifuge tubes with caps
- Analytical balance capable of weighing to two decimal places
- Soil checks (pulverized, homogenous soil as laboratory reference samples)
- Adjustable bottle-top dispensers fitted to a bottle of deionized water and calibrated to deliver 18.0 and 49.5 mL
- Adjustable 10-mL and 100–1,000-µL pipettor and tips

- Oscillating (or horizontal) shaker capable of at least 180 oscillations per minute
- Timer capable of tracking time for 2- and10-min intervals

Sample quantification:

- Clear polystyrene flat-bottom cell culture 96-well plates
- Adjustable 30–300-μL pipettor and tips
- Spectrophotometer capable of reading absorbance at 550 nm

## 0.2 M $KMnO_4$ Stock Solution Preparation (Makes 1 L)

1) Weigh 147 g of $CaCl_2$ and place in a 1,000-mL beaker. Add approximately 900 mL of deionized water. Add a stir bar to the beaker, place on a magnetic stir plate, and stir until completely dissolved.
2) Transfer to a 1,000-mL volumetric flask or graduated cylinder. Bring to volume with deionized water.
3) Weigh 31.60 g of $KMnO_4$ into a 1,000-mL beaker and add approximately 900 mL of the $CaCl_2$ solution. Place on the magnetic stir plate with gentle heat and stir until dissolved completely. Note: Dissolution may be slow and, due to the dark color of this solution, it may be necessary to decant some of the solution to check for undissolved $KMnO_4$.
4) The original protocol (Weil et al., 2003) included a step to adjust the pH to 7.2. Gruver (2015) reported a rapid drop in stock solution pH within days after preparing and reported no differences in POXC values with manipulated pH levels of stock solutions. Our observations are consistent with Gruver (2015) and therefore we no longer recommend pH adjustments to the $KMnO_4$ stock solution (see procedural notes below).
5) Pour the solution into a 1,000-mL volumetric flask or graduated cylinder and bring the volume to 1,000 mL with the $CaCl_2$ solution.
6) Transfer to a brown glass bottle and store in a dark place. This stock solution can be used for up to 6 mo.
7) The amount of $KMnO_4$ solution prepared may be adjusted depending on the total number of samples analyzed. One soil sample will use 2.0 mL of 0.2 M $KMnO_4$.

## Standard Preparation

Four solution standards (0.005, 0.01, 0.015, and 0.02 M) are prepared from the $KMnO_4$ stock solution. The standard preparation involves first making a standard stock solution and then diluting to a final working solution standard.

**Part 1. Standard Stock Solutions**
Use the table below to prepare standard stock solutions. These stock solutions can be prepared in centrifuge tubes or in small brown glass bottles and used for 3 d (in glass and in the dark) to prepare working standards.

| Concentration | Volume of $KMnO_4$ stock solution | Volume of deionized water |
|---|---|---|
| M | mL | mL |
| 0.005 | 0.25 | 9.75 |
| 0.01 | 0.5 | 9.5 |
| 0.015 | 0.75 | 9.25 |
| 0.02 | 1.0 | 9.0 |

**Part 2. Dilution Step**
Dilute each standard stock solution to a working standard by adding 0.5 mL of each stock solution to 49.5 mL of deionized water in 50-mL centrifuge tubes. These tubes now contain the working standards and should be prepared fresh daily.

## Sample Reaction

1) Label two 50-mL centrifuge tubes for each sample. One will be the *reaction tube*, the other the *dilution tube*. Weigh 2.50 g (±0.05 g) of air-dried soil into the reaction tube (may be done in advance). Place the dilution tubes aside.
2) Soil checks should be prepared in the same manner as the unknown soils and serve as laboratory reference samples.
3) Add 18.0 mL of deionized water to each of the reaction tubes containing the soil. Using the 1.0–10.0-mL pipettor, add 2.0 mL of 0.2 M $KMnO_4$ stock solution to each tube.
4) Working quickly, cap tubes tightly and place horizontally on a shaker at 180 oscillations per minute for 2 min.
5) After 2 min, remove samples from the shaker and invert the tubes vigorously to ensure that there is no soil clinging to the sides of the tube. Next, remove the caps to avoid further disturbance of the soil after settling. Allow the soil to settle for 10 min. Settling time is a critical step, so a timer is essential.
6) It is important that the timing of each step be consistent, particularly the shaking and settling times. The permanganate will continue to react as long as it remains in contact with the soil. Hence, working quickly with small batches of 10 or fewer samples is advised.

## Sample Dilution

1) While samples are settling, add 49.5 mL of deionized water to the dilution tubes (may be done in advance).
2) Once the 10-min settling period has passed, quickly transfer 0.5 mL of supernatant (avoiding any particulate matter) from the reaction tube to the corresponding dilution tube containing 49.5 mL of water. Note: This step should be performed as quickly as possible because the permanganate will continue to react with the soil as long as it remains in contact.
3) Cap the dilution tubes and invert to mix. These are the final sample solutions for analysis. They are stable for up to 24 h if stored in the dark.

## Sample Quantification

1) This method has been shown to perform well on both single cuvette machines and 96-well plate reading spectrophotometers ($F_{1,10} = 1.7$, $p = .225$; Wade, Maltais-Landry, et al., 2020). If available, a 96-well plate reader is recommended to save time (as outlined below).
2) It is recommended to replicate all standards on a plate, including deionized water blanks. Running each standard two or more times and taking the average typically yields good results.
3) Dispense 200 µL of each standard and unknown sample into each well of the 96-well plate.
4) Determine and record the absorbance (optical density) of the standards and unknowns at 550 nm using spectrophotometer software. Sample absorbance has a broad spectrum, and reports of using different wavelengths (e.g., 540 nm) have yielded results consistent with 550 nm.
5) Subtract the average of the deionized water blanks from all absorbance values. The intercept of the standard curve should be very close to zero.

## Calculating Mass of POXC for Unknown Soil Samples

1) The amount of C oxidized is a function of the quantity of permanganate reduced. Consequently, the higher the POXC values, the lower the absorbance (intensity of the color of the solution).
2) Use the following equation to determine POXC (after Weil et al., 2003):

$$\mathrm{POXC}\left(\mathrm{mg\ kg^{-1}\ soil}\right) = \left[0.02\ \mathrm{mol\ L^{-1}}\left(a + b \times \mathrm{Abs}\right)\right] \times \left(9{,}000\ \mathrm{mg\ C\ mol^{-1}}\right) \times \left(\frac{0.02\ \mathrm{L\ reaction\ solution}}{\mathrm{Wt}}\right)$$

where 0.02 mol $L^{-1}$ is the initial solution concentration, *a* is the intercept of the standard curve, *b* is the slope of the standard curve, Abs is the absorbance of an unknown, 9,000 mg is the C oxidized by 1 mol of $MnO_4$ changing from $Mn^{7+} \rightarrow Mn^{4+}$, 0.02 L is the volume of stock solution reacted, and Wt is the weight of the air-dried soil sample (in kg).

### Example Calculation

Construct standard curve with the following values:

| | | | | |
|---|---|---|---|---|
| *y* axis (molarity of stock $KMnO_4$ standards) [a] | 0.005 | 0.01 | 0.015 | 0.02 |
| *x* axis ((Abs values from spectrophotometer) | 0.1000 | 0.1984 | 0.3034 | 0.3966 |

[a] The standard curve should use the molarity of the stock standards, and not the working standards, since the stock standards represent the actual concentration (0.02 M $KMnO_4$) used to react with the soil.

This produces the regression line $y = 0.0502x - 0.00004$; $R^2 = .999$. For an unknown sample absorbance of 0.3087 and unknown sample soil weight of 2.48 g:

$$\text{POXC} = \{0.02\ \text{M} - [-0.00004 + (0.0502 \times 0.3087)]\}$$
$$\times (9000\ \text{mg C mol}^{-1}) \times \left(\frac{0.02\ \text{L}}{0.00248\ \text{kg}}\right)$$
$$= 329.75\ \text{mg POXC kg}^{-1}\ \text{soil}$$

## Cleanup and Disposal

Leaving the centrifuge tubes capped but on the benchtop for a week or more will allow the permanganate to completely react with the soil and lose all purple pigmentation. The liquid can then be safely disposed of down the sink, and tubes with soil can be thrown out or cleaned and reused. The second dilution of samples and standards contains very little $KMnO_4$ and may be safely flushed down the drain with copious amounts of water. However, check with your environmental health and safety department to ensure compliance with your institution's procedures.

## Procedural Notes

### POXC Soil Mass

Recently, soil mass effects were evaluated and 2.5 g was found to be robust for soils with <10% SOC (Wade, Maltais-Landry, et al., 2020). At higher SOC contents, 2.5 g may saturate the $KMnO_4$, resulting in an underestimation of POXC. Therefore, for studies that include high SOC contents (>10% SOC), a mass of 0.75 g is recommended. To facilitate comparisons with other studies, the conversion between 0.75 and 2.5 g (at <2-mm sieve size) is

$$\mathrm{POXC}_{2.5\mathrm{g}} = 0.633\mathrm{POXC}_{0.75\mathrm{g}} + 56.40\left(R^2 = .954, \mathrm{RMSE} = 55.4\ \mathrm{mg\ POXC\ kg}^{-1}\right)$$

**Repeatability of POXC with Time**

Wade, Maltais-Landry, et al. (2020) robustly assessed intra- and interlaboratory variability of POXC across 12 different laboratories with soils representing all 12 USDA soil orders. Table 9.1 shows the repeatability of POXC in a research and training laboratory during the course of 5 yr. The reference soils were taken from archived soils from three Ohio State University research farms characterized by Fulford and Culman (2018). The soils were air dried, ground to <2 mm with a hammer flail grinder and mixed thoroughly.

During the 5 yr, there was a total of 133 96-well microtiter plates analyzed during the course of 94 separate days using many batches of $KMnO_4$ stock solution. There was a total of 23 technicians who ran POXC including undergraduate student researchers, graduate students, full-time research technicians, and post-doctoral scholars. (Numerous technicians ran POXC analyses over multiple years.) Nearly all technicians had not run POXC before they came to the lab, and some undergraduates had no previous laboratory experience.

**Table 9.1** Permanganate-oxidizable C (POXC) coefficients of variation (CV) of three internal reference soils run in the Culman Lab at Ohio State University from 2015 to 2019.

| | | | | CV% | | |
|---|---|---|---|---|---|---|
| **Year** | **No. of plates** | **No. of days** | **No. of technicians** | **Reference Soil 1** | **Reference Soil 2** | **Reference Soil 3** |
| 2015 | 21 | 14 | 4 | 11.1 | 4.3 | 7.9 |
| 2016 | 27 | 20 | 6 | 15.0 | 7.8 | 8.6 |
| 2017 | 25 | 21 | 11 | 12.8 | 7.2 | 8.5 |
| 2018 | 31 | 16 | 11 | 14.0 | 7.3 | 14.8 |
| 2019 | 29 | 23 | 3 | 7.1 | 7.2 | 7.8 |
| Total | 133 | 94 | 23 | 12 | 6.8 | 9.5 |

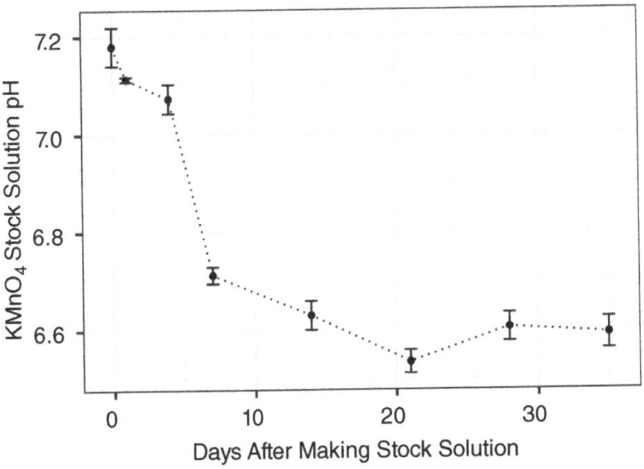


**Figure 9.4** Declines in potassium permanganate ($KMnO_4$) stock solution pH that was initially adjusted to 7.2. Error bars represent standard deviations of the mean ($n$ = 4).

The coefficients of variation (CVs) suggest that some soils are inherently more variable than others (e.g., Reference Soil 1 relative to Reference Soil 2). In addition, CVs typically increased as more technicians ran POXC within a year, probably reflecting that many of these technicians (or students) were being trained in the method. In 2019, there were only three technicians, all of whom were full time and experienced, who ran POXC. This year recorded some of the lowest CVs across soils and probably represents variation that might be more typically encountered in a commercial soil testing laboratory.

**Reductions in pH of $KMnO_4$ Stock Solution**

Gruver (2015) reported a rapid drop in stock solution pH days after preparing the solution and reported no differences of POXC values with manipulated pH levels of stock solutions. However, data were not shown to document this drop. Here, we present data that corroborate the findings of Gruver (2015) (Figure 9.4). We made up stock solution in four batches and measured the pH from all four batches during the course of 5 wk. Within 1 wk after making the stock solution (0.2 M $KMnO_4$), pH levels dropped nearly 0.5, and after 2–3 wk the pH stabilized. This demonstrates that the $KMnO_4$ stock solution is weakly buffered and that adjusting the solution to a final pH of 7.2 is ineffective and therefore not recommended.

## References


Anderson, J. P. E., & Domsch, K. H. (1978). A physiological method for the quantitative measurement of microbial biomass in soils. *Soil Biology and Biochemistry*, 10, 215–221. https://doi.org/10.1016/0038-0717(78)90099-8

Arachchige, P. (2016). Understanding of coupled physicochemical and mineralogical mechanisms controlling soil carbon storage and preservation (Doctoral dissertation). Retrieved from https://krex.k-state.edu/dspace/handle/2097/32921

Awale, R., Emeson, M. A., & Machado, S. (2017). Soil organic carbon pools as early indicators for soil organic matter stock changes under different tillage practices in Inland Pacific Northwest. *Frontiers in Ecology and Evolution*, 5. https://www.frontiersin.org/article/10.3389/fevo.2017.00096

Barré, P., Plante, A. F., Cécillon, L., Lutfalla, S., Baudin, F., Bernard, S., … Chenu, C. (2016). The energetic and chemical signatures of persistent soil organic matter. *Biogeochemistry Letters*, 130, 1–12. https://doi.org/10.1007/s10533-016-0246-0

Bell, M. J., Moody, P. W., Connolly, R. D., & Bridge, B. J. (1998). The role of active fractions of soil organic matter in physical and chemical fertility of Ferrosols. *Australian Journal of Soil Research*, 36, 809–820. https://doi.org/10.1071/S98020

Blair, G., Lefroy, R., & Lisle, L. (1995). Soil carbon fractions based on their degree of oxidation, and the development of a carbon management index for agricultural systems. *Australian Journal of Agricultural Research*, 46, 1459–1466. https://doi.org/10.1071/AR9951459

Blanco-Canqui, H., & Benjamin, J. G. (2015). Impacts of soil organic carbon on soil physical behavior. In S. Logsdon, M. Berli, & R. Horn (Eds.), *Quantifying and modeling soil structure dynamics* (Advances in Agricultural Systems Modeling 3, pp. 11–40). Madison, WI: SSSA.

Bongiorno, G., Bünemann, E. K., Oguejiofor, C. U., Meier, J., Gort, G., Comans, R., … de Goede, R. (2019). Sensitivity of labile carbon fractions to tillage and organic matter management and their potential as comprehensive soil quality indicators across pedoclimatic conditions in Europe. *Ecological Indicators*, 99, 38–50. https://doi.org/10.1016/j.ecolind.2018.12.008

Bundy, L. G., & Bremner, J. M. (1973). Determination of ammonium N and nitrate N in acid permanganate solution used to absorb ammonia, nitric oxide, and nitrogen dioxide evolved from soils. *Communications in Soil Science and Plant Analysis*, 4(3), 179–184. https://doi.org/10.1080/00103627309366435

Buyer, J. S., & Sasser, M. (2012). High throughput phospholipid fatty acid analysis of soils. *Applied Soil Ecology*, 61, 127–130. https://doi.org/10.1016/j.apsoil.2012.06.005

Byrnes, R. C., Eastburn, D. J., Tate, K. W., & Roche, L. M. (2018). A global meta-analysis of grazing impacts on soil health indicators. *Journal of Environmental Quality*, 47, 758–765. https://doi.org/10.2134/jeq2017.08.0313

Calderón, F., Culman, S., Six, J., Franzluebbers, A. J., Schipanski, M., Beniston, J., … Kong, A. Y. Y. (2017). Quantification of soil permanganate oxidizable C (POXC) using infrared spectroscopy. *Soil Science Society of America Journal*, 81, 277–288. https:/doi.org/10.2136/sssaj2016.07.0216

Carski, T. H., & Sparks, D. L. (1987). Differentiation of soil nitrogen fractions using a kinetic approach. *Soil Science Society of America Journal*, 51, 314–317. https://doi.org/10.2136/sssaj1987.03615995005100020010x

Cotrufo, M. F., Wallenstein, M. D., Boot, C. M., Denef, K., & Paul, E. (2013). The Microbial Efficiency–Matrix Stabilization (MEMS) framework integrates plant litter decomposition with soil organic matter stabilization: Do labile plant inputs form stable soil organic matter? *Global Change Biology*, 19, 988–995. https://doi.org/10.1111/gcb.12113

Culman, S., Snapp, S., Freeman, M., Schipanski, M., Beniston, J., Lal, R., … Wander, M. M. (2012). Permanganate oxidizable carbon reflects a processed soil fraction that is sensitive to management. *Soil Science Society of America Journal*, 76, 494–504. https://doi.org/10.2136/sssaj2011.0286

Culman, S., Snapp, S., Green, J., & Gentry, L. (2013). Short- and long-term labile soil carbon and nitrogen dynamics reflect management and predict corn agronomic performance. *Agronomy Journal*, 105, 493–502. https://doi.org/10.2134/agronj2012.0382

de Moraes Sá, J. C., Tivet, F., Lal, R., Briedis, C., Hartman, D. C., Zuffo dos Santos, J., & Burkner dos Santos, J. (2014). Long-term tillage systems impacts on soil C dynamics, soil resilience and agronomic productivity of a Brazilian Oxisol. *Soil and Tillage Research*, 136, 38–50. https://doi.org/10.1016/j.still.2013.09.010

Drinkwater, L. E., Wagoner, P., & Sarrantonio, M. (1998). Legume-based cropping systems have reduced carbon and nitrogen losses. *Nature*, 396, 262–265. https://doi.org/10.1038/24376

Fierer, N., Schimel, J. P., & Holden, P. A. (2003). Variations in microbial community composition through two soil depth profiles. *Soil Biology and Biochemistry*, 35, 167–176. https://doi.org/10.1016/S0038-0717(02)00251-1

Fine, A. K., van Es, H. M., & Schindelbeck, R. R. (2017). Statistics, scoring functions, and regional analysis of a comprehensive soil health database. *Soil Science Society of America Journal*, 81, 589–601. https://doi.org/10.2136/sssaj2016.09.0286

Fulford, A., & Culman, S. (2018). Over-fertilization does not build soil test phosphorus and potassium in Ohio. *Agronomy Journal*, 110, 56–65. https:/doi.org/10.2134/agronj2016.12.0701

Geisseler, D., & Scow, K. M. (2014). Long-term effects of mineral fertilizers on soil microorganisms: A review. *Soil Biology and Biochemistry*, 75, 54–63. https:/doi.org/10.1016/j.soilbio.2014.03.023

Gregorich, E. G., Monreal, C. M., Carter, M. R., Angers, D. A., & Ellert, B.H. (1994). Towards a minimum data set to assess soil organic matter quality in agricultural soils. *Canadian Journal of Soil Science*, 74, 367–385. https://doi.org/10.4141/cjss94-051

Gruver, J. (2015). Evaluating the sensitivity and linearity of a permanganate-oxidizable carbon method. *Communications in Soil Science and Plant Analysis*, 46, 490–510. https://doi.org/10.1080/00103624.2014.997387

Haney, R. L., Haney, E. B., Smith, D. R., Harmel, R. D., & White, M. J. (2018). The soil health tool: Theory and initial broad-scale application. *Applied Soil Ecology*, 125, 162–168. doi:10.1016/j.apsoil.2017.07.035

Haynes, R. J. (2005). Labile organic matter fractions as central components of the quality of agricultural soils: An overview. *Advances in Agronomy*, 85, 221–268. https://doi.org/10.1016/S0065-2113(04)85005-3

Hurisso, T., Culman, S., Horwath, W., Wade, J., Cass, D., Beniston, J. W., …Ugarte, C. M. (2016). Comparison of permanganate-oxidizable carbon and mineralizable carbon for assessment of organic matter stabilization and mineralization. *Soil Science Society of America Journal*, 80, 1352–1364. https://doi.org/10.2136/sssaj2016.04.0106.

Hurisso, T., Culman, S., & Zhao, K. (2018). Repeatability and spatiotemporal variability of emerging soil health indicators relative to routine soil nutrient tests. *Soil Science Society of America Journal*, 82, 939–948. https://doi.org/10.2136/sssaj2018.03.0098

Islam, K. R., & Weil, R. R. (1998). Microwave irradiation of soil for routine measurement of microbial biomass carbon. *Biology and Fertility of Soils*, 27, 408–416. https://doi.org/10.1007/s003740050451

Jaffrain, J., Gérard, F., Meyer, M., & Ranger, J. (2007). Assessing the quality of dissolved organic matter in forest soils using ultraviolet absorption spectrophotometry. *Soil Science Society of America Journal*, 71, 1851–1858. https://doi.org/10.2136/sssaj2006.0202

Jenkinson, D., & Powlson, D. (1976). The effects of biocidal treatments on metabolism in soil: V. A method for measuring soil biomass. *Soil Biology and Biochemistry*, 8, 209–213. https://doi.org/10.1016/0038-0717(76)90005-5

Jensen, J. L., Schjønning, P., Watts, C.,W., Christensen, B.T., Obour, P.B., Munkholm, L. J. (2020). Soil degradation and recovery: Changes in organic matter fractions and structural stability. *Geoderma*, 364, 114181. doi:10.1016/j.geoderma.2020.114181

Jensen, J. L., Schjønning, P., Watts, C. W., Christensen, B. T., Peltre, C., & Munkholm, L. J. (2019). Relating soil C and organic matter fractions to soil structural stability. *Geoderma*, 337, 834–843. https://doi.org/10.1016/j.geoderma.2018.10.034

Joergensen, R. G. (1996). The fumigation–extraction method to estimate soil microbial biomass: Calibration of the $k_{EC}$ value. *Soil Biology and Biochemistry*, 28, 25–31. https://doi.org/10.1016/0038-0717(95)00102-6

Kallenbach, C. M., Frey, S. D., & Grandy, A. S. (2016). Direct evidence for microbial-derived soil organic matter formation and its ecophysiological controls. *Nature Communications*, 7, 13630. https://doi.org/10.1038/ncomms13630

Kim, N., Zabaloy, M. C., Guan, K., & Villamil, M. B. (2020). Do cover crops benefit soil microbiome? A meta-analysis of current research. *Soil Biology and Biochemistry*, 142, 107701. https://doi.org/10.1016/j.soilbio.2019.107701

Kleber, M., Sollins, P., & Sutton, R. (2007). A conceptual model of organo-mineral interactions in soils: Self-assembly of organic molecular fragments into zonal structures on mineral surfaces. *Biogeochemistry*, 85, 9–24. https://doi.org/10.1007/s10533-007-9103-5

Kopittke, P. M., Hernandez-Soriano, M. C., Dalal, R. C., Finn, D., Menzies, N. W., Hoeschen, C, & Mueller, C. W. (2018). Nitrogen-rich microbial products provide new organo-mineral associations for the stabilization of soil organic matter. *Global Change Biology*, 24, 1762–1770. https://doi.org/10.1111/gcb.14009

Lavallee, J. M., Soong, J. L., & Cotrufo, M. F. (2020). Conceptualizing soil organic matter into particulate and mineral-associated forms to address global change in the 21st century. *Global Change Biology*, 26, 261–273. https://doi.org/10.1111/gcb.14859

Lefroy, R. D. B., Blair, G. J., & Strong, W. M. (1993). Changes in soil organic matter with cropping as measured by organic carbon fractions and $^{13}C$ natural isotope abundance. *Plant and Soil*, 155–*156*, 399–402. https://doi.org/10.1007/BF00025067

Lehmann, J., & Kleber, M. (2015). The contentious nature of soil organic matter. *Nature*, 528, 60–68. https://doi.org/10.1038/nature16069

Li, P., & Hur, J. (2017). Utilization of UV-Vis spectroscopy and related data analyses for dissolved organic matter (DOM) studies: A review. *Critical Reviews in Environmental Science and Technology*, 47, 131–154. https://doi.org/10.1080/10643389.2017.1309186

Liang, C., Schimel, J. P., & Jastrow, J. D. (2017). The importance of anabolism in microbial control over soil carbon storage. *Nature Microbiology*, 2, 17105. https://doi.org/10.1038/nmicrobiol.2017.105

Loginow, W., Wisniewski, W., Gonet, S., & Ciescinska, B. (1987). Fractionation of organic C based on susceptibility to oxidation. *Polish Journal of Soil Science*, 20, 47–52.

Lucas, S. T., & Weil, R. R. (2012). Can a labile carbon test be used to predict crop responses to improve soil organic matter management? *Agronomy Journal*, 104, 1160–1170. https://doi.org/10.2134/agronj2011.0415

Majumder, B., Mandal, B., Bandyopadhyay, P. K., & Chaudhury, J. (2007). Soil organic carbon pools and productivity relationships for a 34 year old rice–wheat–jute agroecosystem under different fertilizer treatments. *Plant and Soil*, 297, 53–67. https://doi.org/10.1007/s11104-007-9319-0

Margenot, A. J., Calderón, F. J., Magrini, K. A., & Evans, R. J. (2017). Application of DRIFTS, $^{13}C$ NMR, and py-MBMS to characterize the effects of soil science oxidation assays on soil organic matter composition in a Mollic Xerofluvent. *Applied Spectroscopy*, 71, 1506–1518. https://doi.org/10.1177/0003702817691776

Marschner, B., & Kalbitz, K. (2003). Controls of bioavailability and biodegradability of dissolved organic matter in soils. *Geoderma*, 113, 211–235. https://doi.org/10.1016/S0016-7061(02)00362-2

Matsuda, K., & Schnitzer, M. (1972). The permanganate oxidation of humic acids extracted from acid soils. *Soil Science*, 114, 185–193. https://doi.org/10.1097/00010694-197209000-00005

McDaniel, M. D., Tiemann, L. K., & Grandy, A. S. (2014). Does agricultural crop diversity enhance soil microbial biomass and organic matter dynamics? A meta-analysis. *Ecological Applications*, 24, 560–570. https://doi.org/10.1890/13-0616.1

Moebius-Clune, B. N., Moebius-Clune, D. J., Gugino, B. K., Idowu, O. J., Schindelbeck, R. R., Ristow, A. J., … Abawi, G.S. (2017). Comprehensive assessment of soil health: The Cornell Framework manual (3.1 ed.). Ithaca, NY: Cornell University.

Morrow, J. G., Huggins, D. R., Carpenter-Boggs, L. A., & Reganold, J. P. (2016). Evaluating measures to assess soil health in long-term agroecosystem trials. *Soil Science Society of America Journal*, 80, 450–462. https://doi.org/10.2136/sssaj2015.08.0308

Nunes, M. R., Karlen, D. L., Veum, K. S., Moorman, T. B., & Cambardella, C. A. (2020). Biological soil health indicators respond to tillage intensity: A US meta-analysis. *Geoderma*, 369, 114335. https://doi.org/10.1016/j.geoderma.2020.114335

Oldfield, E. E., Bradford, M. A., & Wood, S. A. (2019). Global meta-analysis of the relationship between soil organic matter and crop yields. *SOIL*, 5, 15–32. https://doi.org/10.5194/soil-5-15-2019

Poeplau, C., Don, A., Six, J., Kaiser, M., Benbi, D., Chenu, C., … Nieder, R. (2018). Isolating organic carbon fractions with varying turnover rates in temperate agricultural soils: A comprehensive method comparison. *Soil Biology and Biochemistry*, 125, 10–26. https://doi.org/10.1016/j.soilbio.2018.06.025

Pribyl, D. W. (2010). A critical review of the conventional SOC to SOM conversion factor. *Geoderma*, 156, 75–83. https://doi.org/10.1016/j.geoderma.2010.02.003

Ramírez, P. B., Calderón, F. J., Fonte, S. J., Santibáñez, F., & Bonilla, C. A. (2020). Spectral responses to labile organic carbon fractions as useful soil quality indicators across a climatic gradient. *Ecological Indicators*, 111, 106042. https://doi.org/10.1016/j.ecolind.2019.106042

Rasmussen, C., Heckman, K., Wieder, W. R., Keiluweit, M., Lawrence, C. R., Berhe, A. A., … Wagai, R. (2018). Beyond clay: Towards an improved set of variables for predicting soil organic matter content. *Biogeochemistry Letters*, 137, 297–306. https://doi.org/10.1007/s10533-018-0424-3

Rinot, O., Osterholz, W. R., Castellano, M. J., Linker, R., Liebman, M., & Shaviv, A. (2018). Excitation–emission–matrix fluorescence spectroscopy of soil water extracts to predict nitrogen mineralization rates. *Soil Science Society of America Journal*, 82, 126–135. https://doi.org/10.2136/sssaj2017.06.0188

Robertson, G. P., Paul, E. A., & Harwood, R.R. (2000). Greenhouse gases in intensive agriculture: Contributions of individual gases to the radiative forcing of the atmosphere. *Science*, 289, 1922–1925. https://doi.org/10.1126/science.289.5486.1922

Romero, C. M., Engel, R. E., D'Andrilli, J., Chen, C., Zabinski, C., Miller, P. R., & Wallander, R. (2018). Patterns of change in permanganate oxidizable soil organic matter from semiarid drylands reflected by absorbance spectroscopy and Fourier transform ion cyclotron resonance mass spectrometry. *Organic Geochemistry*, 120, 19–30. https://doi.org/10.1016/j.orggeochem.2018.03.005

Rousk, J., & Jones, D. L. (2010). Loss of low molecular weight dissolved organic carbon (DOC) and nitrogen (DON) in $H_2O$ and 0.5 M $K_2SO_4$ soil extracts. *Soil Biology and Biochemistry*, 42, 2331–2335. https://doi.org/10.1016/j.soilbio.2010.08.017

Schimel, D. S., Braswell, B. H., Holland, E. A., McKeown, R., Ojima, D. S., Painter, T. H., . . .Townsend, A. R. (1994). Climatic, edaphic, and biotic controls over storage and turnover of carbon in soils. *Global Biogeochemistry Cycles*, 8, 279–293. https://doi.org/10.1029/94GB00993

Schmidt, M. W. I., Torn, M. S., Abiven, S., Dittmar, T., Guggenberger, G., Janssens, I. A., . . . Trumbore, S. E. (2011). Persistence of soil organic matter as an ecosystem property. *Nature*, 478, 49–56. https://doi.org/10.1038/nature10386

Sherrod, L. A., Vigil, M. F., & Stewart, C. E. (2019). Do fulvic, humic, and humin carbon fractions represent meaningful biological, physical, and chemical carbon pools? *Journal of Environmental Quality*, 48, 1587–1593. https://doi.org/10.2134/jeq2019.03.0104

Six, J., Conant, R. T., Paul, E. A., & Paustian, K. (2002). Stabilization mechanisms of soil organic matter: Implications for C-saturation of soils. *Plant and Soil*, 241, 155–176. https://doi.org/10.1023/A:1016125726789

Six, J., Elliott, E. T., Paustian, K., & Doran, J. W. (1998). Aggregation and soil organic matter accumulation in cultivated and native grassland soils. *Soil Science Society of America Journal*, 62, 1367–1377. https://doi.org/10.2136/sssaj1998.03615995006200050032x

Sprunger, C. D., Culman, S. W., Palm, C. A., Thuita, M., & Vanlauwe, B. (2019). Long-term application of low C:N residues enhances maize yield and soil nutrient pools across Kenya. *Nutrient Cycling in Agroecosystems*, 114, 261–276. https://doi.org/10.1007/s10705-019-10005-4

Sprunger, C. D., Martin, T., & Mann, M. (2020). Systems with greater perenniality and crop diversity enhance soil biological health. *Agricultural & Environmental Letters*, 5(1), e20030. https://doi.org/10.1002/ael2.20030

Stiles, C. A., Hammer, R. D., Johnson, M. G., Ferguson, R., Galbraith, J., O'Geen, T., … Miles, R. (2011). Validation testing of a portable kit for measuring an active soil carbon fraction. *Soil Science Society of America Journal*, 75, 2330–2340. https://doi.org/10.2136/sssaj2010.0350

Stine, M. A., & Weil, R. R. (2002). The relationship between soil quality and crop productivity across three tillage systems in south central Honduras. *American Journal of Alternative Agriculture*, 17, 2–8. https://doi.org/10.1079/AJAA20011

Stott, D. E. (2019). *Recommended soil health indicators and associated laboratory procedures (Soil Health Technical Note 450-03)*. Washington, DC: USDA–NRCS.

Tirol-Padre, A., & Ladha, J. K. (2004). Assessing the reliability of permanganate-oxidizable carbon as an index of soil labile carbon. *Soil Science Society of America Journal*, 68, 969–978.

Treseder, K. K. (2008). Nitrogen additions and microbial biomass: A meta-analysis of ecosystem studies. *Ecology Letters*, 11, 1111–1120. https://doi.org/10.1111/j.1461-0248.2008.01230.x

Vance, E. D., Brookes, P. C., & Jenkinson, D. S. (1987). An extraction method for measuring soil microbial biomass C. *Soil Biology and Biochemistry*, 19, 703–707. https://doi.org/10.1016/0038-0717(87)90052-6

Vieira, F. C. B., Bayer, C., Zanatta, J. A., Dieckow, J., Mielniczuk, J., & He, Z. L. (2007). Carbon management index based on physical fractionation of soil organic matter in an Acrisol under long-term no-till cropping systems. *Soil and Tillage Research*, 96, 195–204. https://doi.org/10.1016/j.still.2007.06.007

von Haden, A. C., Kucharik, C. J., Jackson, R. D., & Marín-Spiotta, E. (2019). Litter quantity, litter chemistry, and soil texture control changes in soil organic carbon fractions under bioenergy cropping systems of the North Central U.S. *Biogeochemistry*, 143, 313–326. https://doi.org/10.1007/s10533-019-00564-7

von Lützow, M., Kögel-Knabner, I., Ekschmitt, K., Flessa, H., Guggenberger, G., Matzner, E., & Marschner, B. (2007). SOM fractionation methods: Relevance to functional pools and to stabilization mechanisms. *Soil Biology and Biochemistry*, 39, 2183–2207. https://doi.org/10.1016/j.soilbio.2007.03.007

Wade, J., Culman, S. W., Logan, J. A. R., Poffenbarger, H., Demyan, M. S., Grove, J. H., … West, J. R. (2020a). Improved soil biological health increases corn grain yield in N fertilized systems across the Corn Belt. *Scientific Reports*, 10, 3917. https://doi.org/10.1038/s41598-020-60987-3

Wade, J., Culman, S. W., Sharma, S., Mann, M., Demyan, M. S., Mercer, K. L., & Basta, N. T. (2019). How does phosphorus restriction impact soil health parameters in midwestern corn–soybean systems? *Agronomy Journal*, 111, 1682–1692. https://doi.org/10.2134/agronj2018.11.0739

Wade, J., Maltais-Landry, G., Lucas, D. E., Bongiorno, G., Bowles, T. M., Calderón, … Margenot, A. J. (2020b). Assessing the sensitivity and repeatability of permanganate oxidizable carbon as a soil health metric: An interlab comparison across soils. *Geoderma*, 366, 114235. https://doi.org/10.1016/j.geoderma.2020.114235

Wander, M. (2004). Soil organic matter fractions and their relevance to soil function. In F. Magdoff and R. Weil (Eds.), *Soil organic matter in sustainable agriculture* (pp. 67–102). Boca Raton, FL: CRC Press.

Wander, M. M., & Drinkwater, L. E. (2000). Fostering soil stewardship through soil quality assessment. *Applied Soil Ecology*, 15, 61–73. https://doi.org/10.1016/S0929-1393(00)00072-X

Wardle, D. A. (1992). A comparative-assessment of factors which influence microbial biomass carbon and nitrogen levels in soil. *Biological Reviews, Cambridge Philosophical Society*, 67, 321–358. https://doi.org/10.1111/j.1469-185X.1992.tb00728.x

Weil, R. R., Islam, K. R., Stine, M. A., Gruver, J. B., & Sampson-Liebig, S. E. (2003). Estimating active carbon for soil quality assessment: A simplified method for laboratory and field use. *American Journal of Alternative Agriculture*, 18, 3–17. https://doi.org/10.1079/AJAA200228

Yost, M. A., Veum, K. S., Kitchen, N. R., Sawyer, J. E., Camberato, J. J., Carter, P. R., … Nafziger, E. D. (2018). Evaluation of the Haney Soil Health Tool for corn nitrogen recommendations across eight Midwest states. *Journal of Soil and Water Conservation*, 73, 587–592. https://doi.org/10.2489/jswc.73.5.587

Zuber, S. M., & Villamil, M. B. (2016). Meta-analysis approach to assess effect of tillage on microbial biomass and enzyme activities. *Soil Biology and Biochemistry*, 97, 176–187. https://doi.org/10.1016/j.soilbio.2016.03.011

# 10

## Is Autoclaved Citrate-Extractable (ACE) Protein a Viable Indicator of Soil Nitrogen Availability?

*Tunsisa T. Hurisso and Steve W. Culman*

## Introduction

Soil N is an important indicator of soil health, reflective of how well several ecosystem services are functioning. It is also the most common growth-limiting nutrient in temperate agroecosystems. Root zone soil N is present in several inorganic and organic forms, but with regard to soil health assessment, bioavailable N, defined as a form that can be readily absorbed by plant roots (Keeney, 1982), is a more meaningful indicator than either $NH_4$– or $NO_3$–N. Bioavailable N in crop production systems is derived through several pathways, including mineral fertilizers, biological $N_2$ fixation, and mineralization of organically bound N added to soils through crop residues and amendments such as manure or compost (Yan, Pan, Lavallee, & Conant, 2019). In most agricultural production regions, a significant amount of bioavailable N (<20 to >200 kg ha$^{-1}$) can be mineralized during a growing season (Cabrera, Kissel, & Vigil, 1994; Curtin & Campbell, 2008). The exact quantity of N that becomes available depends on the size of the mineralizable pool, microbial community diversity and activity, and environmental conditions such as soil temperature and moisture content. Because crop N uptake and removal rates are higher than what many agricultural systems can provide (especially if they have been degraded), the use of nitrogenous mineral fertilizers has increased worldwide and, perhaps as an unintended environmental consequence, resulted in a loss of reactive N (Galloway et al., 2008; Gardner & Drinkwater, 2009). Therefore, to

*Soil Health Series: Volume 2 Laboratory Methods for Soil Health Analysis*, First Edition.
Edited by Douglas L. Karlen, Diane E. Stott, and Maysoon M. Mikha.

fully assess biological soil health, an accurate assessment of soil N availability is an extremely important component of cost-effective, environmentally sound agricultural best management practices.

## Nitrogen Availability Indices

Numerous chemical and biological laboratory indices have been developed to measure soil N (Table 10.1). They can be grouped into three broad categories: (a) incubation-based, direct measurements of mineralized N, (b) total soil N by combustion, and (c) chemically extractable labile fractions. Despite decades of research focused on the development and evaluation of soil N tests, the scientific community has yet to converge on an accepted method (or set of methods) to measure the soil N status (Curtin et al., 2017; Schomberg et al., 2009). This chapter presents details on autoclaved citrate-extractable (ACE) soil protein that, albeit operationally defined, we believe is a biologically meaningful, soil organic-N fraction that is amenable to the high-throughput framework necessary for adoption into commercial soil testing laboratories as a soil health indicator (Hurisso, Moebius-Clune, et al., 2018; Moebius-Clune et al., 2016; Schindelbeck et al., 2016). The citrate-extractable and four alternative soil N measurement methods are briefly described in Table 10.1, but providing in-depth evaluations is beyond the scope of this chapter. Therefore, we provide only a brief discussion in the following subsections to illustrate what we believe are important principles as well as advantages and disadvantages associated with each method.

### Aerobic Incubation

Long-term (usually ≥20-wk) aerobic incubations of soil under optimum laboratory temperature and moisture conditions are often used as a reference method to determine N mineralization potential (Stanford & Smith, 1972). While this procedure provides results that more accurately reflect N mineralized during the course of a growing season, its application is limited to a research setting because it is time consuming and hence operationally not suitable as a routine test (Curtin & Campbell, 2008). A shortened version of the aerobic incubation method of Sanford and Smith (1972) was developed by Keeney and Bremner (1966) to provide a measure of mineralizable soil N from samples incubated anaerobically for 7 d at 40°C. Compared with aerobic incubation, the anaerobic incubation technique has significant operational advantage because the same volume of water is added to all soils regardless of water-holding capacity (Curtin & Campbell, 2008). However,

**Table 10.1** Advantages and disadvantages of selected soil N methods and potential usefulness as soil health indicators.

| Assessment metric | Total soil N | Organic matter | Inorganic N ($NO_3$ and $NH_4$) | Mineralizable N | Organic labile N fractions |
|---|---|---|---|---|---|
| Method | dry combustion[a] | loss on ignition[b] | 2 M KCl extraction[c] | aerobic >28-d incubations, anaerobic 7-d incubation | soil protein or hydrolyzable amino sugars |
| Relative sensitivity to soil and crop management? | slowly, several years | slowly, several years | very rapidly, within days | intermediate, within 1–3 yr | intermediate, within 1–3 yr |
| High-throughput potential? | no | yes | yes | no | yes |
| Rapid and inexpensive? | no | yes | yes | no | yes |
| Primary soil function reflected by the test | total pool of soil N | total pool of C and N | immediately plant-available N pool | mineralizable pool of soil N | mineralizable pool of soil N |
| Overall potential as a soil health indicator | expensive as a routine soil test, and management-induced changes occur very slowly | management-induced changes occur very slowly; low analytical precision; not an ideal indicator of nutrient availability | ephemeral, large changes possible over days, therefore not robust as a soil health indicator | expensive, time consuming, and not suitable for high throughput | relatively new methods, but promising for practical and routine soil health testing |

*Note:* Adapted from Hurisso, Moebius-Clune, et al. (2018). "Relative sensitivity to soil and crop management" refers to the length of time it takes to detect management-induced changes using each soil test method.

[a] Nelson & Sommers (1996).

[b] Ball (1964); Combs & Nathan (1998).

[c] Keeney & Nelson (1982).

for high-throughput commercial soil testing laboratories, even a relatively short 7-d incubation is often not practical.

## Inorganic and Total Soil Nitrogen

Inorganic soil N (i.e., $NH_4$– and $NO_3$–N) is typically determined via 2 M KCl extraction (Keeney & Nelson, 1982) and can provide an index of N available for plant uptake. However, this pool is too ephemeral to be a reliable indicator of plant-available N during the growing season because of losses from denitrification, volatilization, and leaching. Furthermore, microbial immobilization and plant uptake can lead to significant time scale changes ranging from days to weeks (Culman, Snapp, Green, & Gentry, 2013; Sela et al., 2017). In contrast, total soil N via dry combustion (Nelson & Sommers, 1996) is a very stable N pool but not very sensitive or effective for predicting seasonal N availability. It can take numerous years to decades for total soil organic matter pools (evaluated on a total organic C or total soil N basis) to reflect management practice effects (Brock, Knies-Deventer, & Leithold, 2011; Ruffo, Bollero, Hoeft, & Bullock, 2005; Smith, 2004). Also, even though most soil-testing laboratories have dry combustion analyzers for total C/N analysis, the high operating expenses for high-throughput soil health analysis justifies the development of alternative bioavailable N assessment methods.

## Labile Soil Nitrogen Fractions

Labile organic N fractions, unlike inorganic N, are subject to fewer loss pathways and therefore potentially available for plant uptake throughout the growing season. For decades, there has been an ongoing research effort to identify assays or tests that would enable quantification of labile soil organic N fractions on a routine basis and in a cost-effective manner. One such method is the Illinois Soil N Test (ISNT), developed to estimate amino sugar N as a soil organic N fraction that is potentially mineralizable throughout the growing season (Khan, Mulvaney, & Hoeft, 2001; Mulvaney, Khan, Hoeft, & Brown, 2001). The ISNT has shown promise for high-throughput commercial soil-test laboratories, but results have been mixed, with limited success at predicting N mineralization (Barker, Sawyer, Al-Kaisi, & Lundvall, 2006; Marriott & Wander, 2006; Osterhaus, Bundy, & Andraski, 2008; Spargo, Alley, Thomason, & Nagle, 2009). More recently, Ros, Temminghoff, and Hoffland (2011) reviewed numerous extraction-based methods suggested as ways to rapidly assess biologically labile soil organic N fractions. Their extensive review revealed that labile N fractions measured using different extractants were positively correlated with mineralizable N but explained only a small proportion (47% on average) of the overall variability. Ros et al. (2011) clearly emphasized the need for continued work toward the development and evaluation of methods for routine measurement of soil N in production fields. In response to that recommendation, another labile soil N test that has attracted attention is the ACE soil protein assay. Therefore, for soil health assessment in a high-throughput laboratory setting, we suggest that ACE soil protein shows promise for rapid, sensitive quantification of functionally relevant soil N pools (Hurisso, Moebius-Clune, et al., 2018).

## Autoclaved Citrate-Extractable Protein

Soil organic matter is comprised of numerous compounds, including cellulose, hemicellulose, lignin, and proteins. Of these compounds, proteins represent by far the largest pool (30–40%) of organically bound soil N (Gillespie et al., 2011, and others therein; Németh, Bartels, Vogel, & Mengel, 1988; Matsumoto, Ae, & Yamagata, 2000). Recent studies have also documented that soil N mineralization is limited by depolymerization (i.e., amino acid production by proteases) (Fujii, Yamada, Hayakawa, Nakanishi, & Funakawa, 2020; Jan, Roberts, Tonheim, & Jones, 2009; Mooshammer et al., 2012; Nannipieri & Paul, 2009; Schimel & Bennett, 2004; Weintraub & Schimel, 2005). The size of the soil protein pool thus has important implications not only for soil health assessment but also for plant nutrition, N storage, and synchrony between soil N release and plant demand. Soil protein assays are therefore an important metric for measuring labile, organically bound N pools available for depolymerization and subsequent contributions to plant-available N uptake during a growing season (Hurisso, Moebius-Clune, et al., 2018; Moebius-Clune et al., 2016).

A large body of literature exists demonstrating the utility of soil protein as an important soil N pool that is sensitive to management practices (Hurisso, Moebius-Clune, et al., 2018; Moebius et al., 2007; Moebius-Clune et al., 2008; Roper, Osmond, Heitman, Wagger, & Reberg-Horton, 2017). Reduced tillage intensity and crop rotational diversity have both been shown to increase soil protein content (Borie et al., 2006; Emran, Gispert, & Pardini, 2012; Liebig, Carpenter-Boggs, Johnson, Wright, & Barbour, 2006; Nichols & Millar, 2013; Nunes, Karlen et al., 2020; Wright, Green & Cavigelli, 2007). Soil protein had the strongest relationship with corn (Zea mays L.; $r^2 = .45–.88$) and soybean [Glycine max (L.) Merr.; $r^2 = .55$] yields compared with a suite of other soil health indicators (Roper, Osmond, & Heitman, 2019; van Es & Karlen, 2019). Geisseler, Miller, Leinfelder-Miles, & Wilson (2019) found that soil protein was not better at predicting 10-wk aerobic N mineralization than total soil N in California soils. In North Carolina, Caudle et al. (2020) found that soil protein was significantly ($p < .05$) correlated with total soil N and soil organic C ($r^2 > .88$) and with a number of enzyme assays including β-glucosidase and β-glucosiaminidase ($r^2 > .67$). Furthermore, positive correlations of soil protein with aggregate stability ($r^2$ = .28–.71) have been reported by a number of investigators (Rillig, Wright, Kimball, & Leavitt, 2001; Wright & Anderson, 2000; Wright, Starr, & Paltineanu, 1999; Wright & Upadhyaya, 1998; Wu, Cao, Zou, & He, 2014).

The procedure for quantifying soil protein as a potential soil health indicator was outlined by Hurisso, Moebius-Clune, et al. (2018). The procedure, details outlined below, is a modified version of the Wright and Upadhyaya (1996) method developed originally to quantify a particular protein known as glomalin or glomalin-related soil protein (GRSP), which is generally believed to be produced by arbuscular mycorrhizal fungi. Previous studies have demonstrated that the Wright and Upadhyaya (1996) method based on a neutral sodium citrate buffer solution extracts proteins from a wide range of sources, not just glomalin or GRSP (Gillespie et al., 2011; Hurisso, Moebius-Clune, et al., 2018; Purin & Rillig, 2007; Rosier, Hoye, & Rillig, 2006; Schindler, Mercer, & Rice, 2007). Hurisso, Moebius-Clune, et al. (2018) also called on the scientific community to adopt a more accurate name—autoclaved citrate-extractable (ACE) protein—rather than glomalin or GRSP. The latter are not only scientifically imprecise terminologies but also could limit the utility, interpretation, and application of the method. The suggested ACE protein procedure is operationally suitable for rapid quantification in high-throughput laboratories and is already included as an indicator in the commercially available Cornell Assessment of Soil Health framework (Idowu et al., 2009; Moebius-Clune et al., 2016; Schindelbeck et al., 2016).

In the ACE protein method, soil is exposed to high temperature and pressure conditions in an autoclave, followed by clarification and quantification steps. The most substantial change compared with the original Wright and Upadhyaya (1996) method is to use more appropriate chemistry in the quantification step. Various techniques exist by which the concentration of proteins in soil extracts can be quantified, including the Lowry assay (Lowry, Rosebrough, Farr, & Randall, 1951), Bradford assay (Bradford, 1976), and bicinchoninic-acid (BCA) assay (Smith et al., 1985). The latter assay is generally preferred for the ACE protein method. First, the BCA assay replaces the Folin–Ciocalteu reagent (phosphomolybdate and phosphotungstate) in the Lowry method with bicinchoninic acid, which results in a protein assay that is less susceptible to interferences from humic substances than other dye-binding protein quantification methods (Noble & Bailey, 2009; Smith et al., 1985). Second, the BCA reaction is temperature dependent and displays less protein-to-protein variability when run at elevated temperatures (e.g., ≥60°C; Noble & Bailey, 2009; Smith et al., 1985; Walker, 2002). The BCA assay is based on the reduction of $Cu^{2+}$ to $Cu^{+}$ by proteins in an alkaline medium (known as the biuret reaction) and colorimetric detection of cuprous ($Cu^{+}$) peptide complex using a unique reagent containing bicinchoninic acid (Smith et al., 1985). The purple-colored reaction product that is formed when bicinchoninic acid reacts with the reduced cuprous cation exhibits a strong absorbance at the 562-nm wavelength.

In the extraction step, the ACE protein method involves the use of a neutral sodium citrate buffer solution (pH 7.0). The sodium and citrate ions contribute to soil dispersal, which is enhanced through mechanical agitation by shaking the mixture. Soil is disaggregated or slaked and thereby exposed to the extractant solution. The mixture is heated in an autoclave to increase protein solubilization. The concentration of dissolved protein in clarified extracts is determined by reaction in bicinchoninic-acid assay according to Smith et al. (1985) and quantified against a bovine serum albumin (BSA) standard curve by colorimetry using a spectrophotometric plate reader. The ACE protein assay has been optimized for use in micro-well plates, typically 96-well plates, to enhance speed and throughput and to lower sample and reagent usage.

## Soil Sampling Considerations

Consistent with all soil health indicator measurements, critical factors affecting the accuracy and reproducibility of bioavailable soil N measurements include both the time of sampling and the number of samples (i.e., sampling density) taken within a field. Chapter 2 (this volume) addresses field sampling for soil health assessment, but once again very little attention has been given to spatiotemporal variability and repeatability of soil protein assays. Hurisso, Culman, and Zhao (2018) conducted repeated grid soil sampling throughout the entire corn growing season at three Ohio fields. They found spatiotemporal variability and repeatability of soil protein similar to that expected for Mehlich-3 extractable nutrients. To minimize ACE protein temporal variability, they suggested sampling at the same time each year, as currently recommended for routine soil test measurements. They also concluded that soil protein measurements should not require greater soil sampling densities than used for routine nutrient testing and thus recommended a composite of five to eight cores per field (Hurisso, Culman & Zhao, 2018). With consistent sampling time and density, Hurisso, Culman, Zone, and Sharma (2018) found very little variation (≤3% CV) among ACE protein analytical replicates using soils ground to pass a 2-mm sieve or hand sieved to <2 mm compared with 8-mm sieved soils. Based on those studies, they recommended the use of soils ground or sieved to <2 mm to reduce analytical error and maximize repeatability (precision) of the results. They also stressed the need for further research on interlaboratory variability (i.e., reproducibility) and quantifying the relationship of ACE protein with other methods of soil N availability and plant functions (e.g., crop yield).

# Methods and Materials for Protein Quantification

## Materials and Reagents

### Sodium Citrate Buffer and BCA Working Reagent Solutions

1) Reagent-grade tribasic sodium citrate dihydrate ($Na_3C_6H_5O_7{\cdot}2H_2O$; formula weight = 294.1 g $mol^{-1}$)
2) Reagent-grade citric acid ($C_6H_8O_7$; formula weight = 192.1 g $mol^{-1}$)
3) 20-L carboy, 250-ml beaker, and 1-L Erlenmeyer flask
4) pH meter
5) BCA Reagents A and B (Pierce BCA protein assay kits, ThermoScientific, Product no. P123225)
6) Pipetting reservoir for multichannel pipette
7) 50-ml disposable polypropylene centrifuge tube with cap (Falcon tube)
8) Adjustable 100–1000-µl pipette and disposable tips
9) 25-ml graduated cylinder

### Sample Extraction and Clarification

1) Glass extraction tubes with caps
2) Metal racks for extraction tubes
3) Analytical balance capable of weighing to two decimal places
4) Adjustable bottle-top dispenser fitted to a bottle of sodium citrate and calibrated to deliver 24 ml
5) Oscillating (or horizontal) shaker capable of at least 180 oscillations per minute
6) Autoclave
7) Adjustable 100–1000-µl pipette and disposable tips
8) Microcentrifuge tubes, 2 ml
9) Centrifuge with microtiter plate inserts to handle 2-ml microcentrifuge tubes
10) Microtiter tubes (1.1-ml open-top tubes in strips of eight) and caps

### Sample Quantification

1) BSA protein standards (Pierce prediluted protein assay standard set, ThermoScientific, Product no. P123208)
2) Adjustable 1–10- and 30–300-µl multichannel pipettes and disposable tips
3) Clear polystyrene, flat-bottom, chimney style 96-well plates
4) Plate sealing tape and roller
5) Block heater (plate incubator)
6) Spectrophotometer capable of reading absorbance at 562 nm

## Reagent Preparation

**Sodium citrate stock solution (pH 7.0):** This solution is prepared first by making a solution of 0.02 mol $L^{-1}$ sodium citrate and another 0.02 mol $L^{-1}$ solution of citric acid. Then, the sodium citrate solution is adjusted to pH 7.0 with the citric acid solution.

**BCA working reagent (WR) solution:** The WR solution is prepared by mixing 50 parts of BCA Reagent A with one part of BCA Reagent B to form a green solution (Reagent A/B ratio = 50:1). Reagent A is a clear mixture of sodium carbonate, sodium bicarbonate, bicinchoninic acid, and sodium tartrate in 0.1 M sodium hydroxide. Reagent B is a blue-green copper sulfate solution.

Note: A cloudiness that appears when Reagent B is first added to Reagent A is normal and quickly disappears on mixing to yield a clear, green WR solution. Prepare sufficient volume of WR based on the number of samples to be assayed. For example, to make enough for a 96-well plate, mix 25 ml of Reagent A and 500 μl (0.5 ml) of Reagent B in a 50-ml Falcon tube.

**BSA standards:** Prepare standards for multichannel use by pipetting aliquots (~750 μl) from each of the vials into corresponding and labeled microtiter tubes. The BSA standards should include the following concentrations: 125, 250, 500, 750, 1000, 1500, and 2000 μg $ml^{-1}$.

**Blank:** Prepare a blank of sodium citrate stock solution.

## Procedure

The ACE protein procedure involves three steps: (a) extraction, (b) clarification, and (c) quantification (Figure 10.1).

### Extraction

1) Weigh 3.00 g of <2-mm air-dry soil, or soil ground to pass a 2-mm sieve, into a labeled 50-mL screw-cap glass extraction tube.
2) Using a bottle-top dispenser, add 24 ml of 0.02 mol $L^{-1}$ sodium citrate stock solution (pH 7.0) to each tube containing the soil.
3) Cap tubes tightly and place on metal racks configured to protect glass tubes from breaking while on the shaker.
4) Shake contents of the tube at 180 oscillations $min^{-1}$ for 5 min ("low" setting on an Eberbach reciprocal shaker).
5) After 5 min, remove tubes from shaker and invert a couple times to ensure that there is no soil clinging to the sides of the tube. Next, loosen the caps so they are not air tight but still placed on the tubes to protect contents. Note: In the event that a sample boils over, the sample will need to be rerun.

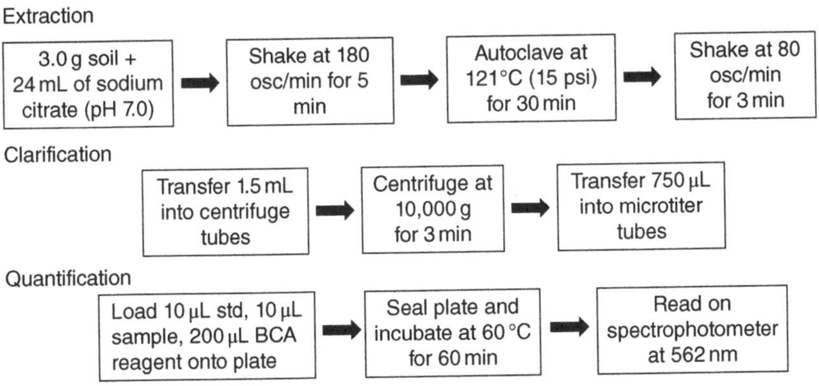


**Figure 10.1** Flow chart of the ACE protein extraction, clarification and quantification.

6) Place tubes in metal rack in an autoclave at 121°C (103 kPa or 15 psi) for 30 min.
7) Remove tubes from the autoclave after 30 min and set aside to cool to room temperature before clarification. Note: Wear heat-protective gloves. Autoclave surfaces are very hot.

**Clarification**

1) Tighten caps on extraction tubes again and place on shaker for 3 min at 180 oscillations $min^{-1}$ to resuspend the soil in the glass extraction tube. Ensure that there is no soil clinging to the sides of the tube.
2) Using an adjustable 100–1000-µl pipette, withdraw 1.5 ml of tube contents and transfer into a clean, labeled 2-ml microcentrifuge tube and close the microcentrifuge cap.
3) Place labeled microcentrifuge tubes in microcentrifuge rotor slots, distributing to ensure rotor balance if less than a full rotor is loaded.
4) Centrifuge contents at 10,000 × g for 3 min. Note: Make sure that the settings are for 10,000 gravities, not 10,000 rpm. On occasion, samples with high clay content will not separate completely. If this happens, simply repeat this step until complete separation occurs.
5) Using a 1000-µl pipette and tips, transfer 750 µl of the clarified extract liquid layer to a microtiter tube in a 96-well format rack. Note: Use a fresh pipette tip for each tube and avoid dislodging the pellet of solids at the bottom of the tube. Clarified extracts can be stored in a refrigerator overnight if not quantified on the same day, but they should be allowed to equilibrate to room temperature before quantification.

**Quantification**

1) Ready a 96-well plate, ensuring bottom is free from scratches, debris, or lint that can affect light transmission.
2) Using the adjustable 1–10-µl multichannel pipette and tips, load 10 µl of each BSA standard, sodium citrate blank and unknown samples into each well of the 96-well plate. Note: well-to-well contamination should be avoided by using a fresh pipette tip for each sample and reagent. Keep careful track of which samples are where on the plate. It is recommended to replicate all standards and sodium citrate blanks on each plate and taking the average typically yields good results.
3) Transfer the BCA working reagent to a clean, dry pipetting reservoir.
4) When all samples including standards and blanks have been placed in the appropriate wells of the reaction plate (Figure 10.2), add 200 µl of the BCA working reagent to each well, preferably using the 30–300-µl multichannel pipette and tips to minimize reaction times between the first and last loaded wells.
5) When the plate is filled, use the roller to seal the plate with a tape seal to prevent evaporative loss of the sample.
6) Once the plate is sealed, place the plate gently on the preheated block heater to incubate at 60°C for 60 min.
7) After 60 min, remove the reaction plate from block heater and allow to cool for at least 5 min. Then, remove the sealing tape.
8) Measure and record the absorbance of each sample on the spectrophotometer at 562 nm.

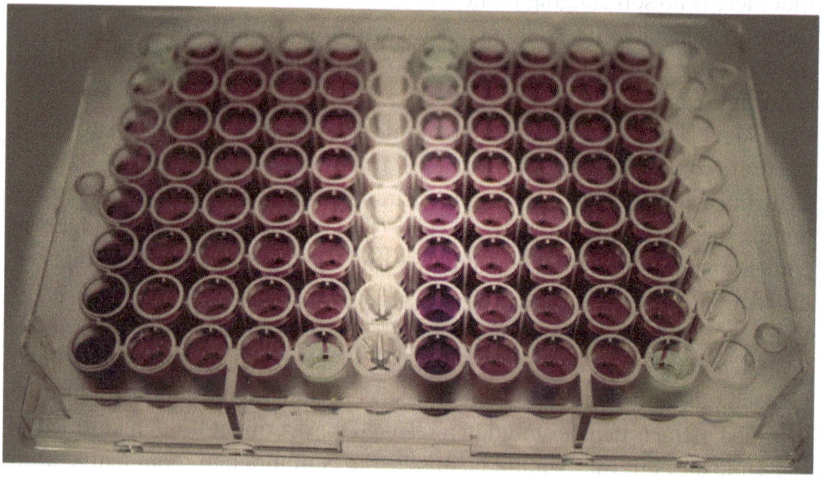

**Figure 10.2** A 96-well reaction plate loaded with samples, blanks and standards for quantification of extracted soil protein through absorbance.

## Calculations

1) First, subtract the average absorbance measurement of the blank replicates from the absorbance values of all other standards and unknown samples.
2) Construct a standard curve by plotting the average blank-corrected 562-nm absorbance measurement for each BSA standard versus its concentration (in μg $ml^{-1}$). This will produce a parabolic second-order response curve:

$$y = ax^2 + bx + c \tag{10.1}$$

where $y$ is absorbance, $x$ is concentration, $a$ and $b$ are coefficients of the standard curve, and $c$ is the intercept of the standard curve. Unknown protein concentrations can be calculated from this curve.

3) Determine the unknown protein concentration of each sample as

$$\text{ACE protein}\left(\mu\text{g ml}^{-1}\right) = \frac{-b + \sqrt{b^2 - 4a\left(c - \text{Abs}\right)}}{2a} \tag{10.2}$$

where the values of a, b, c, and Abs (the absorbance at 562 nm) are obtained from Equation 10.1.

4) Determine the unknown protein content of each sample as

$$\text{ACE protein}\left(\text{mg g}^{-1}\text{ soil}\right) = \frac{\text{Concentration} \times \text{Vol.extractant}}{\text{Weight} \times 1000} \tag{10.3}$$

where Concentration (μg $ml^{-1}$) is the concentration of ACE protein obtained using Equation 10.2, Vol.extractant (ml) is the volume of sodium citrate solution used, and Weight (g) is the mass of air-dry soil used.

## Comments

1) We found very small well-to-well variation (3.2% CV) with the ACE-protein method when quantifying the BSA standard across multiple plates. Therefore, well-level replication is probably not necessary; however, well-level variability is laboratory and instrument dependent, and it is worth examining the repeatability of measurements to determine if well-level replication of the same soil extract is required.
2) Cleanup and disposal: microcentrifuge tubes and microtiter tubes can be discarded after each use. The reaction plate and reagent reservoir can be placed in a fume hood or allowed to dry on the counter. These can be discarded once the reagent has completely evaporated. Extraction tubes and caps including the Falcon tube that was used to prepare the BCA working reagent should be washed using phosphate-free detergent and finally rinsed with deionized water.

## Summary and Conclusions

Soil N is an important indicator of soil health, but despite decades of research, the soil health assessment community has yet to converge on an accepted method or set of measurement methods. In the search for a laboratory method that is simple, rapid, and reproducible, this chapter focuses on autoclaved citrate-extractable (ACE) soil protein, which, albeit operationally defined, we believe is a biologically meaningful soil organic-N measurement that is amenable to the high-throughput framework necessary for adoption into commercial soil testing laboratories as a soil health indicator. Furthermore, ACE protein appears to be a valid index of soil N availability, although additional research is needed to confirm its effectiveness as a universal soil health indicator.

## References

Ball, D. F. 1964. Loss-on-ignition as an estimate of organic matter and organic carbon in noncalcareous soils. *Journal of Soil Science*, 15, 84–92. https://doi.org/10.1111/j.1365-2389.1964.tb00247.x

Barker, D. W., Sawyer, J. E., Al-Kaisi, M. M., & Lundvall, J. R. (2006). Assessment of the amino sugar-nitrogen test on Iowa soils: II. Field correlation and calibration. *Agronomy Journal*, 98, 1352–1358. https://doi.org/10.2134/agronj2006.0034

Borie, F., Rubio, R., Rouanet, J. L., Morales, A., Borie, G., & Rojas, C. (2006). Effects of tillage systems on soil characteristics, glomalin and mycorrhizal propagules in a Chilean Ultisol. *Soil Tillage Research*, 88, 253–261. https://doi.org/10.1016/j.still.2005.06.004

Bradford, M. M. (1976). Rapid and sensitive method for quantitation of microgram quantities of protein utilizing principle of protein–dye binding. *Analytical Biochemistry*, 72, 248–254. https://doi.org/10.1016/0003-2697(76)90527-3

Brock, C., Knies-Deventer, H., & Leithold, G. (2011). Assessment of cropping system impact on soil organic matter levels in short-term field experiments. *Journal of Plant Nutrition and Soil Science*, 174, 867–870. https://doi.org/10.1002/jpln.201100147

Cabrera, M. L., Kissel, D. E., & Vigil, M. F. (1994). Potential nitrogen mineralization: Laboratory and field evaluation. In J. L. Havlin and J. S. Jacobsen (Eds.), *Soil testing: Prospects for improving nutrient recommendations* (pp. 15–30). SSSA Special Publication 40 Madison, WI: SSSA and ASA.

Caudle, C., Osmond, D., Heitman, J., Ricker, M., Miller, G., & Wills, S. (2020). Comparison of soil health metrics for a Cecil soil in the North Carolina Piedmont. *Soil Science Society of America Journal* https://doi.org/10.1002/saj2.20075

Combs, M., & Nathan, M. V. (1998). Soil organic matter. In J. R. Brown (Ed.), *Recommended chemical soil test procedures for the North Central Region* (revised, pp. 53–58). North Central Regional Research Publication 221. Columbia, MO: Missouri Agricultural Experiment Station.

Culman, S. W., Snapp, S. S., Green, J. M., & Gentry, L. E. (2013). Short- and long-term labile soil carbon and nitrogen dynamics reflect management and predict corn agronomic performance. *Agronomy Journal*, 105, 493–502 [erratum: 105, 874]. https://doi.org/10.2134/agronj2012.0382

Curtin, D., Beare, M. H., Lehto, K., Tregurtha, C., Qiu, W., Tregurtha, R., & Peterson, M. (2017). Rapid assays to predict nitrogen mineralization capacity of agricultural soils. *Soil Science Society of America Journal*, 81, 979–991. https://doi.org/10.2136/sssaj2016.08.0265

Curtin, D., & Campbell, C. A. (2008). Mineralizable nitrogen. In M. R. Carter and E. G. Gregorich (Eds.), *Soil sampling and methods of analysis* (pp. 599–606). Boca Raton, FL: CRC Press.

Emran, M., Gispert, M., & Pardini, G. (2012). Patterns of soil organic carbon, glomalin and structural stability in abandoned Mediterranean terraced lands. *European Journal of Soil Science*, 63, 637–649. https://doi.org/10.1111/j.1365-2389.2012.01493.x

Fujii, K., Yamada, T., Hayakawa, C., Nakanishi, A., & Funakawa, S. (2020). Decoupling of protein depolymerization and ammonification in nitrogen mineralization of acidic forest soils. *Applied Soil Ecology*, 153, 103572. https://doi.org/10.1016/j.apsoil.2020.103572

Galloway, J. N., Townsend, A. R., Erisman, J. W., Bekunda, M., Cai, Z., Freney, J. R., … Sutton, M. A. (2008). Transformation of the nitrogen cycle: Recent trends, questions, and potential solutions. *Science*, 320, 889–892. https://doi.org/10.1126/science.1136674

Gardner, J. B., & Drinkwater, L. E. (2009). The fate of nitrogen in grain cropping systems: A meta-analysis of $^{15}N$ field experiments. *Ecolological Applications*, 19, 2167–2184. https://doi.org/10.1890/08-1122.1

Geisseler, D., Miller, K., Leinfelder-Miles, M., & Wilson, R. (2019). Use of soil protein pools as indicators of soil nitrogen mineralization potential. *Soil Science Society of America Journal*, 83, 1236–1243. https://doi.org/10.2136/sssaj2019.01.0012

Gillespie, A. W., Farrell, R. E., Walley, F. L., Ross, A. R. S., Leinweber, P., Eckhardt, K. U., … Blyth, R. I. R. (2011). Glomalin-related soil protein contains non-mycorrhizal-related heat-stable proteins, lipids and humic materials. *Soil Biology and Biochemistry*, 43, 766–777. https://doi.org/10.1016/j.soilbio.2010.12.010

Hurisso, T. T., Culman, S. W., & Zhao, K. (2018a). Repeatability and spatiotemporal variability of emerging soil health indicators relative to routine soil nutrient tests. *Soil Science Society of America Journal*, 82, 939–948. https://doi.org/10.2136/sssaj2018.03.0098

Hurisso, T. T., Culman, S. W., Zone, P., & Sharma, S. (2018b). Absolute values and precision of emerging soil health indicators as affected by soil sieve size.

*Communications in Soil Science and Plant Analysis*, 49, 1934–1942. https://doi.org/10.1080/00103624.2018.1492597

Hurisso, T. T., Moebius-Clune, D. J., Culman, S. W., Moebius-Clune, B. N., Thies, J., & van Es, H. M. (2018c). Soil protein as a rapid soil health indicator of potentially available organic nitrogen. *Agricultural and Environmental Letters*, 3, 180006. https://doi.org/10.2134/ael2018.02.0006

Idowu, O. J., van Es, H. M., Abawi, G. S., Wolfe, D. W., Schindelbeck, R. R., Moebius-Clune, B. N., & Gugino, B. K. (2009). Use of an integrative soil health test for evaluation of soil management impacts. *Renewable Agriculture and Food Systems*, 24, 214–224. https://doi.org/10.1017/S1742170509990068

Jan, M. T., Roberts, P., Tonheim, S. K., & Jones, D. L. (2009). Protein breakdown represents a major bottleneck in nitrogen cycling in grassland soils. *Soil Biology and Biochemistry*, 41, 2272–2282. https://doi.org/10.1016/j.soilbio.2009.08.013

Keeney, D. R. (1982). Nitrogen availability indices. In A. L. Page, R. H. Miller, & D. R. Keeney (Eds.), *Methods of soil analysis, Part 2: Chemical and microbiological properties* (2nd ed., pp. 711–733). Agronomy Monograph 9. Madison, WI: SSSA and ASA.

Keeney, D. R., and Bremner, J. M. (1966). Comparison and evaluation of laboratory methods of obtaining an index of soil nitrogen availability. *Agronomy Journal*, 58, 498–503. https://doi.org/10.2134/agronj1966.00021962005800050013x

Keeney, D. R., & Nelson, D.W. (1982). Nitrogen inorganic forms. In A. L. Page, R. H. Miller, & D. R. Keeney (Eds.), *Methods of soil analysis, Part 2: Chemical and microbiological properties* (2nd ed., pp. 643–698). Agronomy Monograph 9. Madison, WI: SSSA and ASA.

Khan, S., Mulvaney, R. L., & Hoeft, R.G. (2001). A simple soil test for detecting sites that are nonresponsive to nitrogen fertilization. *Soil Science Society of America Journal*, 65, 1751–1760. https://doi.org/10.2136/sssaj2001.1751

Liebig, M., Carpenter-Boggs, L., Johnson, J. M. F., Wright, S., & Barbour, N. (2006). Cropping system effects on soil biological characteristics in the Great Plains. *Renewable Agriculture and Food Systems* 21:36–48. https://doi.org/10.1079/RAF2005124

Lowry, O. H., Rosebrough, N. J., Farr, A. L., & Randall, R. J. (1951). Protein measurement with the Folin phenol reagent. *Journal of Biological Chemistry*, 193, 265–275.

Marriott, E. E., & Wander, M. M. (2006). Total and labile soil organic matter in organic and conventional farming systems. *Soil Science Society of America Journal*, 70, 950–959. https://doi.org/10.2136/sssaj2005.0241

Matsumoto, S., Ae, N., & Yamagata, M. (2000). Extraction of mineralizable organic nitrogen from soils by a neutral phosphate buffer solution. *Soil Biology and Biochemistry*, 32, 1293–1299. https://doi.org/10.1016/S0038-0717(00)00049-3

Moebius, B. N., van Es, H. M., Schindelbeck, R. R., Idowu, O. J., Thies, J. E., & Clune, D. J. (2007). Evaluation of laboratory-measured soil properties as indicators of soil physical quality. *Soil Science*, 172, 895–912. https://doi.org/10.1097/ss.0b013e318154b520

Moebius-Clune, B. N., van Es, H. M., Idowu, O. J., Schindelbeck, R. R., Moebius-Clune, D. J., Wolfe, D. W., … Lucey, R. (2008). Long-term effects of harvesting maize stover and tillage on soil quality. *Soil Science Society of America Journal*, 72, 960–969.

Moebius-Clune, B. N., Moebius-Clune, D. J., Gugino, B. K., Idowu, O. J., Schindelbeck, R. R., Ristow, A. J., … Abawi, G. S. (2016). *Comprehensive assessment of soil health: The Cornell framework* (3rd ed.). Ithaca, NY: Cornell University.

Mooshammer, M., Wanek, W., Schnecker, J., Wild, B., Leitner, S., Hofhansl, F., … Richter, A. (2012). Stoichiometric controls of nitrogen and phosphorus cycling in decomposing beech leaf litter. *Ecology*, 93, 770–782. https://doi.org/10.1890/11-0721.1

Mulvaney, R. L., Khan, S. A., Hoeft, R. G., & Brown, H. M. (2001). A soil organic nitrogen fraction that reduces the need for nitrogen fertilization. *Soil Science Society of America Journal*, 65, 1164–1172. https://doi.org/10.2136/sssaj2001.6541164x

Nannipieri, P., & Paul, E. (2009). The chemical and functional characterization of soil N and its biotic components. *Soil Biology and Biochemistry*, 41, 2357–2369. https://doi.org/10.1016/j.soilbio.2009.07.013

Nelson, D. W., & Sommers, L. E. (1996). Total carbon, organic carbon, and organic matter. In D. L. Sparks (Ed.), *Methods of soil analysis. Part 3* (pp. 961–1010). SSSA Book Series 5. Madison, WI: SSSA. https://doi.org/10.2136/sssabookser5.3.c34

Németh, K., Bartels, H., Vogel, M., & Mengel, K. (1988). Organic nitrogen compounds extracted from arable and forest soils by electro-ultrafiltration and recovery rates of amino-acids. *Biology and Fertility of Soils*, 5, 271–275. https://doi.org/10.1007/BF00262130

Nichols, K. A., & Millar, J. (2013). Glomalin and soil aggregation under six management systems in the northern Great Plains, USA. *Open Journal of Soil Science*, 3, 374–378. https://doi.org/10.4236/ojss.2013.38043

Noble, J. E., & Bailey, M. J. A. (2009). Quantitation of protein. *Methods in Enzymology*, 463, 73–95. https://doi.org/10.1016/S0076-6879(09)63008-1

Nunes, M. R., Karlen, D. L., Veum, K. S., Moorman, T. B., & Cambardella, C. A. (2020). Biological soil health indicators respond to tillage intensity: A US meta-analysis. *Geoderma*, 369, 114335. https://doi.org/10.1016/j.geoderma.2020.114335

Osterhaus, J. T., Bundy, L. G., & Andraski, T.W. (2008). Evaluation of the Illinois soil nitrogen test for predicting corn nitrogen needs. *Soil Science Society of America Journal*, 72, 143–150. https://doi.org/10.2136/sssaj2006.0208

Purin, S., & Rillig, M.C. (2007). The arbuscular mycorrhizal fungal protein glomalin: Limitations, progress, and a new hypothesis for its function. *Pedobiologia*, 51, 123–130. https://doi.org/10.1016/j.pedobi.2007.03.002

Rillig, M. C., Wright, S. F., Kimball, B. A., & Leavitt, S. W. (2001). Elevated carbon dioxide and irrigation effects on water stable aggregates in a sorghum field: A possible role for arbuscular mycorrhizal fungi. *Global Change Biology*, 7, 333–337. https://doi.org/10.1046/j.1365-2486.2001.00404.x

Roper, W. R., Osmond, D. L., & Heitman, J. L. (2019). A response to "Reanalysis validates soil health indicator sensitivity and correlation with long-term crop yield." *Soil Science Society of America Journal*, 83, 1842–1845. https://doi.org/10.2136/sssaj2019.06.0198

Roper, W. R., Osmond, D. L., Heitman, J. L., Wagger, M. G., & Reberg-Horton, S. C. (2017). Soil health indicators do not differentiate among agronomic management systems in North Carolina soils. *Soil Science Society of America Journal*, 81, 828–843. https://doi.org/10.2136/sssaj2016.12.0400

Ros, G. H., Temminghoff, E. J. M., & Hoffland, E. (2011). Nitrogen mineralization: A review and meta-analysis of the predictive value of soil tests. *European Journal of Soil Science*, 62, 162–173. https://doi.org/10.1111/j.1365-2389.2010.01318.x

Rosier, C. L., Hoye, A. T., & Rillig, M. C. (2006). Glomalin-related soil protein: Assessment of current detection and quantification tools. *Soil Biology and Biochemistry*, 38, 2205–2211. https://doi.org/10.1016/j.soilbio.2006.01.021

Ruffo, M. L., Bollero, G. A., Hoeft, R. G., & Bullock, D. G. (2005). Spatial variability of the Illinois Soil Nitrogen Test: Implications for soil sampling. *Agronomy Journal*, 97, 1485–1492. https://doi.org/10.2134/agronj2004.0323

Schimel, J. P., & Bennett, J. (2004). Nitrogen mineralization: Challenges of a changing paradigm. *Ecology*, 85, 591–602.

Schindelbeck, R. R., Moebius-Clune, B. N., Moebius-Clune, D. J., Kurtz, K. S., & van Es, H. M. (2016). *Cornell Soil Health Laboratory: Comprehensive assessment of soil health standard operating procedures*. Ithaca, NY: Cornell University.

Schindler, F. V., Mercer, E. J., & Rice, J. A. (2007). Chemical characteristics of glomalin-related soil protein (GRSP) extracted from soils of varying organic matter content. *Soil Biology and Biochemistry*, 39, 320–329. https://doi.org/10.1016/j.soilbio.2006.08.017

Schomberg, H. H., Wietholter, S., Griffin,T. S., Reeves, D. W., Cabrera, M. L., Fisher, D. S., … Tyler, D. D. (2009). Assessing indices for predicting potential nitrogen mineralization in soils under different management systems. *Soil Science Society of America Journal*, 73, 1575–1586. https://doi.org/10.2136/sssaj2008.0303

Sela, S., van Es, H. M., Moebius-Clune, B. N., Marjerison, R., Moebius-Clune, D., Schindelbeck, R., … Young, E. (2017). Dynamic model improves agronomic and environmental outcomes for maize nitrogen management over static approach. *Journal of Environmental Quality*, 46, 311–319. https://doi.org/10.2134/jeq2016.05.0182

Smith, P. (2004). How long before a change in soil organic carbon can be detected? *Global Change Biology*, 10, 1878–1883. https://doi.org/10.1111/j.1365-2486.2004.00854.x

Smith, P. K., Krohn, R. I., Hermanson, G. T., Mallia, A. K., Gartner, F. H., Provenzano, M. D., … Klenk, D. C. (1985). Measurement of protein using bicinchoninic acid. *Analytical Biochemistry*, 150, 76–85. https://doi.org/10.1016/0003-2697(85)90442-7

Spargo, J. T., Alley, M. M., Thomason, W. E., & Nagle, S. M. (2009). Illinois soil nitrogen test for prediction of fertilizer nitrogen needs of corn in Virginia. *Soil Science Society of America Journal*, 73, 434–442. https://doi.org/10.2136/sssaj2007.0437

Stanford, G., & Smith, S. J. (1972). Nitrogen mineralization potentials of soils. *Soil Science Society of America Journal*, 36, 465–472. https://doi.org/10.2136/sssaj1972.03615995003600030029x

van Es, H. M., & Karlen, D. L. (2019). Reanalysis validates soil health indicator sensitivity and correlation with long-term crop yields. *Soil Science Society of America Journal*, 83, 721–732. https://doi.org/10.2136/sssaj2018.09.0338

Walker, J. M. (2002). The bicinchonic acid (BCA) assay for protein quantitation. In J.M. Walker (Ed.), *The protein protocols handbook* (2nd ed., pp. 11–14). Totowa, NJ: Humana Press. https://doi.org/10.1385/1-59259-169-8:11

Weintraub, M. N., & Schimel, J. P. (2005). Seasonal protein dynamics in Alaskan arctic tundra soils. *Soil Biology and Biochemistry*, 37, 1469–1475. https://doi.org/10.1016/j.soilbio.2005.01.005

Wright, S. F., & Anderson, R. L. (2000). Aggregate stability and glomalin in alternative crop rotations for the central Great Plains. *Biology and Fertility of Soils*, 31, 249–253. https://doi.org/10.1007/s003740050653

Wright, S. F., Green, V. S., & Cavigelli, M. A. (2007). Glomalin in aggregate size classes from three different farming systems. *Soil and Tillage Research*, 94, 546–549. https://doi.org/10.1016/j.still.2006.08.003

Wright, S. F., Starr, J. L., & Paltineanu, I. C. (1999). Changes in aggregate stability and concentration of glomalin during tillage management transition. *Soil Science Society of America Journal*, 63, 1825–1829. https://doi.org/10.2136/sssaj1999.6361825x

Wright, S. F., & Upadhyaya, A. (1996). Extraction of an abundant and unusual protein from soil and comparison with hyphal protein of arbuscular mycorrhizal fungi. *Soil Science*, 161, 575–586. https://doi.org/10.1097/00010694-199609000-00003

Wright, S. F., & Upadhyaya, A. (1998). A survey of soils for aggregate stability and glomalin, a glycoprotein produced by hyphae of arbuscular mycorrhizal fungi. *Plant and Soil*, 198, 97–107. https://doi.org/10.1023/A:1004347701584

Wu, Q. S., Cao, M. Q., Zou, Y. N., & He, X. (2014). Direct and indirect effects of glomalin, mycorrhizal hyphae, and roots on aggregate stability in rhizosphere of trifoliate orange. *Science Reports*, 4, 5823. https://doi.org/10.1038/srep05823

Yan, M., Pan, G., Lavallee, J. M., & Conant, R. T. (2019). Rethinking sources of nitrogen to cereal crops. *Global Change Biology*, 26, 191–199. https://doi.org/10.1111/gcb.14908

# 11

# Metabolic Activity – Enzymes

*Verónica Acosta-Martínez, Lumarie Pérez-Guzmán, Kristen S. Veum, Márcio R. Nunes, and Richard P. Dick*

## Introduction

The microbial component of the soil biological community produces hundreds of enzymes that mediate biochemical reactions associated with decomposition processes in soil (Figure 11.1). Many soil microorganisms have dual purposes, producing enzymes to gain energy and releasing products that other organisms can metabolize. Soils vary in the amount, type, and distribution of enzymes due to their inherent soil organic matter (SOM) content, chemistry, texture, and predominant microbial community (Figure 11.1a). In addition, several management factors also influence enzyme stability, fate, and overall activity making enzymes useful soil ecosensors.

Soil enzymes can exist intracellularly (inside the cytoplasmic membrane) where they contribute to cellular life processes, or within the soil matrix as extracellular (*abiontic*) entities (Figure 11.1b). Both types contribute to potential enzyme activity in soils (Burns et al., 1972; Burns, 1978; Skujiņš, 1976). Extracellular enzymes, as defined by Skujiņš (1976), are those exclusive of live cells and excreted to the soil by microbes to: (1) hydrolyze substances that are too large or insoluble to be taken up directly by cells, (2) detoxify the surrounding environment, and/or (3) create a favorable environment for survival of the organism. The abiontic enzymes identified by McLaren et al. (1957, 1962) using soil sterilized by $\gamma$ radiation was devoid of viable cells but still hydrolyzed urease and phosphatase substrates. Functionality of abiontic

*Soil Health Series: Volume 2 Laboratory Methods for Soil Health Analysis*, First Edition.
Edited by Douglas L. Karlen, Diane E. Stott, and Maysoon M. Mikha.

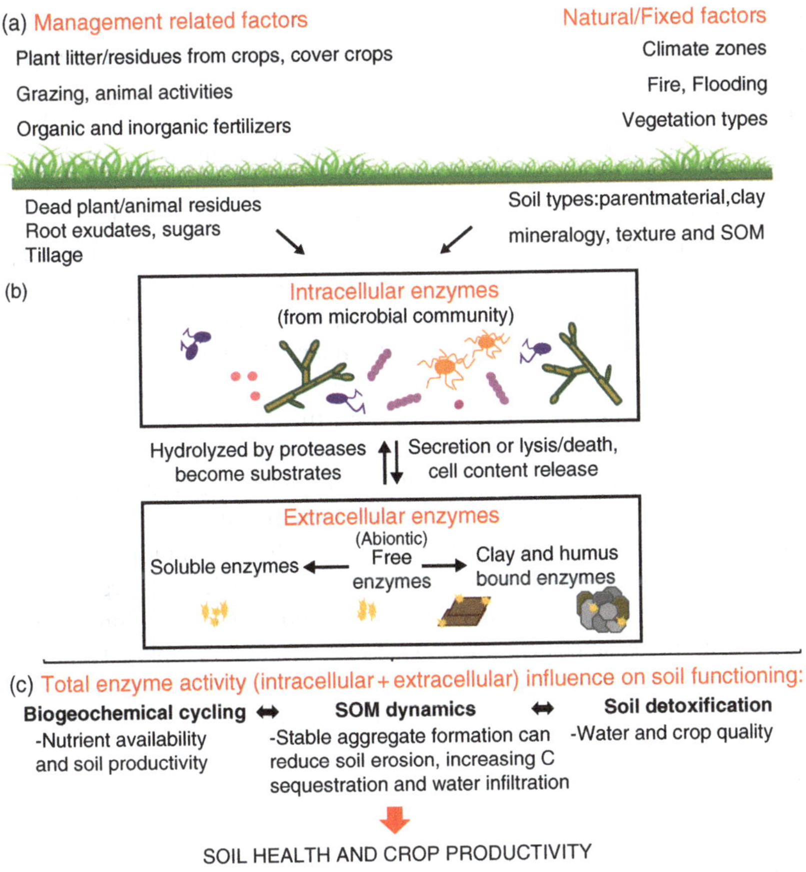


**Figure 11.1** Enzymatic activity in soil can be affected by management related factors and natural/fixed factors (a), and by their distribution within intracellular or extracellular locations (b), which will have an influence in overall soil health and crop productivity (c).

enzymes was subsequently confirmed using chloroform fumigation (Klose & Tabatabai, 1999; Klose & Tabatabai, 2002) and microwave irradiation (Speir et al., 1986; Knight & Dick, 2004) to sterilize the soil. The importance of these enzymes to soil health was documented by Knight and Dick (2004) who also found that abiontic β-glucosidase activity was able to detect land management impacts. This is important because abiontic enzymes accumulate along a steady trajectory, significantly reducing short-term variability (Bandick & Dick, 1999; Ndiaye et al., 2000). This characteristic is what provides the potential for calibrating EAs as soil health indicators by reducing in-season variability due to short term climatic variation or a recent soil management event.

Enzyme activities generally change more quickly (1 to 3 years) than soil physical or chemical properties (*e.g.*, bulk density, aggregate stabilty, soil organic matter) thus providing an early indication of changes in soil function due to biogeochemical cycling, C sequestration, or soil detoxification (*e.g.*, Dick, 1997; Wilson et al., 2009; Stott et al., 2013; Frene et al., 2018) (Figure 11.1c). EAs have also been correlated to shifts in microbial community composition, especially toward greater saprophytic fungal populations, which are more efficient in C transformation than bacteria (Bailey et al., 2002; Zhang et al., 2014). Such changes increase enzyme production and create a more diverse pool capable of degrading more complex substrates (Acosta-Martínez et al., 2011, 2014b). EAs have also been positively correlated with beneficial arbuscular mycorrhizal fungi (AMF) biomass, which is important for alleviating plant water stress in semiarid lands (Cotton et al., 2013; Davinic et al., 2013; Li et al., 2018). This means that soils which have been managed to promote soil health (*e.g.*, minimum tillage, organic amendments, cropping system diversification) should have larger microbial communities and greater enzyme concentrations bound within the soil matrix. Furthermore, since abiontic enzymes are generally complexed and protected within soil humic or clay complexes for many years, EAs provide an integrative soil biological index reflecting past soil management (Bandick & Dick, 1999).

Efforts to relate soil EA to plant productivity have produced mixed results. Early work showed no close relationships to either crop yield or soil nutrient status (Koepf, 1954; Drobnik, 1957; Galstyan, 1960; Haban, 1967). Other studies, however, showed that certain EAs (*e.g.*, phosphatase, invertase, β-glucosidase, and urease) correlate with the productivity of several crops including sugar beet (*Beta vulgaris*) and winter wheat (*Triticum aestivum*) (Verstraete & Voets, 1977), corn and soybean (de Castro Lopes et al., 2013), and cotton (Acosta-Martínez et al., 2011). In managed systems, other factors may confound or override relationships between soil EA and plant productivity since both are influenced by weather, pests, and many nuances of crop management. Recent multi-disciplinary studies have established ecological linkages involving soil EAs (Jian et al., 2016; Zuber & Villamil, 2016; Xiao et al., 2018). For example, Bhandari et al. (2018) found higher EAs in soils where a drought tolerant forage was planted. The ecological benefit was a suppression of fire ants and their associated impact on grass establishment. With regard to soil health, interest in combining EAs with other biological indicators (*e.g.*, nematodes, earthworms) is also increasing (García-Ruíz et al., 2008; Kizilkaya et al., 2010).

Enzyme activity assays were developed by soil enzymology pioneers to quantify the distribution, role and location of soil enzymes across a wide range of soil types (Nannipieri et al., 2018). An enzyme assay measures the rate of reaction, mostly according to product release, that reflects the amount of a specific enzyme in a soil (Tabatabai & Dick, 2002), but they are only valid and reproducible across investigations if the reaction occurs under optimal conditions. This includes excess substrate concentrations and optimum temperature, buffer, pH, and other possible co-factors. Therefore, soil EAs measure *potential* not in situ activity and cannot distinguish between viable cells (intracellular or cell surface) or extracellular enzyme activities (Figure 11.1b). However, assays under optimal conditions have provided a foundation for the current understanding of enzymatic responses to soil and crop management as well as their fate and persistence mechanisms (Skujiņš, 1967, 1976, 1978; Kiss et al., 1975; McLaren, 1975; Burns, 1978, 1982; Nannipieri et al., 1974, 1988; Tabatabai, 1994; Dick, 1994).

Among the most commonly evaluated EAs in agricultural soils, four hydrolases have been selected by soil health initiatives based on their importance to C (β-glucosidase), C and N (β-glucosaminidase), P (acid and alkaline phosphatase), and S (arylsulfatase) cycling (Figure 11.2a, 11.2b, 11.2c, and 11.2d, respectively). These enzyme assays have been vetted and optimized by M.A. Tabatabai and coworkers (Tabatabai, 1994; R. Dick, 2011). The assays are relatively simple, using

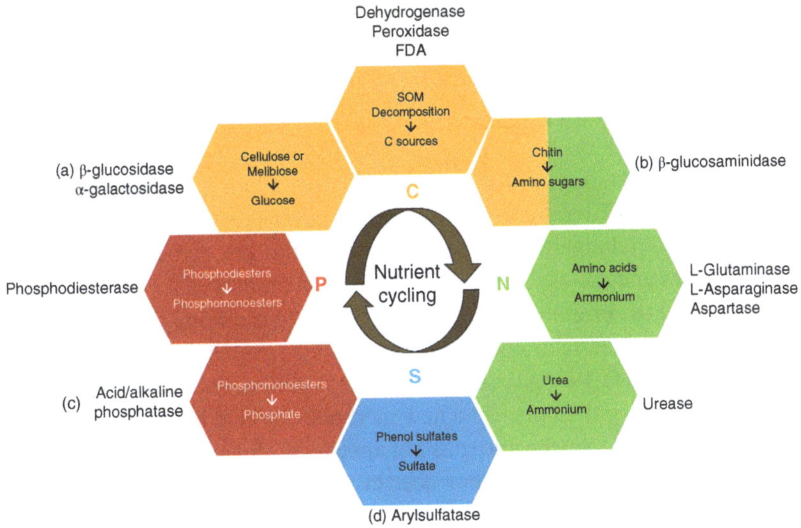


**Figure 11.2** Enzyme activities commonly evaluated in agricultural soils include: β-glucosidase (a), β-glucosaminidase (b), acid and alkaline phosphatase (c), and arylsulfatase (d) as the four selected by soil health initiatives based on their key role to C, C and N, P, and S cycling, respectively.

one gram of soil or less under buffered conditions where a synthetic substrate forms the product, *p*-nitrophenol (PNP) that can be measured by spectrophotometry. Among these four enzymes, the β-glucosidase assay is generally the most informative and capable of distinguishing among various management scenarios. It is a valuable assay because this enzyme is involved in cellulose (most abundant natural polysaccharide) degradation, which releases glucose as a primary energy source for heterotrophic microorganisms (Eivazi & Tabatabai, 1988). β-glucosidase was also the first EA incorporated as an indicator of C cycling in the Soil Management Assessment Framework (SMAF) (Stott et al., 2010). The activity of β-glucosaminidase provides information about C and N cycling and has been shown to be an index of N mineralization in soils since this enzyme hydrolyzes chitin releasing amino sugars (Ekenler & Tabatabai, 2004). Chitin is the second most abundant polysaccharide in nature and thus a major source of mineralized N in soils as it is found in the exoskeleton of most organisms (*e.g.*, insects, fungi, yeast, and algae). Acid or alkaline phosphatase enzymes are involved in SOM dynamics and in the transformation of P, which is globally the second most limiting nutrient for agricultural production. Phosphorus is an essential component of macromolecules such as lipids and DNA. Finally, arylsulfatase is involved in sulfur mineralization and the release of inorganic S, which is often another limiting nutrient for plant and microbial synthesis of enzymes and for some amino acids. Although sulfatases are produced by both bacteria and fungi, this enzyme has been related to fungal biomass because fungi and not bacteria accumulate ester sulfate which is the substrate for arylsulfatase (Saggar et al., 1981). Thus, arylsulfatase is typically correlated with fungal biomass. This is important because fungi are critical in many biogeochemical processes such as decomposition and for aggregation.

This chapter focuses on β-glucosidase, β-glucosaminidase, acid and alkaline phosphatase, and arylsulfatase activities as indicators of biological soil health. We begin by reviewing the literature and discussing how enzyme activities (EAs) can quite rapidly detect relatively recent soil biological changes, across diverse management practices and soil types. We provide recommendations for soil sampling and handling, and the original, accepted method for determining those EAs individually using the vetted methodology developed by Tabatabai and his research group (Tabatabai, 1994). Recognizing the goal for this book series is to be applicable for several years, we then present a new methodology that has been published for a limited number of site-specific soils but appears useful and capable of reducing analytical time and cost if further peer evaluated research confirms what's been observed to date. The combined analysis for simultaneous determination of all four EAs has therefore been included to encourage national and international evaluation by the soil health assessment community and to explore the potential regarding its suitability for commercial application. Finally, for individually determined EAs, we provide recommendations for using those EAs to assess soil biological health and to encourage further interpretation research as well. Our final goal is to discuss the applicability of EAs as a sensitive indicator of soil biological health and how they might be adopted by commercial and Agency service labs as a part of complete soil health assessment.

## Enzymes as Soil Health Indicators

A conceptual model for enzyme assays as indicators of soil quality was first articulated by Dick (1994) with the following four criteria, which are met by the four enzyme activities discussed in this chapter: (1) temporal sensitivity to negative or positive effects of external conditions within a reasonable time frame (*i.e.*, a few years, not decades) so that they will be useful for guiding sustainable soil management, (2) short-term stability without high in-season variability (to avoid obscuring the status of the soil sampled), (3) adaptability for commercial operations, in particular simplicity and high throughput capability, and (4) interpretability that is independent of soil type. This section summaries how EAs can help land managers implement better soil conservation programs by detecting early changes due to tillage frequency and intensity, crop diversity, inorganic fertilizer applications, organic (manure) amendments and other materials (Table 11.1) as well as increasing climate variability.

### Tillage and Crop Residues

Long-term intensive tillage practices (*e.g.*, plowing, disking, and weeding) can decrease SOM by exposing older, physically-protected SOC, decrease soil nutrient retention, and alter soil physical conditions, ultimately impacting the soil microbial community. Tillage also affects the soil microbial community, especially saprophytic (free-living) fungi, by breaking hyphal networks and decreasing both substrate availability and surface area for microbial enzyme stabilization. Gupta and Germida (1988) showed that 69 years of cultivating an Orthic Brown Chernozemic native prairie grassland decreased microbial biomass and the activities of arylsulfatase and acid phosphatase compared to native prairie soils. Alternatively, adoption of less intensive tillage practices can improve overall health of degraded soils, primarily by maintaining crop residues on the soil surface. This increases both content and quality of SOM, independent of soil texture (Nunes et al., 2018), and can positively affect soil EA. For example, a study conducted in India showed that the adoption of no-tillage significantly increased the activity of several enzymes (*i.e.*, β-glucosidase [54.5%], urease [88.8%], acid phosphatase [97.4%], and alkaline phosphatase [85.3%]) in a sandy clay loam soil when compared with conventional tillage (Sharma et al., 2013). In another long-term

**Table 11.1** Enzyme activities from selected studies across management and regions in the United States. C = corn; S = soybean; So = sorghum; B = barley; A = alfalfa; O = oats; P = peanut; Co = Cotton; R = rye; CC = cover crop.

| Soil and Crop Management | State | Soil Textural Class | Depth | SOC | β-Glucosidase | β-Glucosaminidase | Acid Phosphatase | Alkaline Phosphatase | Arylsulfatase | Notes and Other Factors | References |
|---|---|---|---|---|---|---|---|---|---|---|---|
| | | | cm | $g\ kg^{-1}$ | $mg\ PNP\ kg^{-1}\ h^{-1}$ | | | | | | |
| **Annual Crops** | | | | | | | | | | | |
| *No-till C-S-W with CC* | Missouri | Silt Loam | 0–5 | 22 | 236 | * | * | * | * | CEAP/ SMAF | Veum et al. (2015) |
| *No-till C-S with CC* | Missouri | Loam | 0–10 | 19 | 96 | 52 | * | * | * | On-farm | Rankoth et. al (2019) |
| *No-till A-C-S-W* | Minnesota | Clay Loam | 0–10 | 30–46 | 116–163 | 19–24 | 163–190 | * | 175–276 | combined assay | Acosta-Martinez et al. (2018, 2019) |
| *No-till C-S-B* | Colorado | Clay Loam | 0–10 | 12 | 130 | * | * | * | * | N rate | Zobeck et al. (2008) |
| *No-till C-S* | Missouri | Silt Loam | 0–5 | 17 | 167–222 | * | * | * | * | CEAP/ SMAF | Veum et al. (2015) |
| *No-till C-S* | Missouri | Silt Loam | 0–10 | 13 | 118 | 71 | * | * | * | On-farm | Alagele et al. (2019) |
| *No-till C-S* | Colorado | Clay Loam | 0–10 | 12 | 151 | * | * | * | * | N rate | Zobeck et al. (2008) |
| *No-till C-S* | Minnesota | Clay Loam | 0–10 | 37–48 | 80–120 | 12–21 | 145–171 | * | 183–214 | combined assay | Acosta-Martinez et al. (2018, 2019) |

| | | | | | | | | | | | |
|---|---|---|---|---|---|---|---|---|---|---|---|
| *No-till C-S (5 years)* | Minnesota | Clay Loam | 0–5 | 27 | 178–180 | 32–35 | 366–464 | * | * | stover removal | Johnson et al. (2013) |
| *No-till C-S (21 years)* | Minnesota | Clay Loam | 0–5 | 27–29 | 222–258 | 38–45 | 388–405 | * | * | stover removal | Johnson et al. (2013) |
| *No-till C-B* | Colorado | Clay Loam | 0–10 | 12 | 167 | * | * | * | * | N rate | Zobeck et al. (2008) |
| *No-till C (mono)* | Colorado | Clay Loam | 0–10 | 13 | 174 | * | * | * | * | N rate | Zobeck et al. (2008) |
| *Conservation-till Co-W* | Texas | Loamy Sand | 0–5 | 2.8 | 46 | 7 | 32 | 46 | 9 | irrigation | Acosta-Martinez et al. (2003) |
| *Conservation-till Co-W* | Texas | Sandy Loam | 0–5 | 5.8 | 78 | 13 | 47 | 147 | 13 | irrigation | Acosta-Martinez et al. (2003) |
| *Conservation-till Co-W* | Texas | Sandy Clay Loam | 0–5 | 8.9 | 192 | 15 | 98 | 174 | 8 | irrigation | Acosta-Martinez et al. (2003) |
| *Conservation-till Co-So* | Texas | Sandy Clay Loam | 0–5 | 7.7 | 63 | 8 | 37 | 119 | 23 | irrigation | Acosta-Martinez et al. (2003) |
| *Mulch-till C-S-W with manure* | Missouri | Silt Loam | 0–5 | 21 | 186 | * | * | * | * | CEAP/SMAF | Veum et al. (2015) |
| *Mulch-till C-S* | Missouri | Silt Loam | 0–5 | 14–16 | 132–181 | * | * | * | * | CEAP/SMAF | Veum et al. (2015) |
| *Vertical-till C-S* | Missouri | Loam | 0–10 | 17 | 103 | 46 | * | * | * | On-farm | Rankoth et. al (2019) |

*(Continued)*

**Table 11.1** (Continued)

| Soil and Crop Management | State | Soil Textural Class | Depth | SOC | β-Glucosidase | β-Glucosaminidase | Acid Phosphatase | Alkaline Phosphatase | Arylsulfatase | Notes and Other Factors | References |
|---|---|---|---|---|---|---|---|---|---|---|---|
| | | | cm | $g\ kg^{-1}$ | $mg\ PNP\ kg^{-1}\ h^{-1}$ | | | | | | |
| *Chisel plow C-S-W with CC* | Iowa | Loam | 0–10 | 8.2–9.6 | 117–153† | 47-69† | 188-400† | * | * | seasonal | McDaniel and Grandy (2016) |
| *Chisel plow C-C-O-A* | Iowa | Loam | 0–15 | 21–25 | 153–277 | * | * | * | * | rotation phase, N rate | Dodor and Tabatabi (2005); SOC from Moore et al. (2000) |
| *Chisel plow C-C-O-A* | Iowa | Clay Loam | 0–15 | 29–38 | 129–209 | * | * | * | * | rotation phase, N rate | Dodor and Tabatabi (2005); SOC from Moore et al. (2000) |
| *Chisel plow C-S-W* | Iowa | Loam | 0–10 | 7.0–7.6 | 102–118† | 33–54† | 152–280† | * | * | seasonal | McDaniel and Grandy (2016) |
| *Chisel plow C-S* | Iowa | Loam | 0–15 | 17–18 | 158–199 | * | * | * | * | rotation phase, N rate | Dodor and Tabatabi (2005); SOC from Moore et al. (2000) |

| | | | | | | | | | | | |
|---|---|---|---|---|---|---|---|---|---|---|---|
| *Chisel plow C-S* | Iowa | Clay Loam | 0–15 | 30–32 | 114–172 | * | * | * | * | rotation phase, N rate | Dodor and Tabatabi (2005); SOC from Moore et al. (2000) |
| *Chisel plow C-S* | Iowa | Loam | 0–10 | 7.6–7.8 | 107–110† | 20–55† | 129–291† | * | * | seasonal | McDaniel and Grandy (2016) |
| *Chisel plow C-S* | Minnesota | Clay loam | 0–5 | 24–26 | 137–138 | 23 | 269–289 | * | * | stover removal | Johnson et al. (2013) |
| *Chisel plow S (mono)* | Iowa | Loam | 0–15 | 17 | 117 | * | * | * | * | rotation phase, N rate | Dodor and Tabatabi (2005); SOC from Moore et al. (2000) |
| *Chisel plow C (mono)* | Iowa | Loam | 0–15 | 23–25 | 137–172 | * | * | * | * | rotation phase, N rate | Dodor and Tabatabi (2005); SOC from Moore et al. (2000) |
| *Chisel plow C (mono)* | Iowa | Clay Loam | 0–15 | 28–35 | 123–137 | * | * | * | * | rotation phase, N rate | Dodor and Tabatabi (2005); SOC from Moore et al. (2000) |
| *Chisel plow C (mono)* | Iowa | Loam | 0–10 | 8.1–7.9 | 94–111† | 27–67† | 133–270† | * | * | seasonal | McDaniel and Grandy (2016) |
| *Conventional till P-Co-Co* | Texas | Sand | 0–12 | 14 | 8–30 | 1–8 | 1–13 | 12–49 | 1–4 | * | Acosta-Martinez et al. (2004) |

(*Continued*)

**Table 11.1** (Continued)

| Soil and Crop Management | State | Soil Textural Class | Depth | SOC | β-Glucosidase | β-Glucosaminidase | Acid Phosphatase | Alkaline Phosphatase | Arylsulfatase | Notes and Other Factors | References |
|---|---|---|---|---|---|---|---|---|---|---|---|
| | | | cm | g kg$^{-1}$ | mg PNP kg$^{-1}$ h$^{-1}$ | | | | | | |
| *Conventional-till R-Co-W-Fallow* | Texas | Clay Loam | 0–5 | 13–16 | 138–352 | 23–59 | 59–172 | 200–478 | 28–67 | grazing | Acosta-Martinez et al. (2010) |
| *Conventional-till Veg-Fallow* | Oregon | Silt Loam | 0–20 | 23 | 53 | * | * | * | 34 | field moist, storage, seasonal, residue | Bandick and Dick (1999) |
| *Conventional-till Veg-CC* | Oregon | Silt Loam | 0–20 | 20–24 | 82 | * | * | * | 13–34 | field moist, storage, seasonal, residue | Bandick and Dick (1999) |
| *Conventional-till W-Fallow* | Oregon | Silt Loam | 0–20 | 13–17 | 42–76 | * | * | * | 15–39 | field moist, storage, seasonal, residue, amendment | Bandick and Dick (1999) |
| *Conventional-till Co (mono)* | Texas | Clay Loam | 0–5 | 12 | 67–144 | 9–19 | 31–58 | 122–189 | 18–21 | grazing | Acosta-Martinez et al. (2010) |
| *Conventional-till Co (mono)* | Texas | Loamy Sand | 0–5 | 2.8–3.4 | 23 | 5 | 9–10 | 28–29 | 2 | irrigation | Acosta-Martinez et al. (2003) |
| *Conventional-till Co (mono)* | Texas | Sandy Loam | 0–5 | 3.1–3.6 | 23–28 | 4–7 | 36–49 | 45-47 | 9–11 | irrigation | Acosta-Martinez et al. (2003) |
| *Conventional-till Co (mono)* | Texas | Sandy Clay | 0–5 | 3.1–6.9 | 31–57 | 4–8 | 12–38 | 38–93 | 3–13 | irrigation | Acosta-Martinez et al |

| | | | | | | | | | | | |
|---|---|---|---|---|---|---|---|---|---|---|---|
| *Conventional-till Co (mono)* | Texas | Silty Clay Loam | 0–10 | 5.5–6.0 | 78–83 | 10–17 | 60–65 | * | 8–13 | combined assay | Acosta-Martinez et al. (2018; 2019) |
| *Conventional-till C (mono)* | Missouri | Silt Loam | 0–5 | 19 | 131 | * | * | * | * | CEAP/ SMAF | Veum et al. (2015) |
| *Conventional-till C (mono)* | Colorado | Clay Loam | 0–10 | 12 | 103 | * | * | * | * | N rate | Zobeck et al. (2008) |
| *Conventional till P (mono)* | Texas | Sand | 0–12 | 20 | 14–71 | 2–15 | 2–17 | 17–58 | 1–5 | * | Acosta-Martinez et al. (2004) |
| *Corn (unknown mgmt)* | Ohio | Loamy sand | A horizon | 28 | 211 | 59 | 437 | 324 | 98 | field moist, storage | Lee et al. (2007) |
| *Corn (unknown mgmt)* | Ohio | Clay | A horizon | 21 | 137 | 50 | 429 | 314 | 142 | field moist, storage | Lee et al. (2007) |
| *Various tillage and veg. rotations* | Washington | Sandy Loam | 0–10 | 19–24 | 71–144 | * | * | * | * | organic, seasonal, amendments | Pritchett et al. (2011) |
| **Perennial** | | | | | | | | | | | |
| *Cool-season CRP* | Missouri | Silt Loam | 0–5 | 28 | 272 | * | * | * | * | CEAP/ SMAF | Veum et al. (2015) |
| *Warm-season CRP* | Missouri | Silt Loam | 0–5 | 21 | 197 | * | * | * | * | CEAP/ SMAF | Veum et al. (2015) |
| *Hay and Forage* | Missouri | Silt Loam | 0–5 | 24–26 | 211–248 | * | * | * | * | CEAP/ SMAF | Veum et al. (2015) |
| *Fescue pasture* | Missouri | Silt Loam | 0–5 | 36 | 534 | * | * | * | * | CEAP/ SMAF | Veum et al. (2015) |

*(Continued)*

Table 11.1 (Continued)

| Soil and Crop Management | State | Soil Textural Class | Depth | SOC | β-Glucosidase | β-Glucosaminidase | Acid Phosphatase | Alkaline Phosphatase | Arylsulfatase | Notes and Other Factors | References |
|---|---|---|---|---|---|---|---|---|---|---|---|
| | | | cm | g $kg^{-1}$ | mg PNP $kg^{-1}$ $h^{-1}$ | | | | | | |
| *Fescue* | Oregon | Silt Loam | 0–20 | unknown | 58 | * | * | * | 59 | field moist, storage, seasonal | Bandick and Dick (1999) |
| *Grass pasture* | Oregon | Silt Loam | 0–20 | 22 | 202 | * | * | * | 86 | field moist, storage, seasonal | Bandick and Dick (1999) |
| *Mixed pasture* | Missouri | Silt Loam | 0–5 | 27 | 181 | * | * | * | * | CEAP/SMAF | Veum et al. (2015) |
| *Restored prairie* | Missouri | Silt Loam | 0–5 | 20 | 139 | * | * | * | * | CEAP/SMAF | Veum et al. (2015) |
| *Grass buffer* | Missouri | Silt Loam | 0–10 | 15 | 130 | 75 | * | * | * | On-farm | Alagele et al. (2019) |
| *Fescue waterway* | Missouri | Silt Loam | 0–10 | 24 | 219 | 121 | * | * | * | On-farm | Alagele et al. (2019) |
| *Biomass crop (switchgrass)* | Missouri | Silt Loam | 0–10 | 16 | 138 | 80 | * | * | * | On-farm | Alagele et al. (2019) |
| *Native prairie* | Kansas | Silty Clay Loam | 0–5 | 30–35 | 21–45 | * | 46–92 | 35–82 | 30–49 | burn, N rate, seasonal | Ajwa et al. (1999) |
| *Warm-season pasture* | Texas | Clay Loam | 0–5 | 13–14 | 112–396 | 28–120 | 92–239 | 159–318 | 51–90 | grazing | Acosta-Martinez et al. (2010) |

| | | | | | | | | | | | |
|---|---|---|---|---|---|---|---|---|---|---|---|
| *Warm-season CRP* | Texas | Loam | 0–10 | 18–22 | 180–201 | 46–63 | 370–476 | * | 64–79 | combined assay | Acosta-Martinez et al. (2018, 2019) |
| *Warm-season grass (lovegrass)* | Texas | Silty Clay Loam | 0–10 | 8.2–8.4 | 124–128 | 14–16 | 70–76 | * | 18–19 | combined assay | Acosta-Martinez et al. (2018 2019) |
| *Warm-season grass* | Minnesota | Clay Loam | 0–10 | 38–44 | 70–102 | 23–27 | 152–167 | * | 200–241 | combined assay | Acosta-Martinez et al. (2018, 2019) |
| *Warm-season CRP* | Texas | Loamy Sand | 0–5 | 4 | 49 | 9 | 63 | 120 | 25 | * | Acosta-Martinez et al. (2003) |
| *Warm-season CRP* | Texas | Sandy Loam | 0–5 | 11 | 131 | 18 | 58 | 181 | 25 | * | Acosta-Martinez et al. (2003) |
| *Warm-season CRP* | Texas | Sandy Clay Loam | 0–5 | 11–13 | 144–231 | 26–32 | 73–270 | 238–256 | 15–53 | * | Acosta-Martinez et al. (2003) |
| *Native Rangeland* | Texas | Sandy Clay Loam | 0–5 | 11–19 | 191–198 | 27–34 | 154–185 | 251–308 | 13–58 | * | Acosta-Martinez et al. (2003) |
| *Native Rangeland* | Texas | Sandy Loam | 0–5 | 14 | 124 | 40 | 345 | 190 | 34 | * | Acosta-Martinez et al. (2003) |
| *Agroforestry -Oak* | Missouri | Silt Loam | 0–10 | 17 | 162 | 87 | * | * | * | On-farm | Alagele et al. (2019) |
| *Forest - deciduous* | Ohio | Silt Loam | A horizon | 34 | 200 | 175 | 738 | 463 | 694 | field moist, storage | Lee et al. (2007) |

† Data presented in nmol $g^{-1}$ $h^{-1}$.

study conducted on a silt loam soil, no-till significantly increased activities of phenol oxidase, peroxidase, dehydrogenase, and β-glucosaminidase compared with conventional tillage (Chu et al., 2016). Similarly, in a corn–soybean rotation, Veum et al. (2015) also found that no-till alone (without cover crops) demonstrated 19% greater β-glucosidase activity than that associated with a mulch-till system.

Deng and Tabatabai (1997) investigated the effect of different tillage practices (no-till, chisel plow, and moldboard plow) in combination with corn residue placement on activities of phosphatases and arylsulfatase. The greatest EAs were found when no-till practices were combined with residue return, whereas the lowest activities were found under no-till with bare soil or with moldboard plowing after crop residues were removed. In another study, Wegner et al. (2015) reported that during the soybean phase of a corn–soybean rotation microbial activity was up to 31% greater in a soil with a low rate of corn residue removal than with high corn residue removal, but there were no significant EA differences during the corn phase. In a later study, Wegner et al. (2018) concluded that multiple years of cover cropping can partially mitigate potential negative effects of excessive crop residue removal. Additional results from a U.S. meta-analysis across different soils, cropping systems, and weather conditions, showed that reducing tillage intensity from moldboard plow to no-till or perennial system increased β-glucosidase activity within the topsoil (0-15 cm) by 1.81 and 2.11 times, respectively (Nunes et al., 2020). These studies emphasize the importance of residue return in conjunction with no-tillage practices.

In general, EAs are concentrated in surface soils and decrease with depth due to decreased subsoil organic matter content. This stratification of microbial and biochemical activities impacts soil forming processes, biogeochemical cycling, plant productivity, and pollutant transformation and fate. Depth stratification is magnified in no-till systems where organic matter and nutrients are typically concentrated at or just beneath the soil surface (Luo et al., 2010). This results in significantly greater EAs in surface soils under no-till management (*e.g.*, Dick 1983; Veum et al., 2015; Rankoth et al., 2019). In contrast, tillage incorporates organic matter and nutrients by mixing plant residues into the soil and thus decreases depth stratification. Soil physical properties are often also affected by no-till, leading to increased water content and reduced pore space (Mielke et al., 1986), which in turn can suppress microbial activity at depth (Linn & Doran, 1984; Torbert & Wood, 1992). This was observed in Missouri claypan soils, where a diversified, corn-soybean-wheat rotation under no-till with a cover crop had greater EA in the 0- to 5-cm depth relative to a corn-soybean mulch-till system, but the reverse was observed within the 5- to 15-cm depth layer, due depth stratification (Veum et al., 2015). These seemingly subtle differences are important because the soil–atmosphere interface is where multiple ecosystem services are regulated. Enhanced surface soil function is beneficial for maintaining both agricultural productivity and environmental quality.

It is also important to recognize that the response of different enzymes to tillage intensity varies depending on soil type, crop sequence, climatic zone, and the specific enzyme of interest. For example, a long-term study comparing no-tillage and conventional tillage for corn showed a greater response for arylsulfatase than either acid or alkaline phosphatase activity (Eivazi et al., 2003). Similarly, another long-term (21 years) experiment comparing no-tillage and conventional tillage under three different cropping systems (soybean/wheat; maize/wheat; and cotton/wheat) within a subtropical climate on a clay soil (85% clay) showed that independent of crop rotation arylsulfatase response was greater than either acid or alkaline phosphatase activity (Balota et al., 2004). Adoption of no-tillage increased acid phosphatase up to 46%, alkaline phosphatase 61%, and arylsulfatase 219% within the 0- to 5-cm depth as compared to conventional tillage. Similar trends were observed at the 5- to 10- and 10- to 20-cm depths. In contrast, a 10-year study on sandy soils in the Southern High Plains showed no response in EA, probably due to soil water differences and/or because overall biomass incorporation from cotton was lower (Acosta-Martínez et al., 2011).

The benefits of no-tillage compared to conventional tillage on EAs and the processes they mediate within the soil ecosystem are well known, but additional studies are needed to elucidate their response to emerging practices such as strip tillage, vertical tillage, or rotational tillage. Those practices disturb only a small portion of the soil profile directly below the seed row; lightly till the soil while cutting, mixing, and anchoring a small portion of the residue into the upper few inches of soil; or combine different tillage practices and depths across a period of years. All, however, strive to leave large quantities of residue on the soil surface.

## Crop Diversity

### Crop Rotations

Compared to monocultures, benefits of diverse crop rotations, including leguminous crops, have been well documented (Zak et al., 2003; Karlen et al., 2006). Extended crop rotations generally increase EAs compared to monocropping because organic residue inputs are more diversified, the growing season is often extended, and rhizosphere activity, root exudates, and root density are often greater (Bandick & Dick, 1999; Deng et al., 2000; Klose et al., 1999; Eivazi et al., 2003). However, EA response to crop diversification is not easily predicted because of its dependence on other factors including soil type, climate, and the specific rotation. For example, Zhang et al. (2014) did not find differences in EAs between a corn–soybean rotation and monoculture corn on a clay loam soil under a humid continental climate. The lack of detectable EA effects may reflect the: (1) short-term (*i.e.*, 2 year) duration of the experiment, (2) lack of a cover crop, and/or (3) history of monoculture corn with conventional tillage for over 50 years before the Zhang et al. (2014) experiment was established. A study by Acosta-Martínez et al. (2011) showed that it took at least 5 years for measurable changes in soil microbial biomass C (MBC) and EAs to be detectable in sandy Southern Plains soils under a cotton-sorghum rotation as compared to continuous cotton. This likely reflects cotton's low residual biomass, and the weather patterns, since the soils being studied in this semiarid region were experiencing more frequent growth-limiting droughts that further reduced SOM accumulation. Overall, Acosta-Martínez et al. (2011) reported a 32 to 36% increase in biogeochemical cycling (depending on the enzyme). This is ecologically and environmentally important because that sandy soil has less than 1% of SOM and therefore, very little microbial biomass (*i.e.*, ~20 mg C $kg^{-1}$ soil) with continuous cotton.

Each EA may respond differently to crop rotation. For example, Chu et al. (2016) reported that dehydrogenase activity was greater under a corn-wheat-clover (*Trifolium partense*) rotation as compared to monoculture soybean, wheat (*Triticum aestivum*) or corn (*Zea mays*), but β-glucosaminidase activity was higher in soil from continuous corn than from the rotation. Those findings indicate that microorganisms producing the different enzymes have an optimum root zone environment and that plant species are important for selective development of soil microorganisms (Chu et al., 2016). Although sometimes considered a rotation effect (*i.e.,* cropping phase or rotation year), EAs also differ due to the amount, quality, and timing of crop residue input, as well as differences in root exudates and turnover among crops. Once again, those effects tend to be most evident in near-surface soil layers (Campbell et al., 1992).

### Cover Crops

Cover crops, sometimes referred to as green manure, have received a lot of recent attention for their potential to mitigate environmental challenges while simultaneoulsy boosting production and economic benefits for enhanced sustainability (Teasdale, 1996; Delgado & Gantzer, 2015; Lal, 2015; Dunn et al., 2016). Multiple studies have demonstrated improved soil physical, chemical, and biological properties due to cover crops (*e.g.*, Ding et al., 2006; Doran & Zeiss, 2000; Fageria et al., 2005; Veum et al., 2015; Rankoth et al., 2019). With regard to EAs, cover crops have been shown to increase several enzymes on a variety of soils when compared to simple rotations or monoculture cash crops (*e.g.*, Miller & Dick, 1995; Chander et al., 1997; Bandick & Dick 1999; Ndiaye et al., 2000; Debosz et al., 1999; Hoorman, 2009; Chu et al., 2016; Fernández et al., 2016; Jain et al., 2018). Cover crops increase EAs and improve soil health by providing cover to bare soil, reducing erosion potential, increasing the time that living plants and roots are present within agroecosystems, and providing greater root exudates, plant biomass and residue to the system. Those factors increase both the quantity and quality of SOM available to stimulate soil microbial activity (Kumar et al., 2017). Diversifying cropping systems by including cover crops in rotations can also increase EA resilience to seasonal climatic effects when compared to cropping systems without cover crops (Miller & Dick, 1995).

Several other factors can influence cover crop effects on agroecosystems and EAs. For example, Liang et al. (2014) reported crop species effects by showing that among Austrian winter pea (*Pisum sativum*), hairy vetch (*Vicia villosa*), and crimson clover (*Trifolium incarnatum*), Austrian winter pea had the greatest positive effect on both β-glucosidase and β-glucosaminidase. Chander et al. (1997) showed that using the legume *Sesbania aculeata* as a cover crop increased dehydrogenase and alkaline phosphatase activities, while oilseed crops (*e.g.*, sunflower or mustard) decreased enzyme activities. *Sesbania aculeata* presumably had a positive effect because it can supply high amounts of readily mineralizable C and N, which can stimulate microbial biomass production and result in higher enzyme activity. In contrast, the allelopathic effect of either sunflower or mustard inhibits development of microbial biomass, consequently decreasing enzyme activity in oilseed-based rotations.

Integration of cover crops with other conservation practices can have a synergistic effect on soil health indicators including EAs. In Missouri claypan soils, a diversified, corn-soybean-wheat rotation under no-till with a cover crop resulted in 68% more β-glucosidase activity in the 0- to 5-cm depth relative to a corn-soybean mulch-till system (Veum et al., 2015). They also noted that biological soil function, including EAs, was greatest for annual cropping systems that combined cover crops with no-till or extended rotations. For many of the studies, effects on soil health indicator values were nearly equivalent to perennial systems. Another vegetable system study on sandy soil in California, with combined inputs of yard-waste compost (none or 15.2 Mg $ha^{-1}$ $year^{-1}$), winter cover crops (annual or every fourth year), and different cover crop species (legume-rye, mustard, or rye) showed that β-glucosidase, β-glucosaminidase, alkaline phosphatase, dehydrogenase, and L-asparaginase were lowest without compost, intermediate with compost and infrequent cover cropping, and highest with compost and annual cover cropping (Brennan & Acosta-Martínez, 2019).

## Perennial Systems

Perennial grasses are expected to add organic C and N to soils through large, belowground, organic matter inputs (DuPont et al., 2010), and soils under perennial systems generally demonstrate greater inherent stability relative to cultivated soil (Zheng et al., 2004). Reintroduction of perennial grasslands by U.S. government-funded conservation initiatives, such as the Conservation Reserve Program

(CRP) since 1985, was initially designed to mitigate soil erosion and to enhance soil C sequestration, but CRP has also been important for improving soil health because it combines minimal disturbance, increased biodiversity, and maximization of soil cover and living root systems (USDA–NRCS, 2020). Several comparisons between cropland and CRP across different regions (*e.g.*, Acosta-Martínez et al., 2003; Staben et al., 1997; Tyler et al., 2016; Veum et al., 2015) have shown significant increases in all four EAs discussed in this chapter. For example, β-glucosidase activity was up to four times greater under CRP than under cropland within four different soil types, while β-glucosaminidase, acid phosphatase and arylsulfatase were up to 12 times greater under CRP than cropland. This was interpreted to reflect differences in both the reaction(s) occurring and substrates provided by the various types of grass (Acosta-Martínez et al., 2003). Other studies have reported activities of β-glucosidase and phosphatase to be at least two times greater in CRP areas than in croplands, highlighting the potential for CRP to enhance biological soil function (Acosta-Martínez et al., 2010; Veum et al., 2015; Tyler et al., 2016).

Although EAs under CRP and grasslands are generally greater than under cropland, to understand the soil biological restoration process of those ecosystems, it is important to quantify the temporal changes (Li et al., 2018). For example, Baer et al. (2000) reported that even after 10 years of CRP, microbial biomass was still significantly lower than in native prairies. A study by Li et al. (2017) showed the greatest changes in microbial community and C dynamics during the first 13 to15 years within a chronosequence that ranged from 0 to 28 years. The research team also found that nine EAs within the 0- to10-cm layer, and five EAs within the 10- to 30-cm layer increased linearly as years of CRP increased. Ultimately, there was a three-fold increase after 28 years (C. Li, unpublished data). The perennial systems studied by Veum et al. (2015), which included working grasslands (*i.e.*, pasture, forage and hay production), CRP, grassed waterways, and restored prairie systems, demonstrated enhanced β-glucosidase activity relative to a wide range of annual cropping systems in the 0- to 5-cm layer. Their results emphasize the soil quality/health benefits achieved with permanent vegetative cover and living roots extending deep within the soil profile, and underscore the potential loss of soil function on reversion of CRP lands to annual cropping systems. Information regarding the rate at which CRP benefits are lost when the site is returned to cropland is generally lacking, although a recent study showed rapid decreases in EAs and other soil health indicators within a month after a perennial grass site was tilled four times. This included a 68 and 74% decrease in phosphodiesterase and β-glucosidase activities, respectively (Cotton & Acosta-Martínez, 2018). That study confirmed that SOM accumulations after 30 years of CRP were still very fragile and raised questions regarding the extent of denaturation and hydrolysis by proteases that occur within the accumulated enzyme pool when tillage exposes them to oxidation. These results confirm that additional replicated studies across other geographic areas are needed to determine how soil type, grass species, and climatic affects EAs.

Species composition and growth phenology of perennial vegetation play a key role in microbial enzyme activity. Veum et al. (2015) observed greater β-glucosidase activity under cool-season grasses relative to warm-season systems. This was attributed to the coarser, less fibrous root systems under warm-season grasses, whereas cool-season grasses generally are more fibrous and have significantly more primary and secondary roots than warm-season grasses (Hetrick et al., 1988, 1991). Additionally, warm-season grasses tend to form clumps and leave soil more exposed than cool-season grasses. For example, tall fescue forms a tight, dense sod (Blanco-Canqui & Lal, 2008).

**Biomass Removal– Perennial Systems**

Returning crop residues in annual systems has been shown to impact a wide range of soil health properties and processes. Similarly, removal of perennial crops has the potential to compromise the sustainability of natural resources, particularly soil health and functioning (Franzluebbers, 2015), if not done with appropriate, site-specific, and science-based guidance. Several studies have shown that excessive biomass harvest as feedstock for biofuel or bio-product production or as animal feed can negatively affect soil health. This can occur through increases in soil temperature and/or penetration resistance as well as through decreased soil C and N inputs, soil water dynamics, and nutrient availability (*e.g.*, Doran et al., 1984; Karlen et al., 1994; Blanco-Canqui et al., 2007). Similarly, harvesting herbage for hay or forage production can substantially reduce nutrient levels without replacement fertilization (Veum et al., 2015). Those changes in soil nutrient status and other characteristics can also affect soil EAs.

Effects of biomass removal rates on soil EAs will vary across climate scenarios, cropping sequences, or soil types. For example, in a sandy soil with low SOM (<0.9%) in a semiarid region in Texas had higher MBC and MBN (11–15%) in plots where 50% of the biomass (forage sorghum; *Sorghum bicolor* L. Moench) remained, compared to 100% removal by the second growing season (Cotton et al., 2013). Additionally, the activities of β-glucosidase, β-glucosaminidase and alkaline phosphatase were lower in non-irrigated plots (vs. irrigated) with a 100% biomass removal rate. EAs showed no changes following three cycles of biomass removal from clay loam soils with high SOM in Minnesota, probably because the SOC pools had particulate organic matter (POM) and MBC levels that were more than 10 times higher than in coarse-textured southern soils (Johnson et al., 2013). Furthermore, compared to Texas cooler temperatures throughout most of the year in Minnesota likely slowed residue decomposition and thus minimized changes in microbial biomass, community structure, and EAs. Detecting residue removal effects on EAs may be more challenging in Minnesota soils, but aboveground biomass removal for biofuel feedstock, grazing, or other purposes must still be managed judiciously to avoid detrimental effects on soil health.

## Inorganic Fertilizers, Organic Amendments and other Materials

Soil amendments (*e.g.*, inorganic fertilizer, manure, compost, biochar, or nanomaterials) influence EAs differently because they have highly variable nutrient and organic matter contents. Changes in EAs due to inorganic fertilizer applications were demonstrated decades ago (*e.g.*, Klose et al., 1999; Deng et al., 2000; Saha et al., 2008). Combining fertilizer with other management practices can have an even greater effect on both EAs and overall soil health than either practice alone. For example, in a semiarid Eastern Oregon region, Bandick and Dick (1999) showed that applying a combination of inorganic N (34 kg $ha^{-1}$ $year^{-1}$), green manure or inorganic N (111 kg $ha^{-1}$ $year^{-1}$), and beef manure resulted in higher EAs than either 0 or 90 kg N $ha^{-1}$ $year^{-1}$ alone. A positive effect of combining green manure and inorganic fertilizer on EAs was also observed by Miller and Dick (1995). They reported that applying 280 kg inorganic N $ha^{-1}$ increased the EAs when compared to no N fertilizer (0 kg N $ha^{-1}$), but applying the same N rate with green manure increased EAs even more. Those results were observed during the third year of the experiment, indicating a short-term positive EA effect associated with combining inorganic and organic N fertilizers. Other studies have also reported increased fertilizer use efficiency associated with manure (Sharma et al., 2011, 2013). Sharma et al. (2013) also reported that different combinations of inorganic and organic N sources showed similar EA (*i.e.*, β-glucosidase, urease, and phosphatases) improvement when applied using conservation tillage (no-till) or with an optimum water supply. Their study, conducted on a sandy soil, showed that substitution of 25% N by farmyard manure, green manure, or biofertilizers or using only organic N sources resulted in significant positive EA changes. Positive responses of dehydrogenase and β-glucosidase activities to combinations of inorganic fertilizer and organic amendments were also detected in a paddy-rice ecosystem (Islam et al., 2011). In that study, chemical NPK fertilizers were applied alone or in combination with different rates of composted rice straw. After 53 years, EAs were the lowest in untreated soil and highest in soils treated with both organic and inorganic fertilizers. Among the latter, EAs were generally the highest in soils receiving the maximum amount of compost: NPK + 30 Mg $ha^{-1}$ compost > NPK + 22.5 Mg $ha^{-1}$ compost > NPK + 7.5 Mg $ha^{-1}$ compost > NPK > without fertilizer or compost.

Recently, interest has increased in application of inorganic fertilizers in combination with biochar, an organic material produced by pyrolysis at temperatures between 350 and 600°C, as a practice to improve soil health (Liang et al., 2006; Chan et al., 2007; Lehmann & Joseph, 2009; Novak et al., 2012; Gul et al., 2015). For example, a 3-year study on a silt loam soil showed that applying N, P and K with biochar increased alkaline phosphatase and urease activities in a corn production system when compared to the same soil with N-P-K but without biochar (Bera et al., 2016). The increase in urease activity was suggested to be due to protection of this enzyme by its association with biochar or a synergistic interaction between soil microbes and biochar (Lehmann et al., 2011). This specific combination (NPK plus biochar), however, decreased the activity of β-glucosidase and acid phosphatase. The increase in alkaline phosphatase and decrease in acid phosphatase activity were presumably caused by higher soil pH due to the biochar addition. This may explain how biochar impacts enzymes differently due to the chemistry involved in their stabilization and because of the different roles enzymes have in the decomposition of this material (Albiach et al., 2000; Lehmann et al., 2011).

Biochar can be produced using a wide variety of pyrolysis conditions which results in very diverse characteristics for the same feedstock, and therefore influences the substrates available for EA catalyzed reactions. For example, alkaline phosphatase activity has been shown to increase or decrease depending on the type of biochar. Trupiano et al. (2017) incubated soils with orchard pruning biochar and found that the activities of acid and alkaline phosphatase increased when compared to control (no biochar). Another study showed increased alkaline phosphatase activity in soils incubated with wheat biochar, but significantly lower activity in soils incubated with either rice or corn biochar for 67 d (Purakayastha et al., 2015). Pine-wood biochar increased β-glucosaminidase while it decreased β-glucosidase and phosphatase activities (Foster et al., 2016). They attributed EA increases to higher SOC, increased co-location and stabilization of the C substrates for the enzymes, as well as increased surface area for stabilizing those enzymes in the biochar plots.

Studies have also shown that biochar application rates are an important factor influencing EAs. An incubation study in China reported maximum increases in β-glucosidase activity at 0.5 and 1% application rates but no detectable response at 2% (Demisie et al., 2014). Similarly, Khadem and Raiesi (2017) reported that corn biochar decreased protease activity at a rate of 0.5% when compared to 1% application rate on sandy soil, while an opposite trend was observed for clay soil. Another study with corn stover biochar showed higher β-glucosidase activity in control or low (1 t $ha^{-1}$) treatment plots, while the highest application rate (30 t $ha^{-1}$) significantly increased alkaline phosphatase activity (Jin, 2010). A review by Gul et al. (2015) reported that coarse-textured soils tend to have less aggregation, lower microbial biomass and decreased EAs when amended with slow pyrolysis biochars than with high temperatures (>600°C) materials, even though other soil properties were not affected. Another study reported that plant-based (*e.g.*, pine wood and *Miscanthus* spp.) biochars increased β-glucosidase activity when compared to sludge biochar, with a response being higher for an Acrisol than a Ferralsol (Paz-Ferreiro et al., 2014). That study also showed increased activities of β-glucosaminidase and arylsulfatase in soils with the different biochars. Those results contrasted with other studies that reported no changes in arylsulfatase activity and suggested that the S cycle is unaltered by biochar additions (Paz-Ferreiro et al., 2012; Sun et al., 2014).

Nanomaterials are products engineered to have specific sizes, shapes, properties (*e.g.*, optical, electrical, magnetic, and chemical), and reactivity (Dinesh et al., 2012). Their environmental fate and reactivity are largely unknown (Nowack & Bucheli, 2007) and outcomes associated with their application to agricultural soils are uncertain. Some studies have reported nanoparticles possess antibacterial properties and may directly affect soil microbial communities (Vance et al., 2015) or indirectly affect them through changes in nutrient availability and other interactions (Aruguete & Hochella, 2010). Three studies that evaluated nanomaterials in soil suggest that they do not represent a long-term threat to the enzymatic potential or overall biogeochemical cycling. This includes Eivazi et al. (2018) who reported a decrease in the four EAs discussed in this chapter within 1 h to 1 wk after exposure to silver nanoparticles regardless of the application rate (1600 or 3200 μg Ag $kg^{-1}$ dry soil), but with apparent recovery beyond that time. Another study showed that multi-walled carbon nanotubes decreased activities of acid phosphatase and β-glucosidase by ≤30% for up to 28 d, with no effects detectable after 90 d (Shrestha et al., 2013). The third study with C60 fullerene, a carbon-based nanomaterial, also reported no detectable effects on several enzyme activities during a 6-month study with exposure to a dose of 1000 mg $kg^{-1}$ in soil (Tong et al., 2007), but we suggest further evaluations of nanomaterial applications under different soil, climate, and management scenarios need to be conducted to determine potential environmental outcomes of those practices.

## Climate Change

Almost a decade ago, a soil methodology book emphasized the importance of soil enzyme research to help understand potential consequences of climate change, and described it as the most important challenge soil biologists have ever faced (R. Dick, 2011). To date, this continues to be a challenge for understanting and improving soil health. Most climate change studies have been conducted under laboratory and/or simulated conditions (*e.g.*, Sardans et al., 2008; Bérard et al., 2011; A'Bear et al., 2014; Alster et al., 2013). We did find a few studies addressing the effects of extreme weather events (*e.g.*, record drought, heat waves, record precipitation) on EAs of soils (Hammerl et al., 2019; Acosta-Martinez et al., 2014a, 2014). One of the studies found shifts in microbial community composition and significantly higher EAs (8 out of 10) in agricultural soils during a record drought (*i.e.*, Palmer Drought Severity Index = −7.93) and heat wave in the Texas High Plains. The net results were a decrease in the inherent SOM content of the soils and consistent EA decreases due an exhaustion of SOM substrates in the following sampling times (Acosta-Martínez et al., 2014a, 2014b). A recent continuous cotton study by Pérez-Guzmán et al. (2020) on five soil types in the same agricultural region showed that all sites (regardless of textural class), responded similarly from 2015 to 2016 to decreased precipitation and increased temperature. They reported EAs were reduced by as much as 87% in the summer of 2016 (warmest year in the planet) when compared to the summer of 2015, a year of record regional precipitation (NOAA, 2018). Those extreme weather events caused shifts in microbial community composition illustrated by a 59% decrease in saprophytic fungal markers. The study which evaluated a 5-year cycle, showed increasing EA and fungal population trends between 2016 and 2017, demonstrating remarkable resilience within these fragile (highly erodible, low SOM) soils. However, more research is needed to quantify climate change effects for other regions (*e.g.*, humid, tropical), soil types (*e.g.*, higher clay and SOM content), management practices, and the interactions among those factors. Furthermore, questions regarding composition of recovered microbial communities compared to initial communities prior to the extreme weather event, and how different enzymatic pools, including the loss of abiontic enzymatic pool that had been built for so many years, are altered by extreme weather events remain unanswered.

## Soil Sampling and Handling

Sampling and handling soils for EA assessment, as for other biological measurements, requires careful decisions regarding sampling time, soil type, management history, approach (*e.g.*, sub-reps within a site, field replicates in research plots), and storage (Lorenz & Dick, 2011).

### Field Sampling

Time of sampling is important for every soil health indicator, but for β-glucosidase, arylsulfatase and phosphatase it has been established that a significant amount of the activity is associated with the abiontic enzyme fraction (McLaren et al., 1957; Knight & Dick, 2004). Therefore, since abiontic enzymes are accumulated and stabilized in the soil matrix they exhibit less seasonal variation (Ndiaye et al., 2000). This suggests EA samples can be taken at any time of year, but we recommend a spring sampling since that typically coincides with soil fertility testing (*i.e.*, before planting). Sampling after harvest can also be beneficial for monitoring system stability, but the most important decision for valid, multi-year comparisons is collecting samples at the same time every year.

After determining when to sample, the next step is to know what soil types are present at each sampling site. Local soil surveys (*e.g.*, the U.S. county surveys are available from Natural Resources Conservation Service, online at http://soils.usda.gov/survey/printed_surveys/) provide the best guidance for this question. Printed or web-based (http://websoilsurvey.nrcs.usda.gov/app/HomePage.htm) they identify the soil type(s) and provide general soil physical and chemical properties for each site. However, even though the manuals are valuable resources, field inspections are still recommended.

Following site selection it is important to determine if it will be randomly sampled, divided into subsample units (stratified random sample), systematically sampled, selectively sampled (judgment sampling), or evaluated using a composite sample (Lorenz & Dick, 2011). For agricultural applications, a randomized, composite sampling is generally most appropriate and is typical for routine soil fertility testing. Composite sampling greatly reduces the cost for analysis and integrates spatial variability by homogenizing over the area of interest. The number of samples to take for small uniform areas (<0.5 ha or <1.2 acres) to form a composite sample may be as few as 5 to 10, while for larger areas 25 samples appear to be sufficient as collecting more generally gains little in precision (Webster & Oliver, 1990). Typically, most soil scientists use a 2.5-cm diameter probe and collect several cores to the desired depth of sampling, but for sandy soils, which have less structure, or soils with dense root material, or those under grass, it may be better to use a shovel to take the soil samples, but care must be taken to ensure sampling depth is consistent.

To determine the optimum sampling depth, knowledge regarding management history and the specific questions to be answered need to be known. Influential factors include fertilizer type, quantity and method of application (*e.g.*,banding vs. broadcast), prior tillage, crop rotation, and land leveling or erosion that may have exposed the subsoil. The standard depth for sampling agricultural soils is 0 to 15 cm because that is the typical, historical depth of tillage. For minimum- or no-tillage systems several studies have shown that hot spots for biological properties tend to be in the top 7.5 cm or from 7.5 to 15 cm (Dick & Daniel, 1987), but other studies have used 0 to 10 cm or even more stratified increments of 0 to 5, 5 to 10, and 10 to 15 cm (*e.g.*, Acosta-Martínez et al., 2004; Johnson et al., 2013). For undisturbed or natural sites such as grasslands and forests, we recommend sampling by horizon.

### Sample Handling and Storage

As with other biological measurements, soil samples should be kept fresh in a sealed plastic bag, brought to the laboratory in a cooler, and stored at 4°C until they can be processed. If analyses are to be done on fresh soil samples, they should be analyzed as soon as possible—within days—to obtain results that are most characteristic of the sampling day. An advantage of the enzymes considered in this chapter is that they can be run on air-dried soil, which overcomes the cold storage requirement of –20 to –80°C for PLFA and DNA, rapid analysis for microbial biomass C, and/or the need to correct for water content (Lee et al., 2007). Overall, air-drying has the biggest impact on enzymes associated with the viable microbial population, with less effect on the catalytic fraction stabilized in the soil matrix. This could be an advantage for soil health evaluations as air-drying would reduce the impact of conditions that affect the highly variable microbial component relative to EA. Furthermore, EAs after air-drying are likely to better reflect the true long-term trajectory of a given management practice on soil health. We advise readers to review the literature to understand how EAs vary in field-moist vs. air-dried soil (W. Dick, 2011). Some studies have reported decreases in activity (20 to 50%) for β-glucosidase (Lee et al., 2007; Bandick & Dick, 1999), acid phosphatase (Lee et al., 2007), and arylsulfatase (Bandick and Dick, 1999; Zornoza et al., 2006), while others reported increased activity due to air-drying (Tabatabai & Bremner, 1970; Eivazi & Tabatabai 1977). Those differences simply emphasize the multiple locations and enzyme origins within soils.

For sample handling, we recommend sieving field moist soil (2- to 5-mm mesh) and then air-drying as needed, since, in our experience, activities of β-glucosidase, phosphatase, and arylsulfatase always go down if this protocol is followed but maintains treatment effects. This is desirable for commercial soil testing labs as they are used to handling air dried soils and reduces the urgency for immediate analysis after sampling or storage at –20 or 80°C. Also, experience from Acosta-Martínez's laboratory shows that EAs, especially for air-dried sandy soils, are highly reproducible and that reanalysis of the same sample will result in similar absorbance readings over time regardless of the operator or spectrophotometer (data not presented).

## Method

This section describes two approaches to determine the four EAs discussed in this chapter. The first approach involves running individual assays at bench scale (1-g samples) using a *p*-nitrophenyl derivate substrate at optimal buffer pH (37°C for 1 h) and colorimetric determination of PNP at 400 nm according to Tabatabai (1994) (Figure 11.3a). We recommend this as the standard EA approach for soil health

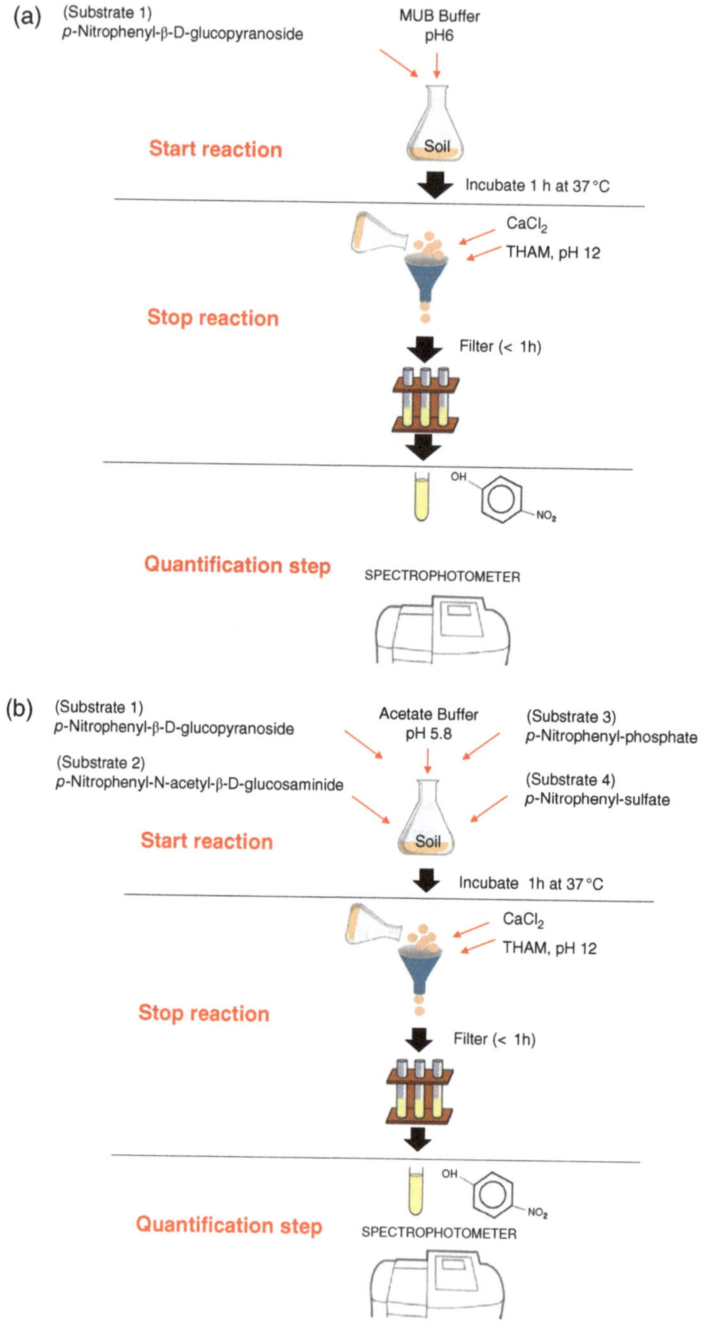


**Figure 11.3** General steps for an individual assay to measure activities of β-glucosidase, β-glucosaminidase, phosphatases or arylsulfatase (a) and for the combined assay to measure these four EAs simultaneously (b). The example of the individual assay is given for β-glucosidase (without toluene).

assessment because the protocols have been vetted to optimize pH, buffer, and other factors. Furthermore, several papers, some since the early 1970s, have been published on these methods and they have been shown to be sensitive for detecting changes due to land management, physical disturbance, and selected contaminants. The second approach is a combined assay for evaluating all four EAs simultaneously using Tabatabai's substrate concentrations and steps (Acosta-Martínez et al., 2019) (Figure 11.3b). This combined analysis appears to be a promising, cost-effective approach for EA measurement in commercial labs, but it still needs to be tested for accuracy and sensitivity across a diverse set of soils and environments. Therefore, we view the combined method as still being in a research mode and strongly encourage the research community to use it alongside the four individual assays to ultimately vet it and develop a database for calibration and interpretation.

Another EA assay for which there has been growing interest is the fluorescence-based microplate method because it is sensitive and has potential for simultaneous and rapid evaluations using the same soil suspension (Deng et al., 2013; Drouillon & Merckx, 2005; Marx et al., 2001; Pritsch et al., 2004; Trap et al., 2012). Recent methylumbelliferone (MUF) microplate results have correlated well with the bench-scale PNP method when using strict protocols (Popova & Deng 2010; Deng et al., 2013; Dick et al., 2013; Dick et al., 2018). Those studies showed consistent rankings among soil types and/or management practices. Although the results are encouraging and a cross lab study confirmed that different operators (using a strict and consistent protocol) could get reasonably similar results with the MUF microplate method (Dick et al., 2018), we do not recommend it as a soil health assay. First, there is no standardized method that has been as fully vetted as the "Tabatabai protocol", even though the MUF protocol (*e.g.*, optimum pH of buffer) is based on PNP results. Second, the micro-scale method has great potential for operator error because of the challenge in having a consistent soil suspension across labs. Factors affecting the suspension include: the type of homogenizer, size of glassware and stir bars, differences in mixing speed, and the time a sample remains in suspension (Dick et al., 2018, 2013). Third, there is also high potential for operator error due to ML volume pipetting and the need for strict temperature control (*i.e.*, reagents need to be preheated and incubation temperature maintained) since MUF fluorescence is very sensitive to temperature (Guilbault, 1990). Finally, although the microplate method is promoted as a high throughput method, it is questionable whether it is really a "high throughput" method because of the need to: (1) create soil suspensions (a labor-intensive step that is not needed for conventional bench assays), (2) provide for high analytical replication (5 or 6 reps/sample), (3) have more controls, and (4) develop a standard curve for every soil sample. Therefore, even though the MUF microplate method (if carefully done) can correlate with the PNP colorimetric method, the reaction rates on a molar basis are different for the same soil samples. This means that adopting the MUF protocol will require an individual calibration based on a local database, and thus would not contribute to the much larger PNP bench-scale database that is already being developed.

## Bench Scale Individual Assays to Measure Activities of β-Glucosidase, β-Glucosaminidase, Phosphatases, and Arylsulfatase

The original protocol for enzyme assays by Tabatabai (1994) incubates one gram of air-dried soil at 37°C for 1 h in an Erlenmeyer flask with 0.2 mL of toluene, 4 mL of the appropriate buffer at optimum pH and 1 mL of the required *p*-nitrophenyl-derivate substrate (Figure 11.3a). Following incubation, 1 mL of 0.5 M $CaCl_2$ is added and followed by 4 mL of the appropriate solution/buffer to stop the reaction. The suspension is filtered (Whatman No.2) and the color intensity of PNP is measured colorimetrically with a spectrophotometer at 400 nm. A modification is possible for determining EAs with half of all solutions and amount of soil without changing the proportions of the original assays. The same absorbance is obtained per sample and the time to perform the assays remains the same. However, it can reduce significantly the amount of soil, resources, and waste generated in the long-term. Another modification is the elimination of toluene as an antiseptic. Multiple studies have demonstrated no effect on certain EAs (e.g., phosphatases) during a 1-h incubation (Eivazi & Tabatabai 1977; Tabatabai, 1994; W. Dick, 2011). Eliminating toluene also reduces safety concerns when performing the assay (*e.g.*, avoid the need to perform assays under the hood) and environmental risks associated with the waste generated (Acosta-Martínez and Tabatabai, 2011). In brief, the approach suggested involves incubation of 0.5 g of air-dried soil with 2 mL of the appropriate buffer (acetate or MUB) and 0.5 mL of the appropriate substrate for 1 h at 37°C. After incubation, 0.5 mL of 0.5 M $CaCl_2$ is added followed by 2 mL of the stop solution (NaOH or THAM buffer).

A list of the equipment and materials needed, along with instructions for preparation of solutions and substrates are described in Tabatabai (1994) and summarized below. Reagents and substrates for each individual EA are shown in Table 11.2.

### Equipment, Materials, and Reagents

Equipment

- Analytical balance (± 0.0001 g sensitivity) and electronic balance (± 0.1 g sensitivity)
- Magnetic stir plate, and stir bars
- Incubator set at 37°C
- Funnel stand to accommodate several glass funnels
- Colorimeter or spectrophotometer set at 400 nm.

**Table 11.2** Description of the enzyme assay procedure and reagents needed for determining enzyme activities in soils when analyzing 0.5 g of soil.

| Enzyme and EC number | Start reaction | | Stop reaction | |
|---|---|---|---|---|
| | Buffer used to start reaction | Substrate | 0.5 M $CaCl_2$ | Solution to stop reaction |
| **Individual Assays** | | | | |
| **β-Glucosidase (3.2.1.21)** | 2 mL MUB pH 6.0 | 0.5 mL *p*-Nitrophenyl-β-D-glucopyranoside (0.05 M) | 0.5 mL | 2 mL 0.1 M THAM pH 12.0 |
| **β-Glucosaminidase (3.2.1.30)** | 2 mL 0.1 M acetate buffer pH 5.5 | 0.5 mL *p*-Nitrophenyl-N-acetyl-β-D-glucosaminide (0.01 M) | 0.5 mL | 2 mL 0.5 N NaOH |
| **Alkaline phosphatase (3.1.3.1)** | 2 mL MUB pH 11.0 | 0.5 mL *p*-Nitrophenyl phosphate (0.05 M) | 0.5 mL | 2 mL 0.5 N NaOH |
| **Acid phosphatase (3.1.3.2)** | 2 mL MUB pH 6.5 | 0.5 mL *p*-Nitrophenyl phosphate (0.05 M) | 0.5 mL | 2 mL 0.5 N NaOH |
| **Arylsulfatase (3.1.6.1)** | 2 mL 0.5 M acetate buffer pH 5.8 | 0.5 mL *p*-Nitrophenyl sulfate (0.05 M) | 0.5 mL | 2 mL 0.5 N NaOH |
| **Combined Assay** | | | | |
| **CNPS activity** | 0.5 mL 0.5 M acetate buffer pH 5.8 | 0.5 mL of each substrate prepared in same acetate buffer with concentration of individual assays | 0.5 mL | 2 mL 0.1 M THAM pH 12.0 |

Materials

- Volumetric flasks (acid washed), 100 mL, 1 L
- Incubation flasks, 25 mL Erlenmeyer flasks (acid washed) fitted with No. 1 stoppers.
- Funnels, long stem
- Filter paper, Whatman 2V
- Test tubes and rack
- Pipettes and tips
- Cuvettes
- Reagents
  - Deionized water (DI $H_2O$)
  - Calcium chloride (0.5 M $CaCl_2$). Prepared by dissolving 73.5 g of $CaCl_2.2H_2O$ (CAS 10035-04-8) in 700 mL of water and dilute to 1 L with water.
  - Modified universal buffer (MUB) stock solution. Prepared by mixing 12.1 g of Tris (hydroxymethyl) aminomethane (THAM), 11.6 g of maleic acid, 14.0 g of citric acid, and 6.3 g of boric acid ($H_3BO_3$) in 500 mL of 1N sodium hydroxide (NaOH) and dilute the solution to 1 L with water. Store it in refrigerator until use.
  - MUB buffer, pH 6. Prepared using 200 mL of the MUB stock solution, adjusting to pH 6 with 0.1 M hydrochloric acid (HCl), and diluting to 1 L with DI water.
  - MUB buffer, pH 6.5. Prepared using 200 mL of the MUB stock solution, adjusting with 0.1M hydrochloric acid (HCl) until pH 6.5 is reached, and diluting to 1 L with DI water.
  - MUB buffer, pH 11. Prepared using 200 mL the MUB stock solution and titrating to pH 11 with 0.1 M NaOH, then diluting to 1 L with DI water.
  - 0.1 M acetate buffer, pH 5.5: Prepared by dissolving 13.6 g of sodium acetate trihydrate in 800 mL of deionized DI water. The pH is adjusted to 5.5 by adding 99% glacial acetic acid and then, bringing the final volume to 1 L with DI water.
  - 0.5 M acetate buffer, pH 5.8. Prepared by dissolving 68 g of sodium acetate trihydrate in 700 mL of deionized (DI) water. The pH is adjusted to 5.8 by adding 99% glacial acetic acid and then, bringing the final volume to 1 L with DI water.
  - Sodium hydroxide (0.5 M NaOH). This solution is prepared by dissolving 20 g of NaOH (CAS 1310-73-2) in 700 mL of water and then adjust volume to 1 L.
  - 0.1 M THAM, pH 12. This solution is prepared by dissolving 12.2 g of THAM (CAS 77-86-1) in 800 mL of water and then adjust with 0.5 M NaOH until pH 12.0, then adjust to 1 L with DI water.
  - Standard *p*-nitrophenol, spectrophotometric grade (CAS 100-02-7): Prepare the standard solution by dissolving 1 g into 1 L of water and follow instructions given in section 4.3.

- Substrates:
  - *p*-Nitrophenyl-β-D-glucopyranoside (0.05 M): Add 1.506 g of this substrate (CAS 2492-87-7) and adjust the volume to 100 mL with start buffer (MUB pH 6.0) for β-glucosidase activity.
  - *p*-Nitrophenyl-N-acetyl-β-D-glucosaminide (0.01 M): Add 0.342 g of this substrate (CAS 3459-18-5), and adjust to 100 mL with start buffer (0.1 M acetate buffer pH 5.5) for β-glucosaminidase activity.
  - *p*-Nitrophenyl phosphate (0.05 M): Add 1.68 g of this substrate (CAS 333338-18-4), and adjust to 100 mL with the start buffer for each phosphomonoesterase (e.g., MUB pH 11.0 for alkaline phosphatase activity or MUB pH 6.5 for acid phosphatase activity)
  - *p*-Nitrophenyl sulfate (0.05 M): Add 1.228 g of this substrate (CAS 6217-68-1), and adjust to 100 mL with the start buffer (0.5 M acetate buffer pH 5.8) for arylsulfatase activity.

## Procedure Using Reduced Soil Amount and Solutions to Half

- Start reaction
  - Label three 25 mL Erlenmeyer flasks with replicates A and B, and use C for the control per sample.
  - Add 0.5 g of air-dried soil to each Erlenmeyer flask (A, B and C).
  - Add 2 mL of start buffer to each Erlenmeyer flask (A, B and C).
  - Add 0.5 mL of substrate to A and B flasks only. Place stopper in each flask.
  - Swirl each flask gently, and place in an incubator for 1 h at 37 °C.
- Stop reaction
  - Remove flasks from the incubator and remove stoppers.
  - Add 0.5 mL $CaCl_2$ to the soil in each flask.
  - Add 2 mL of stop buffer or solution to flasks A and B first, then to C. Swirl gently after each addition.
  - Add 0.5 mL of substrate to flask C only. Then, swirl.
  - Pour into a funnel lined with filter paper, capturing solution in test tubes. Let stand for ~ 30 min until fully filtered.
- Colorimetric reading
  - Remove filter from funnel and discard accordingly.
  - Remove the test tube and place into a rack.
  - Read samples in a spectrophotometer at 400 nm. Dilute if necessary, to get an absorbance value that fits into the calibration curve (section 4.3).
  - All *p*-nitrophenol waste is considered a hazardous waste. Please discard accordingly.

## Combined Assay to Determine the Four EAs Simultaneously (CNPS Activity)

The CNPS combined assay suggested by Acosta-Martínez et al. (2019) uses original steps and optimized conditions of individual assays by Tabatabai (1994) and Parham and Deng (2000) as shown in Figure 11.3. Except that the CNPS assay involves incubation of 0.5 g of air-dried soil for 1 h at 37°C with 0.5 mL of 0.5 M acetate buffer (pH 5.8) and 0.5 mL of each of the four substrates required for individual evaluation of acid phosphatase, β-glucosidase, β-glucosaminidase, and arylsulfatase (see also Figure 11.3b). All the substrates should be prepared according to their specified concentration in the original protocols (Tabatabai, 1994; Parham & Deng, 2000), but using 0.5 M acetate buffer (pH 5.8) instead, because this will provide the volume of buffer of original protocols for reactions to take place during the incubation step. After incubation, the reaction is terminated adding 0.5 mL of 0.5 M $CaCl_2$ followed by 2 mL of 0.1 M THAM pH 12.0. Subsequently, the soil suspension is filtered (Whatman No.2) and the color intensity of PNP is measured colorimetrically with a spectrophotometer at 400 nm. For this combined assay, the reactions are terminated using THAM because NaOH can react with the substrate of β-glucosidase, cause non-enzymatic degradation of PNP (Hayano, 1973; Tabatabai, 1994), and can also extract dissolved organic matter (DOM) which can contribute to absorbance at 400 nm (Klose et al., 2011). Similar to the individual assays, toluene has been eliminated and it is possible to increase the molarity of $CaCl_2$ for soils with higher SOM without affecting the reactions occurring (Margenot et al., 2018).

### Equipment, Materials, and Reagents

The same equipment and materials described on page 218 for the individual assays are used, except for the following reagents described below, which are specific for the combined assay.

Reagents

- 0.5 M acetate buffer, pH 5.8. Prepared by dissolving 68 g of sodium acetate trihydrate in 700 mL of deionized (DI) water. The pH is adjusted to 5.8 by adding 99% glacial acetic acid and then, bringing final volume to 1 L with DI water.
- Calcium chloride (0.5 M $CaCl_2$). Prepared by dissolving 73.5 g of $CaCl_2.2H_2O$ (CAS 10035-04-8) in 700 mL of water and dilute to 1 L with water.
- 0.1 M THAM, pH 12. Prepared by dissolving 12.2 g of THAM (CAS 77-86-1) in 800 mL of water and then adjust with 0.5 M NaOH until pH 12.0, then adjust to 1 L with DI water.
- Standard *p*-nitrophenol, spectrophotometric grade (CAS 100-02-7): Prepare the standard solution by dissolving 1 g into 1 L of water and follow instructions given in section 4.3.

- Substrates. Each substrate is prepared individually and the assay requires adding 0.5 mL of each per sample. However, the four substrate solutions can be mixed together. This will reduce pipetting time by dispensing 2 mL of this combined solution to each sample.
- *p*-Nitrophenyl-β-D-glucopyranoside (0.05 M): Add 1.506 g of this substrate (CAS 2492-87-7) and adjust volume to 100 mL with 0.5 M acetate buffer (pH 5.8).
- *p*-Nitrophenyl-N-acetyl-β-D-glucosaminide (0.01 M): Add 0.342 g of this substrate (CAS 3459-18-5), and adjust to 100 mL with 0.5 M acetate buffer (pH 5.8).
- *p*-Nitrophenyl phosphate (0.05 M): Add 1.68 g of this substrate (CAS 333338-18-4) and adjust volume to 100 mL with 0.5 M acetate buffer (pH 5.8).
- *p*-Nitrophenyl sulfate (0.05 M): Add 1.228 g of this substrate (CAS 6217–68–1), and adjust volume to 100 mL with 0.5 M acetate buffer (pH 5.8).

**Procedure**

- Start reaction
  - Label three 25 mL Erlenmeyer flasks with replicates A and B, and use C for the control
  - Add 0.5 g of soil to each Erlenmeyer flask
  - Add 0.5 mL of 0.5 M acetate buffer, pH 5.8 to flasks A, B and C. Place stopper to the C flasks.
  - Add 0.5mL of each substrate to flasks A and B only.
  - Swirl each flask gently, and place in an incubator at 37°C for 1 h.
- Stop reaction
  - Remove flasks from incubator and remove stoppers.
  - Add 0.5 mL $CaCl_2$ to flasks A, B and C
  - Add 2 mL of 0.1 M THAM, pH 12 to flasks A, B and C, followed by adding the substrates (0.5 mL each) to flask C. Swirl gently after each addition.
  - Pour into funnel lined with filter paper, capturing solution in test tubes. Let stand for ~ 30 min until fully filtered.
- Colorimetric reading
  - Remove filter from funnel and discard accordingly.
  - Remove the test tube and place into a rack.
  - Read samples in a spectrophotometer at 400 nm. Dilute if necessary, to get an absorbance value that fits into the calibration curve (section 4.3).
  - All *p*-nitrophenol waste is considered a hazardous waste. Please discard accordingly.

### Calibration Curve for all Enzyme Assays

The content of PNP in the filtrate can be calculated using a calibration graph plotted from the absorbance readings obtained with *p*-nitrophenol standards containing 0, 10, 20, 30, 40, and 50 μg of PNP. Briefly, take 1 mL of the PNP standard solution (prepared by mixing 1 g of PNP in 1 L of water), and dilute to a total volume of 100 mL with water (10 μg $mL^{-1}$). Then, 0, 1, 2, 3, 4, and 5 mL of the diluted PNP standard solution are pipetted into volumetric flasks, and the volume is adjusted to 5 mL by adding water (*i.e.*, 5-, 4-, 3-, 2-, 1-, and 0-mL, respectively). Finally, 1 mL of 0.5 M $CaCl_2$ and 4 mL of stop solution are added to each flask as done when the assay is terminated (e.g., 0.1 M THAM is used for β-glucosidase and CNPS assays, or 0.5 M NaOH for the other enzymes). We have found no difference in the standard curve (e.g., same slope) obtained with either stop solution. The suspension is then filtered and measured with a spectrophotometer at 400 nm. This calibration curve uses solution amounts as the original assays from Tabatabai (1994) described in section 4.1. However, it is also applicable when the soil and solutions are reduced by half because concentrations and proportions are maintained. The color intensity of PNP can be read between 400nm and 420nm with minimal difference in enzyme activity (mg PNP kg-1 soil h-1) as long as the same wavelength is used for standards and samples.

## Interpretation– Putting Enzyme Measurements into Context

### Use of EAs vs. Other Measurements

A review of the literature documents that EAs have greater sensitivity to management effects than some other soil properties including SOC (Powlson et al., 1987). Studies using soil respiration, C mineralization, and qPCR markers based on genes involved in the four EAs suggested in this chapter, may be as sensitive or comparable to the EAs (Lehman et al., 2015; Sinsabaugh et al., 1991). However, enzyme assays are overall less tedious and do not require sophisticated instrumentation for PNP quantification to represent activity.

Although EAs have been reliable assessments for determining potential activity in soil, they do not provide information about enzyme sources (intracellular vs. extracellular), or the microbial community (size and composition [structure and diversity]) involved in the processes investigated. For the latter, we recommend other measurements to obtain information of the changes in the microbial component (Figure 11.4). Although genomic DNA is not currently considered an accessible soil health indicator, in part due to a lack of standardization of protocols across labs, lipid fatty acid profiling methods (phospholipid fatty acid analysis [PLFA] and ester linked-fatty acid methyl ester analysis [EL-FAME]) have standardized methods available to evaluate the microbial community composition/structure (*e.g.*, Schutter & Dick, 2002; Buyer &

Sasser, 2012). Numerous research papers have associated increased microbial biomass (*e.g.*, estimated from chloroform fumigation methods or total FAMEs), arbuscular mycorrhizal fungi (AMF) FAME marker, and/or fungal:bacteria ratios with changes in several EAs (*e.g.*, Cotton et al., 2013; Li et al., 2018). These microbial measurements are also associated with enhanced soil health and serve as sensitive indicators to short-term changes in management (1–3 years).

Methods based on 16S rRNA analysis such as quantitative PCR (qPCR) for different functions have shown potential to link rRNA genes to key enzymes in the soil. This is an area that should be explored in our search for our next generation of soil health indicators. The use of gene primers for alkaline phosphatase (Ragot et al., 2015) and β-glucosidase (Cañizares et al., 2011) have provided valuable information regarding microbial functioning in soil. Trivedi et al. (2016) measured EAs involved in the C cycle (*e.g.*, β-xylosidase, β-glucosaminidase) using microplates and found strong positive correlations between the relative abundance of the microbial genes and their respective enzyme activities. Their study also found that microbial community structure indirectly regulated the activity of extracellular enzymes via functional genes in different soil types. A recent study showed strong positive correlations between the activities of β-glucosidase, alkaline phosphatase and β-glucosaminidase via PNP assays, and the abundance of related microbial functional genes. Among the enzymes, β-glucosidase had the strongest correlation with the gene encoding that enzyme, and similar trends were observed regardless of sampling time (growing season vs. postharvest), year or soil textural classes that ranged from 17 to 52 % of clay (Perez-Guzman et al., 2021).

Near infrared spectroscopy (NIRS) has been used to predict EAs including three of the four discussed in this chapter (except for arylsulfatase) and other soil properties such as nitrification potential, soil respiration, SOC, and amino sugar N concentrations (Reeves et al., 1999; Zornoza et al., 2012; Dick et al., 2013; Veum et al., 2017). This approach relies on the correlation of EAs with other soil properties that directly produce spectral features, especially SOM and inorganic bonds in clay minerals. However, the relationship between the EAs and the spectrally active soil properties may be decoupled, reducing the efficacy of the model for estimating EAs or other properties by proxy. Currently, NIR models do not perform well when applied outside the region of calibration, and it is not possible to apply models at the regional scale for estimation of EAs.

## Calibration and Interpretation of EAs Independent of Soil Type and Region

Soil health literature suggests that higher EAs represent management systems that promote soil health (*e.g.*, cover cropping, no-tillage, organic amendments) and are lower in poorly managed soils (*e.g.*, intensive tillage, low organic inputs), therefore interpretation generally follows a "more-is-better" model. A major

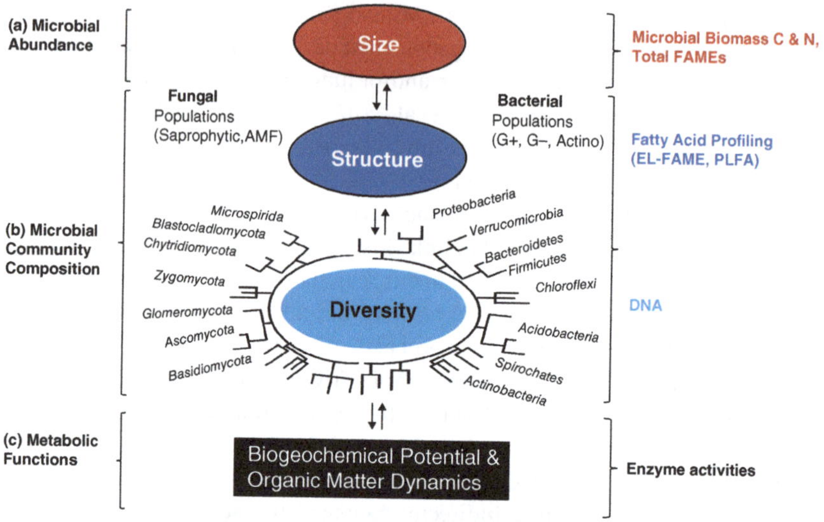


**Figure 11.4** A conceptual overview of the methods used to determine: (a) microbial abundance (e.g., microbial biomass C and N via fumigation techniques or total FAMEs); (b) community composition (Fatty acid profiling via EL-FAME or PLFA and with DNA techniques); and (c) metabolic functions (via enzyme activities).

challenge for calibration of any soil property as a soil health indicator is to have an interpretation independent of soil type. This is because soil properties typically vary much more due to soil type than the effects of soil management within a soil type (*e.g.*, for microbial properties, see Bossio et al., 1998; Schutter et al., 2001). It is possible that a sandy soil, no matter the level of favorable management, could have lower activity than a clayey or loamy soil that has been subjected to poor management. However, on a regional basis with relatively uniform soils, it may be possible to develop threshold values. For example, Mendes et al. (2019) developed thresholds of low, moderate and adequate on air-dried soils on a Typic clayey Dystrophic Red Oxisol, for which β-glucosidase values were <66, 67 to 115 and >116, mg PNP $kg^{-1}$ $h^{-1}$, respectively, and arylsulfatase values were <30, 21 to 70, and >71 mg PNP $kg^{-1}$ $h^{-1}$, respectively. However, there are no universal thresholds or ranges to define relative quality or function potential. One limitation for establishing ranges is that standard procedures have not been uniformly applied. Thus, the first step toward the development of EAs as a soil health indicator is the universal adoption of standardized protocols across laboratories and regions. The protocols for EAs presented in this chapter have been vetted to optimize activity and have been consistent in detecting soil management effects across a wide range of soil types. Interpretation of results from EAs as indicators of soil health can become more consistent across regions by adopting any of the following suggested approaches: (1) Determining percent of change in EAs over time for evaluation of soil management implementation at a producer field scale; (2) Using indexes from several EAs to identify ranges or thresholds in biogeochemical cycling to compare across management and regions; and/or (3) Using calculations or equations including several properties to develop a more comprehensive soil health index to compare across regions (Table 11.3).

The first approach will help producers to monitor percentage increase in EAs (without statistical criteria) after implementing better management practices. This could provide valuable insight at the single farm scale when producers send samples to commercial labs for standard soil pH, organic matter, and nutrient content. This approach is contextual and requires an understanding of the soil agroecosystem such as previous EA values from different soil samplings. Integration with crop and other soil measurements, and standardization of the laboratory method will provide the appropriate context to understand the soil ecosystem dynamics. It should be taken into account that climate variability could possibly mask the comparison of management effects from year to year. A current limitation for implementing this approach is the lack of availability of biological measurements for soil samples in commercial labs. Another alternative is to establish ranges in EAs due to management by defining native grass, unmanaged or CRP land as ideal to represent 100% recovery potential of EAs (*e.g.*, Acosta-Martínez et al., 2003; Li et al., 2018). This would facilitate comparisons of EAs across producer management practices when previous samplings are not available by determining how close the percent recovery is for a system with respect to the native system. However, this requires finding native or undisturbed land with similar soil series for the comparisons to define the 100% value, as EAs vary widely across soil types, and indeed vary much more widely due to soil type than as a result of management practices (Schutter et al., 2001). We propose government entities consider this approach for evaluating the adoption of the four principles to restore soil health (mentioned in section 2).

The second approach for EA data interpretation is the use of indexes from many mediated EAs. We present two alternatives for obtaining a better overview of soil biogeochemical cycling and the reactions involved in SOM transformation. The first is to compare management according to the calculation of a geometric mean from several EAs assayed individually. This is achieved by multiplying the values of several EAs and then taking the root value of the number of EAs (García-Ruíz et al., 2008). This can provide an index of total enzyme activity potential (Cotton et al., 2013; Acosta-Martínez & Cotton, 2017). The second alternative is the simultaneous determination of the four EAs in a single reaction vessel (described in section 4.2 and Figure 11.3b) that could potentially accomplish the same result of the geometric mean of the four separate assays.

The third approach requires an index such as the Soil Management Assessment Framework (SMAF; Andrews et al., 2004), which identifies how EAs relate to other properties related to soil-water relations or SOM (*e.g.*, SOC, bulk density),

**Table 11.3** Approaches for interpreting results from enzyme activities (EAs) as indicators of soil health.

| Approach | Scale | Goal | Examples |
|---|---|---|---|
| 1 | Producer level comparison from values obtained by commercial labs | Monitor changes in EAs without statistical criteria after implementing better management practices | Comparison of percentage increase in EAs using values from previous samplings |
| | | | Establishing ranges in EAs due to management by defining an "ideal" system (e.g., native grass, unmanage, or CRP land) to represent 100% recovery potential of EA |
| 2 | Comparison among systems by research institutions | Use of indexes from several EAs to obtain overview of soil biogeochemical cycling and reactions involved in soil organic matter (SOM) transformation | Compare management via calculation of a geometric mean (GM) from several EAs |
| | | | Value obtained from the combined assay (CNPS cycling) as a direct index |
| 3 | Comparison across regions and systems for national soil health assessments | Compare across regions using calculations or equations that include EAs along with several soil properties | A comprehensive soil quality assesment such as SMAF (Andrews et al., 2004) by including biological, chemical and physical indicators. |

soil structure (*e.g.*, SOC, aggregate stability) and active SOM pools (*e.g.*, MBC, potentially mineralizable N). To date, β-glucosidase activity has been the only enzyme to be included in this soil quality/health index to relate to the soil metabolic capacity and C cycling (Stott et al., 2010). The interpretation algorithm (scoring curve) for β-glucosidase is an S-shaped curve ($y = a/[1 + b^{exp}(-cx)]$), where x is β-glucosidase activity (mg PNP released $kg^{-1}$ soil $h^{-1}$), a and b are constants, and c is a factor modified by soil classification, texture, and climate. This interpretation algorithm was developed based on studies with a wide range in SOC content (247 to 687 g $kg^{-1}$) and β-glucosidase activity (33 to 675 mg $kg^{-1}$ $h^{-1}$). The SMAF scoring index integrates site specific soil inherent properties, climatic factors, and requirements of crops under production to obtain unitless values ranging from 0 to 1 representing fractional attainment of its associated soil functional potential (Andrews et al., 2004). In the development of the β-glucosidase curve, SOC indicator scores ranged from 0.25 to 0.73, while β-glucosidase activity scores varied from 0.17 to 0.93. Their work indicated that by normalizing β-glucosidase activity by the SOC content, the resulting ratio could indicate C sequestration trends, with ratios of 10 to 17 g PNP $kg^{-1}$ SOC $h^{-1}$ showing systems in equilibrium. Ratios > 17 were mostly from recently altered management systems with SOC contents trending upward, while ratios < 10 were obtained from soils expected to continue to lose soil C. Studies that have applied SMAF with β-glucosidase activity were able to discriminate no-till compared to till systems and those under vetch compared to wheat and no cover treatments under cotton production in West Tennessee (Mbuthia et al., 2015), among various cropping systems and management practices in Missouri (Veum et al., 2014, 2015), and showed higher soil quality scores in perennial grass systems, followed by a soybean-dominated rotation, followed by a corn-based rotation in Indiana, but did not differentiate no-till crop production vs. chisel-disk tillage (Hammac et al., 2016). Currently, work is underway to evaluate and potentially refine the SMAF scoring curves for all indicators, including β-glucosidase activity. It is also expected that an expanded suite of EAs will be included in this index, and/or that similar indexes would become available with other combinations of soil properties.

## Importance of Strict EA Protocols and Practical Modifications

One of the main challenges for comparing EA trends across studies and among regions is that different laboratory methods are used (microplates vs. bench scale) or modifications are applied to original assays without validation. To avoid this issue, we recommend the bench scale PNP method as the standard method for the four enzyme assays. To have comparable results, it is essential that the optimized and vetted protocols given in this chapter are followed (*e.g.*, appropriate temperature of incubation, optimum pH and buffer type, and correct solution to stop the reaction). Deviations from each enzyme protocol will affect kinetics and stoichiometry of the reaction. However, there are certain modifications that would help improve the sensitivity of enzyme assays on soil types with extreme pH, low SOM, and/or sandy textures. For example, sandy soils which tend to have low SOM (1% or less), may have activities as low as 1 mg PNP $kg^{-1}$ soil $h^{-1}$ for certain enzymes such as arylsulfatase and/or β-glucosaminidase. This limits the ability to detect management effects on these enzymes, compared with other EAs. Thus, we recommend incubating longer (at least 2 h, with swirling after 1 h) for those EAs as opposed to increasing the amount of soil in the reaction, to avoid changing the proportions of the assay. Problems can also be encountered when soils have high SOM (> 3%), iron mineralogy, or low pH (*e.g.*, < 5), such as tropical soils. For such cases, the molarity of $CaCl_2$ can be increased without interfering with the reactions related to PNP release (Margenot et al., 2018; Acosta-Martínez et al., 2018, 2019).

Other discrepancies in methodology include calibration curves calculated based on molarity (*e.g.*, μmol) vs. concentration (*e.g.*, μg, mg). We recommend the calibration curve be performed following the same steps used in the enzyme assay, which are described in section 4.3, as it better represents the concentrations obtained during an assay. New spectrophotometers allow for higher absorbance readings which reduce the need for dilutions, and it is possible to obtain a linear extrapolation of concentrations with new spectrophotometers that are more precise than previous colorimeters used. For soils with high clay content and SOM, dilutions will still be needed.

## Limitations of EAs as Soil Health Indicators

It is important noting that none of the four EAs suggested can provide direct information on the microbial community size, composition, or activity as affected by management. The assays available for these four EAs provide potential activity and are an index of the total pool of a given enzyme. However, certain precautions should be taken when assessing soil health. Phosphatase activity may not be a good soil quality/health indicator when certain management practices are in use. For example, inorganic phosphate fertilizers naturally suppress this enzyme by a feedback mechanism, as the $PO_4$ product of the reaction is the same form of P found in the fertilizer (Chunderova & Zubers, 1969; Mathur & Rayment, 1977; Spiers & McGill, 1979; Clarholm, 1993). Even though a soil might be of high quality due to cover cropping or no-till, if long-term $PO_4$ fertilizer has been applied, it will suppress phosphatase activity. Nonetheless, phosphatase activity is a valuable enzyme activity for evaluation of the P-cycle in native ecosystems and semiarid agricultural soils where fertilization is not used.

Caution is advised when evaluating the effects of organic amendments such as biochar due to potential absorption of chemicals added for assays, including substrates, buffers, $CaCl_2$, and/or PNP. During pyrolysis, feedstocks undergo chemical and physical changes that may increase surface area and porosity, leading to the stabilization of soil enzymes. For instance, Bailey et al. (2011) reported that substrate sorption by biochar impeded enzyme activity. Later, Jindo et al. (2014) reported that PNP was retained in samples containing biochar and that the retention increased at low pH values. It was recently reported that additions of grass biochar and pine biochar resulted in a decrease in the activities of β-glucosidase and phosphatase due to direct sorption with pine biochar. This resulted in approximately 40% of the enzymes being retained (Foster et al., 2018). Thus, adjustments with a correction factor, if one can be determined, may be needed for evaluation of the effects of organic amendments across soils and amendments types and rates of application.

## Future Research Directions

Establishing standardized protocols for the measurement of EAs is a top priority for national soil conservation and soil health initiatives. Currently, lack of consistent information on the four selected EAs across regions, climates, and soils is limiting our ability to observe trends and determine thresholds for this group of enzymes crucial to C, C and N, P, and S cycling. Although it is expected that EAs should increase with more diversified cropping systems, conservation tillage, and the addition of organic amendments, the complex interactions combined with a lack of data preclude wide-scale interpretation of EAs.

Novel management practices, introduction of new crops, expanded crop rotations, and climatic variability all represent new challenges in understanding and interpreting EAs. Additionally, lack of information in several regions, along with differences in field and laboratory methodologies adds to the complication. Large-scale soil health assessments are underway to compare key land use and soil management practices to begin to address this challenge. The protocol recommendations provided in this chapter and the suggested approaches for interpretation of the four EAs (β-glucosidase, β-glucosaminidase, acid or alkaline phosphatases, and arylsulfatase), will facilitate meta-analyses to provide needed trends and thresholds related to biogeochemical cycling. Databases that integrate these four EAs with other important soil health measurements, environmental and management data will quantify the role of EAs in biogeochemical cycling, soil remediation potential, and soil organic matter dynamics for improved assessment of soil health in the U.S. and globally.

## References


A'Bear, A.D., Jones, T.H., Kandeler, E., and Boddy, L. (2014). Interactive effects of temperature and soil moisture on fungal-mediated wood decomposition and extracellular enzyme activity. *Soil Biol. Biochem.* 70, 151–158. doi:10.1016/j.soilbio.2013.12.017

Acosta-Martínez, V., Burow, G., Zobeck, T.M., and Allen, V.G. (2010). Soil microbial communities and function in alternative systems to continuous cotton. *Soil Sci. Soc. Am. J.* 74, 1181–1192. doi:10.2136/sssaj2008.0065

Acosta-Martínez, V., Cano, A., and Johnson, J.M.F. (2018). Simultaneous determination of multiple soil enzyme activities for soil health-biogeochemical indices. *Appl. Soil Ecol.* 126, 121–128. doi:10.1016/j.apsoil.2017.11.024

Acosta-Martínez, V., and Cotton, J. (2017). Lasting effects of soil health improvements with management changes in cotton-based cropping systems in a sandy soil. *Biol. Fertil. Soils* 53, 533–546. doi:10.1007/s00374-017-1192-2

Acosta-Martínez, V., Lascano, R., Calderón, F., Booker, J.D., Zobeck, T.M., and Upchurch, R. (2011). Dryland cropping systems influence the microbial biomass and enzyme activities in a semiarid sandy soil. *Biol. Fertil. Soils* 47, 655–667. doi:10.1007/s00374-011-0565-1

Acosta-Martínez, V., Moore-Kucera, J., Cotton, J., Gardner, T., and Wester, D. (2014a). Soil enzyme activities during the 2011 Texas record drought/heat wave and implications to biogeochemical cycling and organic matter dynamics. *Appl. Soil Ecol.* 75, 43–51. doi:10.1016/j.apsoil.2013.10.008

Acosta-Martínez, V., Moore-Kucera, J., Cotton, J., Gardner, T., Wester, D., and Cox, S. (2014b). Predominant bacterial and fungal assemblages in agricultural soils during a record drought/heat wave and linkages to enzyme activities of biogeochemical cycling. *Appl. Soil Ecol.* 84, 69–82. doi:10.1016/j.apsoil.2014.06.005

Acosta-Martínez, V., Pérez-Guzmán, L., and Johnson, J.M.F. (2019). Simultaneous determination of β-glucosidase, β-glucosaminidase, acid phosphomonoesterase, and arylsulfatase activities in a soil sample for a biogeochemical cycling index. *Appl. Soil Ecol.* 142, 72–80. doi:10.1016/j.apsoil.2019.05.001

Acosta-Martínez, and M.A. Tabatabai. 2011. *Phosphorous cycle enzymes. In: R.P. Dick, editor,* Methods of soil enzymology. Madison, WI: Soil Sci. Soc. Am. p. 161–183.

Acosta-Martínez, V., Zobeck, T.M., and Allen, V. (2004). Soil microbial, chemical and physical properties in continuous cotton and integrated crop-livestock systems. *Soil Sci. Soc. Am. J.* 68, 1875–1884. doi:10.2136/sssaj2004.1875

Acosta-Martínez, V., Zobeck, T.M., and Gill, T.E. (2003). Enzyme activities in semiarid soils under conservation reserve program, native rangeland, and cropland. *J. Plant Nutr. Soil Sci.* 166, 699–707. doi:10.1002/jpln.200321215

Albiach, R., Canet, R., Pomares, F., and Ingelmo, F. (2000). Microbial biomass content and enzymatic activities after the application of organic amendments to a horticultural soil. *Bioresour. Technol.* 75, 43–48. doi:10.1016/S0960-8524(00)00030-4

Alster, C.J., German, D.O., Lu, Y., and Allison, S.D. (2013). Microbial enzymatic responses to drought and to nitrogen addition in a southern California grassland. *Soil Biol. Biochem.* 64, 68–79. doi:10.1016/j.soilbio.2013.03.034

Andrews, S.S., Karlen, D.L., and Cambardella, C.A. (2004). The soil assessment framework: A quantitative soil quality evaluation method. *Soil Sci. Soc. Am. J.* 68, 1945–1962. doi:10.2136/sssaj2004.1945

Aruguete, D.M., and Hochella, M.F. (2010). Bacteria-nanoparticle interactions and their environmental implications. *Environ. Chem.* 7, 3–9. doi:10.1071/EN09115

Baer, S.G., Rice, C.W., and Blair, J.M. (2000). Assessment of soil quality in fields with short and long term enrollment in the CRP. *J. Soil Water Conserv.* 55, 142–146.

Bailey, V.L., Fansler, S.J., Smith, J.L., and Bolton, H. (2011). Reconciling apparent variability in effects of biochar amendment on soil enzyme activities by assay optimization. *Soil Biol. Biochem.* 43, 296–301. doi:10.1016/j.soilbio.2010.10.014

Bailey, V.L., Smith, Jr., J.L., and Bolton, H. (2002). Fungal-to-bacterial ratios in soils investigated for enhanced C sequestration. *Soil Biol. Biochem.* 34, 997–1008. doi:10.1016/S0038-0717(02)00033-0

Balota, E.L., Kanashiro, M., Colozzi Filho, A., Andrade, D.S., and Dick, R.P. (2004). Soil enzyme activities under long-term tillage and crop rotation systems in subtropical agro-ecosystems. *Braz. J. Microbiol.* 35, 300–306. doi:10.1590/S1517-83822004000300006

Bandick, A., and Dick, R.P. (1999). Field management effects on soil enzyme activities. *Soil Biol. Biochem.* 31, 1471–1479. doi:10.1016/S0038-0717(99)00051-6

Bera, T., Collins, H.P., Alva, A.K., Purakayastha, T.J., and Patra, A.K. 2016. Biochar and manure effluent effects on soil biochemical properties under corn production. *Appl. Soil Ecol.* 107, 360–367. doi:10.1016/j.apsoil.2016.07.011

Bérard, A., Bouchet, T., Sévenier, G., Pablo, A.L., and Gros, R. (2011). Resilience of soil microbial communities impacted by severe drought and high temperature in the context of Mediterranean heat waves. *Eur. J. Soil Biol.* 47, 333–342. doi:10.1016/j.ejsobi.2011.08.004

Bhandari, K., West, C.P., Acosta-Martínez, V., Cotton, J., and Cano, A. (2018). Soil health indicators as affected by diverse forage species and mixtures in semi-arid pastures. *Appl. Soil Ecol.* 132, 179–186. doi:10.1016/j.apsoil.2018.09.002

Blanco-Cangui, H., and R. Lal. 2008. Buffer strips. In: H. Blanco and R. Lal, editors, *Principles of soil conservation and management*. Springer Science, New York. p. 223–240.

Blanco-Canqui, H., Lal, R., Post, W.M., Izaurralde, R.C., and Shipitalo, M.J. (2007). Soil hydraulic properties influenced by corn stover removal from no-till corn in Ohio. *Soil Tillage Res.* 92, 144–155. doi:10.1016/j.still.2006.02.002

Bossio, D.A., Scow, K.M., Gunapala, N., and Graham, K.J. (1998). Determinants of soil microbial communities: effects of agricultural management, season, and soil type on phospholipid fatty acid profiles. *Microb. Ecol.* 36, 1–12.

Brennan, E., and Acosta-Martínez, V. (2019). Cover crops and compost influence soil enzymes during six years of tillage-intensive, organic vegetable production. *Soil Sci. Soc. Am. J.* 83, 624–637. doi:10.2136/sssaj2017.12.0412

Burns, R.G. 1978. *Enzymes in soil: Some theoretical and practical considerations* (p. 295–339). In: Soil enzymes. Academic Press, London, New York, and San Francisco.

Burns, R.G. (1982). Enzyme activity in soil: Location and a possible role in microbial activity. *Soil Biol. Biochem.* 14, 423–427. doi:10.1016/0038-0717(82)90099-2

Burns, R.G., Pukite, A.H., and McLaren, A.D. (1972). Concerning the location and persistence of soil urease. *Soil Sci. Soc. Am. J.* 36(2), 308–311. doi:10.2136/sssaj1972.03615995003600020030x

Buyer, J.S., and Sasser, M. (2012). High throughput phospholipid fatty acid analysis of soils. *Appl. Soil Ecol.* 61, 127–130. doi:10.1016/j.apsoil.2012.06.005

Campbell, C.A., Biederbeck, V.O., R.P., Zentner, Brandt, S.A., and Schnitzer, M. (1992). Effect of crop rotations and rotation phase on characteristics of soil organic matter in a Dark Brown Chernozemic soil. *Can. J. Soil Sci.* 72, 403–416. doi:10.4141/cjss92-034

Cañizares, R., Benítez, E., and Ogunseitan, O.A. (2011). Molecular analyses of β-glucosidase diversity and function in soil. *Eur. J. Soil Biol.* 47, 1–8. doi:10.1016/j.ejsobi.2010.11.002

Chan, K.Y., Van Zwieten, L., Meszaros, I., Downie, A., and Joseph, S. (2007). Agronomic values of greenwaste biochar as a soil amendment. *Aust. J. Soil Res.* 45, 629–634. doi:10.1071/SR07109

Chander, K., Goyal, S., Mundra, M.C., and Kapoor, K.K. (1997). Organic matter, microbial biomass and enzyme activity of soils under different crop rotations in the tropics. *Biol. Fertil. Soils* 24, 306–310. doi:10.1007/s003740050248

Chu, B., Zaid, F., and Eivazi, F. (2016). Long-term effects of different cropping systems on selected enzyme activities. *Commun. Soil Sci. Plant Anal.* 47, 720–730. doi:10.1080/00103624.2016.1146749

Chunderova, A.I., and T. Zubers. (1969). Phosphatase activity in dernopodzolic soils. *Pochvovedeniye* 11, 47–53.

Clarholm, M. (1993). Microbial biomass P, labile P, and acid phosphatase activity in the humus layer of a spruce forest, after repeated additions of fertilizers. *Biol. Fertil. Soils* 16, 287–292. doi:10.1007/BF00369306

Cotton, J., and Acosta-Martínez, V. (2018). Intensive tillage converting grassland to cropland immediately reduces soil microbial community size and organic carbon. *Agric. Environ. Lett.* 3, 1–4.

Cotton, J., Acosta-Martínez, V., Moore-Kucera, J., and Burow, G. (2013). Early changes due to sorghum biofuel cropping systems in soil microbial communities and metabolic functioning. *Biol. Fertil. Soils* 49, 403–413. doi:10.1007/s00374-012-0732-z

Davinic, M., Moore-Kucera, J., Acosta-Martínez, V., Zak, J., and Allen, V. (2013). Soil fungal distribution and functionality as affected by grazing and vegetation components of integrated crop–livestock agroecosystems. *Appl. Soil Ecol.* 66, 61–70. doi:10.1016/j.apsoil.2013.01.013

Debosz, K., Rasmussen, P.H., and Pedersen, A.R. (1999). Temporal variations in microbial biomass C and cellulolytic enzyme activity in arable soils: Effects of organic matter input. *Appl. Soil Ecol.* 13, 209–218. doi:10.1016/S0929-1393(99)00034-7

de Castro Lopes, A.A., Gomes de Sousa, D.M., Montandon Chaer, G., dos Reis, Junior., F.B., Goedert, W.J., and de Carvalho Mendes, I. (2013). Interpretation of microbial soil indicators as a function of crop yield and organic carbon. *Soil Sci. Soc. Am. J.* 77, 461–472. doi:10.2136/sssaj2012.0191

Delgado, J.A., and Gantzer, C.J. (2015). The 4Rs for cover crops and other advances in cover crop management for environmental quality. *J. Soil Water Conserv.* 70, 142A–145A. doi:10.2489/jswc.70.6.142A

Demisie, W., Liu, Z.Y., and Zhang, M.K. (2014). Effect of biochar on carbon fractions and enzyme activity of red soil. *Catena* 121, 214–221. doi:10.1016/j.catena.2014.05.020

Deng, S.P., Moore, J.M., and Tabatabai, M.A. 2000. Characterization of active nitrogen pools in soils under different cropping systems. *Biol. Fertil. Soils* 32, 302–309. doi:10.1007/s003740000252

Deng, S., Popova, I.E., Dick, L., and Dick, R. (2013). Bench scale and microplate format assay of soil enzyme activities using spectroscopic and fluorometric approaches. *Appl. Soil Ecol.* 64, 84–90. doi:10.1016/j.apsoil.2012.11.002

Deng, S.P., and Tabatabai, M.A. (1997). Effect of tillage and residue management on enzyme activities in soil. III. Phosphatases and arylsulfatase. *Biol. Fertil. Soils* 24, 141–146. doi:10.1007/s003740050222

Dick, L.K., Jia, G., Deng, S., and Dick, R.P. (2013). Evaluation of microplate and bench-scale β-glucosidase assays for reproducibility, comparability, kinetics, and homogenization methods in two soils. *Biol. Fertil. Soils* 49, 1227–1236. doi:10.1007/s00374-013-0820-8

Dick, R.P., editor. (2011b). *Methods of soil enzymology. SSSA Book Series 9*. Madison, WI: Soil Sci. Soc. Am. doi:10.2136/sssabookser9

Dick, R.P. (1997). Soil enzyme activities as integrative indicators of soil health. In: Pankhurst, C.E., Doube, B.M., and Gupta, V.V.S.R., (eds.), *Biological indicators of soil health*. Wallingford: CAB International. p. 121–156.

Dick, R.P. 1994. Soil enzyme activities as indicators of soil quality. In: J.W. Doran, D.C. Coleman, D.F. Bezdicek, and B.A. Stewart, editors, *Defining soil quality for a sustainable environment*. Soil Sci. Soc. Am., Special Publication, Madison, WI. p. 107–124. doi:10.2136/sssaspecpub35.c7

Dick, R.P., L.K. Dick, S. Deng, X. Li, E. Kandeler, C. Poll, C. Freeman, T. Graham Jones, M.N. Weintraub, K.A. Esseili, and J. Saxena. 2018. Cross laboratory comparison of fluorimetric microplate and colorimetric bench-scale soil enzyme assays. *Soil Biol. Biochem.* 121:240–248. doi:10.1016/j.soilbio.2017.12.020

Dick, W.A. 2011c. Development of a soil enzyme reaction assay. In: R.P. Dick, editor, *Methods of soil enzymology*. Soil Sci. Soc. Am., Madison, WI. p. 71–84. doi:10.2136/sssabookser9

Dick, W.A. (1983). Organic carbon, nitrogen, and phosphorus concentrations and pH in soil profiles as affected by tillage intensity. *Soil Sci. Soc. Am. J.* 47, 102–107. doi:10.2136/sssaj1983.03615995004700010021x

Dick, W.A., and Daniel, T.C. (1987). Soil chemical and biological properties as affected by conservation tillage: Environmental impacts. In: T.J. Logan, et al., (eds.), *Effects of conservation tillage on groundwater quality: Nitrates and pesticides.* Chelsen, MI: Lewis Publishers, Inc. p. 124–147.

Dinesh, R., Anandaraj, M., Srinivasan,V., and Hamza, S. (2012). Engineered nanoparticles in the soil and their potential implications to microbial activity. *Geoderma* 173-174, 19–27. doi:10.1016/j.geoderma.2011.12.018

Ding, G., Liu, X., Herbert, S., Novak, J., Amarasiriwardena, D., and Xing, B. (2006). Effect of cover crop management on soil organic matter. *Geoderma* 130, 229–239. doi:10.1016/j.geoderma.2005.01.019

Doran, J.W., Wilhelm, W.W., and Power, J.F. (1984). Crop residue removal and soil productivity with no-till corn, sorghum, and soybean. *Soil Sci. Soc. Am. J.* 48, 640–645. doi:10.2136/sssaj1984.03615995004800030034x

Doran, J.W., and Zeiss, M.R. (2000). Soil health and sustainability: Managing the biotic component of soil quality. *Appl. Soil Ecol.* 15, 3–11. doi:10.1016/S0929-1393(00)00067-6

Drobnik, J. (1957). Biological transformations of organic substances in the soil. *Pochvovedeniye* 12, 62–71.

Drouillon, M., and Merckx, R. (2005). Performance of para-nitrophenyl phosphate and 4-methylumbelliferyl phosphate as substrate analogues for phosphomonoesterase in soils with different organic matter content. *Soil Biol. Biochem.* 37, 1527–1534. doi:10.1016/j.soilbio.2005.01.008

Dunn, M., Ulrich-Schad, J.D., Prokopy, L.S., Myers, R.L., Watts, C.R., and Scanlon, K. (2016). Perceptions and use of cover crops among early adopters: Findings from a national survey. *J. Soil Water Conserv.* 71, 29–40. doi:10.2489/jswc.71.1.29

DuPont, S.T., Culman, S.W., Ferris, H., Buckley, D.H., and Glover, J.D. (2010). No-tillage conversion of harvested perennial grassland to annual cropland reduces root biomass, decreases active carbon stocks, and impacts soil biota. *Agric. Ecosyst. Environ.* 137, 25–32. doi:10.1016/j.agee.2009.12.021

Eivazi, F., Bayan, M., and Schmidt, K. (2003). Select soil enzyme activities in the historic Sanborn Field as affected by long-term cropping systems. *Commun. Soil Sci. Plant.* 34, 2259–2275. doi:10.1081/CSS-120024062

Eivazi, F., and Tabatabai, M.A. (1977). Phosphatases in soils. *Soil Biol. Biochem.* 9, 167–172. doi:10.1016/0038-0717(77)90070-0

Eivazi, F., and Tabatabai, M.A. (1988). Glucosidases and galactosidases in soils. *Soil Biol. Biochem.* 20, 601–606. doi:10.1016/0038-0717(88)90141-1

Eivazi, F., Zahra, A., and Elizabeth, J. (2018). Effects of silver nanoparticles on the activities of soil enzymes involved in carbon and nutrient cycling. *Pedosphere* 28, 209–214. doi:10.1016/S1002-0160(18)60019-0

Ekenler, M., and Tabatabai, M.A. (2004). β-Glucosaminidase activity as an index of nitrogen mineralization in soils. *Commun. Soil Sci. Plant* 35, 1081–1094. doi:10.1081/CSS-120030588

Fageria, N.K., Baligar, V.C., and Bailey, B.A. (2005). Role of cover crops in improving soil and row crop productivity. *Commun. Soil. Sci. Plant* 36, 2733–2757. doi:10.1080/00103620500303939

Fernández, A.L., Sheaffer, C.C., Wyse, D.L., Staley, C., Gould, T.J., and Sadowsky, M.J. (2016). Associations between soil bacterial community structure and nutrient cycling functions in long-term organic farm soils following cover crop and organic fertilizer amendment. *Sci. Total Environ.* 566-567, 949–959. doi:10.1016/j.scitotenv.2016.05.073

Foster, E.J., Fogle, E.J., and Cotrufo, M.F. (2018). Sorption to biochar impacts β-glucosidase and phosphatase enzyme activities. *Agriculture* 8, 158. doi:10.3390/agriculture8100158

Foster, E.J., Hansen, N., Wallenstein, M., and Cotrufo, M.F. (2016). Biochar and manure amendments impact soil nutrients and microbial enzymatic activities in a semi-arid maize cropping system. *Agric. Ecosyst. Environ.* 233, 404–414. doi:10.1016/j.agee.2016.09.029

Franzluebbers, A.J. (2015). Farming strategies to fuel bioenergy demands and facilitate essential soil services. *Geoderma* 259-260, 251–258. doi:10.1016/j.geoderma.2015.06.007

Frene, J.P., L.A. Gabbarini, and L.G. Wall. (2018). Efectos del sistema de labranza sobre la actividad y la estructura microbiana a nivel de los microagregados. *Cienc. Suelo* 36, 50–62.

Galstyan, A.S. (1960). Enzyme activities in solonchaks. *Doklady Akademii Nauk Armyanskoi SSR* 30, 61–64.

García-Ruíz, R., Ochoa, V., Hinojosa, M.B., and Carreira, J.A. (2008). Suitability of enzyme activities for the monitoring of soil quality improvement in organic agricultural systems. *Soil Biol. Biochem.* 40, 2137–2145. doi:10.1016/j.soilbio.2008.03.023

Guilbault, G.G. (1990). *Practical fluorescence*. New York: Marcel Dekker, Inc.

Gul, S., Whalen, J.K., Thomas, B.W., Sachedva, V.V., and Deng, V. (2015). Physico-chemical properties and microbial responses in biochar-amended soils: Mechanisms and future directions. *Agric. Ecosyst. Environ.* 206, 46–59. doi:10.1016/j.agee.2015.03.015

Gupta, V.V.S.R., and Germida, J.J. (1988). Distribution of microbial biomass and its activity in different soil aggregate size classes as affected by cultivation. *Soil Biol. Biochem.* 20, 777–786. doi:10.1016/0038-0717(88)90082-X

Haban, L. 1967. Effect of ploughing depth and cultivated crops on the soil microflora and enzyme activity of the soil. V d. Pr. Vysk. Ust. *Rast/Vyroby Pietanoch* 5:159–169.

Hammac, W.A., Stott, D.E., Karlen, D.L., and Cambardella, C.A. (2016). Crop, tillage and landscape effects on near-surface soil quality indices in Indiana. *Soil Sci. Soc. Am. J.* 80, 1638–1652. doi:10.2136/sssaj2016.09.0282

Hammer V.B., Grant, K., Pritsch, K., Jentsch, A., Schloter, M., Beierkuhnlein, C., and Gschwendtner, S. (2019) Environ. Sci. 6: 1–10. doi.org/10.3389/fenvs.2018.00157

Hayano, K. (1973). A method for determination of β-glucosidase activity in soil. *Soil Sci. Plant Nutr.* 19, 103–108. doi:10.1080/00380768.1973.10432524

Hetrick, B.A.D., Wilson, G.W.T., and Leslie, J.F. (1991). Root architecture of warm- and cool-season grasses: Relationship to mycorrhizal dependence. *Can. J. Bot.* 69(1), 112–118. doi:10.1139/b91-016

Hetrick, B.A.D., Kitt, D.G., and Wilson, G.T. (1988). Mycorrhizal dependence and growth habit of warm-season and cool-season tallgrass prairie plants. *Can. J. Bot.* 66(7), 1376–1380. doi:10.1139/b88-193

Hoorman, J.J. (2009). *Using cover crops to improve soil and water quality*. Lima, OH: Ohio State University Extension.

Islam, M.R., Singh Chauhan, P., Kim, Y., Kim, M., and Sa, T. 2011. Community level functional diversity and enzyme activities in paddy soils under different long-term fertilizer management practices. *Biol. Fertil. Soils* 47, 599. doi:10.1007/s00374-010-0524-2

Jain, N.K., Jat, R.S., Meena, H.N., and Chakraborty, K. (2018). Productivity, nutrient, and soil enzymes influenced with conservation agriculture practices in peanut. *Agron. J.* 110, 1165–1172. doi:10.2134/agronj2017.08.0467

Jian, S., Li, J., Chen, J., Wang, G., Mayes, M.A., Dzantor, K.E., Hui, D., and Luo, Y. (2016). Soil extracellular enzyme activities, soil carbon and nitrogen storage under nitrogen fertilization: A meta-analysis. *Soil Biol. Biochem.* 101, 32–43. doi:10.1016/j.soilbio.2016.07.003

Jin, H. (2010). *Characterization of microbial life colonizing biochar and biochar-amended soils. PhD Dissertation*. Ithaca, NY: Cornell University.

Jindo, K., Matsumoto, K., García-Izquierdo, C., Sonoki, T., and Sánchez-Monedero, A.A. (2014). Methodological interference of biochar in the determination of extracellular enzyme activities in composting samples. *Solid Earth* 5, 713–719. doi:10.5194/se-5-713-2014

Johnson, J.M.F., Acosta-Martínez, V., Cambardella, C., and Barbour, N. (2013). Crop and soil responses to using corn stover as a bioenergy feedstock: Observations from the northern US Corn Belt. *Agriculture* 3, 72–89. doi:10.3390/agriculture3010072

Karlen, D.L., Hurley, E.G., Andrews, S.S., Cambardella, C.A., Meek, D.W., Duffy, M.D., and Mallarino, A.P. (2006). Crop rotation effects on soil quality at three northern corn/soybean belt locations. *Agron. J.* 98, 484–495.

Karlen, D.L., Wollenhaupt, N.C., Erbach, D.C., Berry, E.C., Swan, J.B., Eash, N.S., and Jordahl, J.L. (1994). Crop residue effects on soil quality following 10-years of no-till corn. *Soil Tillage Res.* 31, 149–167. doi:10.1016/0167-1987(94)90077-9

Khadem, A., and Raiesi, F. (2017). Influence of biochar on potential enzyme activities in two calcareous soils of contrasting texture. *Geoderma* 308, 149–158. doi:10.1016/j.geoderma.2017.08.004

Kiss, S., Drăgan-Bularda, M., and Rădulescu, D. 1975. *Biological significance of enzymes in soil. Adv. Agron.* 27, 25–87. doi:10.1016/S0065-2113(08)70007-5

Kizilkaya, R., Karaca, A., Turgay, O.C., and Cetin, S.C. (2010). Earthworm interactions with soil enzymes. In A. Karaka, (ed.), *Biology of earthworms.* Soil biology 24. New York: Springer.

Klose, S., S. Bilen, Tabatabai, M.A., and Dick, W.A. (2011). Sulfur cycle enzymes. In R.P. Dick, (ed.), *Methods of soil enzymology.* Madison, WI: SSSA. p. 125–159.

Klose, S., J.M. Moore, and Tabatabai, M.A. (1999). Arylsulfatase activity of microbial biomass in soils as affected by cropping systems. *Biol. Fertil. Soils* 29, 46–54. doi:10.1007/s003740050523

Klose, S., and Tabatabai, M.A. (2002). Response of glycosidases in soils to chloroform fumigation. *Biol. Fertil. Soils* 35, 262–269. doi:10.1007/s00374-002-0463-7

Klose, S., and Tabatabai, M.A. (1999). Arylsulfatase activity of the microbial biomass in soils. *Soil Sci. Soc. Am. J.* 63, 569–574. doi:10.2136/sssaj1999.03615995006300030020x

Knight, T.R., and Dick, R.P. (2004). Differentiating microbial and stabilized β-glucosidase activity relative to soil quality. *Soil Biol. Biochem.* 36, 2089–2096. doi:10.1016/j.soilbio.2004.06.007

Koepf, H. 1954. Investigations on the biological activity in soil. I. Respiration curves of the soil and enzyme activity under the influence of fertilizing and plant growth. Zeitschrift far Acker-und Pf/anzenbau 98:289-312.

Kumar, A., Dorodnikov, M., Splettstößer, T., Kuzyakov, Y., and Pausch, J. (2017). Effects of maize roots on aggregate stability and enzyme activities in soil. *Geoderma* 306, 50–57. doi:10.1016/j.geoderma.2017.07.007

Lal, R. (2015). Soil carbon sequestration and aggregation by cover cropping. *J. Soil Water Conserv.* 70, 329–339. doi:10.2489/jswc.70.6.329

Lee, Y.B., Lorenz, N., Kincaid Dick, L., and Dick, R.P. (2007). Cold storage and pretreatment incubation effects on soil microbial properties. *Soil Sci. Soc. Am. J.* 71, 1299–1305. doi:10.2136/sssaj2006.0245

Lehman, R.M., Acosta-Martínez, V., Buyer, J.S., Cambardella, C.A., Collins, H.P., Ducey, T.F., Halvorson, J.J., Jin, V.L., Johnson, J.M.F., Kremer, R.J., Lundgren, J.G.,

Manter, D.K., Maul, J.E., Smith, J.L., and Stott, D.E. (2015). Soil biology for resilient, healthy soil. *J. Soil Water Conserv.* 70, 12A–18A. doi:10.2489/jswc.70.1.12A

Lehmann, J., and Joseph, S. (2009). Biochar for environmental management: An introduction. In Lehmann, J., and S. Joseph, S., (eds.), *Biochar for environmental management: Science and technology*. London: Earthscan. p. 1–12.

Lehmann, J., Rillig, M., Thies, J., Masiello, C.A., Hockaday, W.C., and Crowley, D. (2011). Biochar effects on soil biota: A review. *Soil Biol. Biochem.* 43, 1812–1836. doi:10.1016/j.soilbio.2011.04.022

Li, C., Fultz, L.M., Moore-Kucera, J., Acosta-Martínez, V., Horita, J., Strauss, R., Zak, J., Calderón, F., and Weindorf, D. (2017). Soil carbon sequestration potential in semi-arid grasslands in the Conservation Reserve Program. *Geoderma* 294, 80–90. doi:10.1016/j.geoderma.2017.01.032

Li, C., Fultz, L.M., Moore-Kucera, J., Acosta-Martínez, V., Kakarla, M., and Weindorf, D.C. (2018). Soil microbial community restoration in Conservation Reserve Program semi-arid grasslands. *Soil Biol. Biochem.* 118, 166–177. doi:10.1016/j.soilbio.2017.12.001

Liang, B., Lehmann, J., Solomon, D., Kinyangi, J., Grossman, J., O'Neill, B., Skjemstad, J.O., Thies, J., Luizao, F.J., Petersen, J., and Neves, E.G. (2006). Black carbon increases cation exchange capacity in soils. *Soil Sci. Soc. Am. J.* 70, 1719–1730. doi:10.2136/sssaj2005.0383

Liang, S., Grossman, J., and Shi, W. (2014). Soil microbial responses to winter legume cover crop management during organic transition. *Eur. J. Soil Biol.* 65, 15–22. doi:10.1016/j.ejsobi.2014.08.007

Linn, D.M., and Doran, J.W. (1984). Effect of water-filled pore space on carbon dioxide and nitrous oxide production in tilled and non-tilled soils. *Soil Sci. Soc. Am. J.* 48, 1267–1272. doi:10.2136/sssaj1984.03615995004800060013x

Lorenz, N., and Dick, R.P. (2011). Sampling and pretreatment of soil before enzyme analysis. In Dick, R.P., (ed.), *Methods of soil enzymology* (p. 85–101). Madison, WI: SSSA. doi:10.2136/sssabookser9.c5

Luo, Z., Wang, E., and Sun, O.J. (2010). Can no-tillage stimulate carbon sequestration in agricultural soils? A meta-analysis of paired experiments. *Agric. Ecosyst. Environ.* 139, 224–231. doi:10.1016/j.agee.2010.08.006

Margenot, A.J., Nakayama, Y., and Parikh, S.J. (2018). Methodological recommendations for optimizing assays of enzyme activities in soil samples. *Soil Biol. Biochem.* 125, 350–360. doi:10.1016/j.soilbio.2017.11.006

Marx, M.C., Wood, M., and Jarvis, S.C. (2001). A microplate fluorimetric assay for the study of enzyme diversity in soils. *Soil Biol. Biochem.* 33, 1633–1640. doi:10.1016/S0038-0717(01)00079-7

Mathur, S.P., and Rayment, A.F. (1977). Influence of trace element fertilization on the decomposition rate and phosphatase activity of a mesic fibrisol. *Can. J. Soil Sci.* 57, 397–408. doi:10.4141/cjss77-045

Mbuthia, L.W., V. Acosta-Martínez, J. DeBruyn, S. Schaeffer, D. Tyler, D. Odoi, M. Mpheshea, F. Walker, and N. Eash. 2015. Long term tillage, cover crop, and fertilization effects on microbial community structure, activity: Implications for soil quality. *Soil Biol. Biochem.* 89, 24–34. doi:10.1016/j.soilbio.2015.06.016

McLaren, A.D. (1975). Soil as a system of humus and clay immobilized enzymes. *Chem. Scr.* 8, 97–99.

McLaren, A.D., Luse, R.A., and Skujiņš, J.J. (1962). Sterilization of soil by irradiation and some further observations on soil enzyme activity. *Soil Sci. Soc. Am. Proc.* 26, 371–377. doi:10.2136/sssaj1962.03615995002600040019x

McLaren, A.D., Reshetko, L., and Huber, W. (1957). Sterilization of soil by irradiation with an electron beam, and some observations on soil enzyme activity. *Soil Sci.* 83, 497–502. doi:10.1097/00010694-195706000-00011

Mendes, I.C., Souza, L.M., Sousa, D.M.G., Lopes, A.A.C., Reis-Junior, F.B., Coelho Lacerda, M.P., Malaquias, J.V. (2019). Critical limits for microbial indicators in tropical Oxisols at post-harvest: The FERTBIO soil sample concept. *Appl. Soil Ecol.* 139, 85–93.

Mielke, L.N., Doran, J.W., and Richards, K.A. (1986). Physical environment near the surface of plowed and no-tilled soils. *Soil Tillage Res.* 7, 355–366. doi:10.1016/0167-1987(86)90022-X

Miller, M., and Dick, R.P. (1995). Dynamics of soil C and microbial biomass on whole soil and aggregates in two cropping systems. *Appl. Soil Ecol.* 2, 253–261. doi:10.1016/0929-1393(95)00060-6

Nannipieri, P., Ceccanti, B., and Bianchi, D. (1988). Characterization of humus phosphatase complexes extracted from soil. *Soil Biol. Biochem.* 20:683–691. doi:10.1016/0038-0717(88)90153-8

Nannipieri, P., Ceccanti, B., Cervelli, S., and Sequi, P. (1974). Use of 0.1 M pyrophosphate to extract urease for podzol. *Soil Biol. Biochem.* 6, 359–362. doi:10.1016/0038-0717(74)90044-3

Nannipieri, P., Trasar-Cepeda, C., and Dick, R.P. (2018). Soil enzyme activity: A brief history and biochemistry as a basis for appropriate interpretations and meta-analysis. *Biol. Fertil. Soils* 54, 11–19. doi:10.1007/s00374-017-1245-6

Ndiaye, E.L., Sandeno, J.M., McGrath, D., and Dick, R.P. (2000). Integrative biological indicators for detecting change in soil quality. *Am. J. Altern. Agric.* 15, 26–36. doi:10.1017/S0889189300008432

National Oceanic and Atmospheric Administration (NOAA). (2018). *Global climate report: Annual 2018*. Washington, D.C: NOAA. https://www.ncdc.noaa.gov/sotc/global/201813

Novak, J.M., Busscher, W.J., Watts, D.W., Amonette, J.E., Ippolito, J.A., Lima, I.M., Gaskin, J., Das, K.C., Steiner, C., Ahmedna, M., Rehrah, D., and Schomberg, H. (2012). Biochars impact on soil-moisture storage in an Ultisol and two Aridisols. *Soil Sci.* 177, 310–320. doi:10.1097/SS.0b013e31824e5593

Nowack, B., and Bucheli, T.D. (2007). Occurrence, behavior and effects of nanoparticles in the environment. *Environ. Pollut.* 150, 5–22. doi:10.1016/j.envpol.2007.06.006

Nunes, M.R., Karlen, D.L., Veum, K.S., Moorman, T.B., & Cambardella, C.A. (2020). Biological soil health indicators respond to tillage intensity: A US meta-analysis. *Geoderma* 369, 114335.

Nunes, M.R., van Es, H.M., Schindelbeck, R., Ristow, A.J., and Ryan, M. (2018). No-till and cropping system diversification improve soil health and crop yield. *Geoderma* 328, 30–43. doi:10.1016/j.geoderma.2018.04.031

Parham, J.A., and Deng, S.P. (2000). Detection, quantification and characterization of β-glucosaminidase activity in soil. *Soil Biol. Biochem.* 32, 1183–1190. doi:10.1016/S0038-0717(00)00034-1

Paz-Ferreiro, J., Fu, S., Méndez, A., and Gascó, G. (2014). Interactive effects of biochar and the earthworm *Pontoscolex corethrurus* on plant productivity and soil enzyme activities. *J. Soils Sediments* 14, 483–494. doi:10.1007/s11368-013-0806-z

Paz-Ferreiro, J., Gascó, G., Gutiérrez, B., and Méndez, A. (2012). Soil biological activities and the geometric mean of enzyme activities after application of sewage sludge and sewage sludge biochar. *Biol. Fertil. Soils* 48, 511–517. doi:10.1007/s00374-011-0644-3

Pérez-Guzmán, L., Acosta-Martínez, Phillips, L., and Mauget, S.A. (2020). Resilience of the microbial communities of semiarid agricultural soils during natural climatic variability events. *Appl. Soil Ecol.* 149. doi:10.1016/j.apsoil.2019.103487

Perez-Guzman, L., Phillips, L. A., Acevedo, M. A., and Acosta-Martinez, V. (2021). Comparing biological methods for soil health assessments: EL-FAME, enzyme activities, and qPCR. Soil Sci. Soc. Am. J. 1–17. doi.org/10.1002/saj2.20211

Popova, I.E., and Deng, S. (2010). A high-throughput microplate assay for simultaneous colorimetric quantification of multiple enzyme activities in soil. *Appl. Soil Ecol.* 45, 315–318. doi:10.1016/j.apsoil.2010.04.004

Powlson, D.S., Brooke, P.C., and Christensen, B.T. (1987). Measurement of soil microbial biomass provides an early indication of changes in total soil organic matter due to straw incorporation. *Soil Biol. Biochem.* 19, 159–164. doi:10.1016/0038-0717(87)90076-9

Pritsch, K., Raidl, S., Marksteiner, E., Blaschke, H., Agerer, R., Schloter, M., and Hartmann, A. (2004). A rapid and highly sensitive method for measuring enzyme activities in single mycorrhizal tips using 4-methylumbelliferone-labelled fluorogenic substrates in a microplate system. *J. Microbiol. Methods* 58, 233–241. doi:10.1016/j.mimet.2004.04.001

Purakayastha, T.J., Kumari, S., and Pathak, H. (2015). Characterization, stability and microbial effects of four biochars produced from crop residues. *Geoderma* 239-240, 293–303. doi:10.1016/j.geoderma.2014.11.009

Ragot, S.A., Kertesz, M.A., and Bünemann, E.K. (2015). *phoD* alkaline phosphatase gene diversity in soil. *Appl. Environ. Microbiol.* 81, 7281–7289. doi:10.1128/AEM.01823-15

Rankoth, L.M., Udawatta, R.P., Veum, K.S., Jose, S., and Alagele, S. (2019). Cover crop influence on soil enzymes and selected chemical parameters for a claypan corn–soybean rotation. *Agriculture* 9, 125. doi:10.3390/agriculture9060125

Reeves, J.B., McCarty, G.W., and Meisinger, J.J. (1999). Near infrared reflectance the analysis of agricultural soils. *JNIRS* 7, 179–193.

Saggar, S.,. Bettany, J.R., and Stewart, J.W.B.B. (1981). Measurement of microbial S in soil. *Soil Biol. Biochem.* 13, 493–498. doi:10.1016/0038-0717(81)90040-7

Saha, S., Prakash, V., Kundu, S., Kumar, N., and Minaet, B.L. (2008). Soil enzymatic activity as affected by long term application of farm yard manure and mineral fertilizer under a rainfed soybean–wheat system in N-W *Himalaya. Eur. J. Soil Biol.* 44, 309–315. doi:10.1016/j.ejsobi.2008.02.004

Sardans, J., Peñuelas, J., and Ogaya., R. (2008). Experimental drought reduced acid and alkaline phosphatase activity and increased organic extractable P in soils in a *Quercus ilex* Mediterranean forest. *Eur. J. Soil Biol.* 44, 509–520. doi:10.1016/j.ejsobi.2008.09.011

Schutter, M.E., and Dick, R.P. (2002). Microbial community profiles and activities among aggregates of winter fallow and cover-cropped soil. *Soil Sci. Soc. Am. J.* 66, 142–153. doi:10.2136/sssaj2002.1420

Schutter, M., J. Sandeno, and R.P. Dick. 2001. Seasonal, soil type, and alternative management influences on microbial communities of vegetable cropping systems. *Biol. Fertil. Soils* 34, 397–410. doi:10.1007/s00374-001-0423-7

Sharma, P., Singh, G., and Singh, R.P. (2013). Conservation tillage and optimal water supply enhance microbial enzyme (glucosidase, urease and phosphatase) activities in fields under wheat cultivation during various nitrogen management practices. *Arch. Agron. Soil Sci.* 59, 911–928. doi:10.1080/03650340.2012.690143

Sharma, P., Singh, G., and Singh, R.P. (2011). Conservation tillage, optimal water and organic nutrient supply enhance soil microbial activities during wheat (*Triticum aestivum* L.) cultivation. *Braz. J. Microbiol.* 42, 531–542. doi:10.1590/S1517-83822011000200018

Shrestha, B., Acosta-Martínez, V., Cox, S.B., Green, M.J., Li, S., and Cañas-Carrell, J.E. (2013). An evaluation of the impact of multiwalled carbon nanotubes on soil microbial community structure and functioning. *J. Hazard. Mater.* 261, 188–197. doi:10.1016/j.jhazmat.2013.07.031

Sinsabaugh, R.L., R.K. Antibus, and A.E. Linkins. 1991. An enzymic approach to the analysis of microbial activity during plant litter decomposition. *Agric. Ecosyst. Environ.* 34, 43–54. doi:10.1016/0167-8809(91)90092-C

Skujiņš, J. 1978. *History of abiontic soil enzyme research.* In: R.G. Burns, editor, Soil enzymes. Academic Press, London, New York, and San Francisco. p. 1–49.

Skujiņš, J. 1976. Extracellular enzymes in soil. *CRC Crit. Rev. Microbiol.* 4, 383–421. doi:10.3109/10408417609102304

Skujiņš, J. 1967. Enzymes in soil. In: A.D. McLaren and G.H. Peterson, editors, *Soil biochemistry*. Marcel Dekker, New York. p. 371–414.

Speir, T.W., S.C. Cowling, and G.P. Sparling. 1986. Effects of microwave radiation on the microbial biomass, phosphatase activity and levels of extractable N and P in a low fertility soil under pasture. *Soil Biol. Biochem.* 18:377–382. doi:10.1016/0038-0717(86)90041-6

Spiers, G.A., and W.B. McGill.(1979). Effects of phosphorus addition and energy supply on acid phosphatase production and activity in soils. *Soil Biol. Biochem.* 11, 3–8. doi:10.1016/0038-0717(79)90110-X

Staben, M.L., D.F. Bezdicek, M.F. Fauci, and J.L. Smith. 1997. Assessment of soil quality in Conservation Reserve Program and wheat-fallow soils. *Soil Sci. Soc. Am. J.* 61, 124–130. doi:10.2136/sssaj1997.03615995006100010019x

Stott, D.E., S.S. Andrews, M.A. Liebig, B.J. Wienhold, and D.L. Karlen. 2010. Evaluation of β-glucosidase activity as a soil quality indicator for the Soil Management Assessment Framework. *Soil Sci. Soc. Am. J.* 74, 107–119. doi:10.2136/sssaj2009.0029

Stott, D.E., D.L. Karlen, C.A. Cambardella, and Harmel, R.D. 2013. A soil quality and metabolic activity assessment after fifty-seven years of agricultural management. *Soil Sci. Soc. Am. J.* 77, 903–913. doi:10.2136/sssaj2012.0355

Sun, Z., Bruun, E.W., Arthur, E., Wollesen de Jonge, L., Moldrup, P., Hauggaard-Nielsen, H., and Elsgaard, L. (2014). Effect of biochar on aerobic processes, enzyme activity and crop yields in two sandy loam soils. *Biol. Fertil. Soils* 50:1087–1097.

Tabatabai, M.A. 1994. Soil enzymes. In: R.W. Weaver, J.S. Angle, and P.S. Bottomley, editors, *Methods of soil analysis: Part 2. Microbiological and biochemical properties*. Vol. 5. Madison, WI: SSSA. p. 775–833.

Tabatabai, M.A., and J.M. Bremner. 1970. Factors affecting soil arylsulfatase activity. *Soil Sci. Soc. Am. J.* 34, 427–429. doi:10.2136/sssaj1970.03615995003400030023x

Tabatabai, M.A., and W.A. Dick. 2002. Enzymes in soil: Research and developments in measuring activities. In Burns, R.G., and Dick, R.P., editors, *Enzymes in the environment: Activity, ecology and applications*. New York: Marcel Dekker. p. 567–596.

Teasdale, J.R. (1996). Contribution of cover crops to weed management in sustainable agricultural systems. *J. Prod. Agric.* 9, 475–479. doi:10.2134/jpa1996.0475

Tong, Z., M. Bischoff, L. Nies, Applegate, B., and Turco, R.F. (2007). Impact of fullerene (C-60) on a soil microbial community. *Environ. Sci. Technol.* 41, 2985–2991. doi:10.1021/es061953l

Torbert, H.A., and Wood, C.W. (1992). Effects of soil compaction and water-filled pore space on soil microbial activity and N losses. *Commun. Soil Sci. Plant* 23, 1321–1331.

Trap, J., Riah, W., Akpa-Vinceslas, M., Bailleul, C., Laval, K., and Trinsoutrot-Gattin, I. (2012). Improved effectiveness and efficiency in measuring soil enzymes as universal soil quality indicators using microplate fluorimetry. *Soil Biol. Biochem.* 45, 98–101. doi:10.1016/j.soilbio.2011.10.010

Trivedi, P., Delgado-Baquerizo, M., Trivedi, C., Hu, H., Anderson, I.C., Jeffries, T.C., Zhou, J., and Singh, B.K. (2016). Microbial regulation of the soil carbon cycle: Evidence from gene-enzyme relationships. *ISME J.* 10, 2593–2604. doi:10.1038/ismej.2016.65

Trupiano, D., Cocozza, C., Baronti, S., Amendola, C., Vaccari, F.P., Lustrato, G., Di Lonardo, S., Fantasma, F., Tognetti, R., and Scippa, G.S. (2017). The effects of biochar and its combination with compost on lettuce (Lactuca sativa L.) growth, soil properties, and soil microbial activity and abundance. *Int. J. Agron.* doi:10.1155/2017/3158207

Tyler, H.L., Locke, M.A., Moore, M.T., and Steinriede, R.W. (2016). Impact of conservation land management practices on soil microbial function in an agricultural watershed. *J. Soil Water Conserv.* 71, 396–403. doi:10.2489/jswc.71.5.396

USDA-NRCS. (2020). *Soil health key points*. Washington, D.C.: USDA-NRCS. https://www.nrcs.usda.gov/wps/portal/nrcs/detail/national/organic/?cid=nrcseprd1363633 (verified 4 June 2020).

Vance, M.E., Kuiken, T., Vejerano, E.P., McGinnis, S.P., Hochella, Jr., M.F., Rejeski, D., and Hull, M.S. (2015). Nanotechnology in the real world: Redeveloping the nanomaterial consumer products inventory. *J. Nanotechnol.* 6, 1769–1780.

Verstraete, W., and Voets, J.P. (1977). Soil microbial and biochemical characteristics in relation to soil management and fertility. *Soil Biol. Biochem.* 9, 253–258. doi:10.1016/0038-0717(77)90031-1

Veum, K.S., Goyne, K.W., Kremer, R.J., Miles, R.J., and Sudduth, K.A. (2014). Biological indicators of soil quality and soil organic matter characteristics in an agricultural management continuum. *Biogeochemistry* 117, 81–99. doi:10.1007/s10533-013-9868-7

Veum, K.S., Kremer, R.J.,, Sudduth, K.A., Kitchen, N.R., Lerch, R.N., Baffaut, C., Stott, D.E., Karlen, D.L., and Sadler, E.J. (2015). Conservation effects on soil quality indicators in the Missouri Salt River Basin. *J. Soil Water Conserv.* 70, 232–246. doi:10.2489/jswc.70.4.232

Veum, K.S., Sudduth, K.A., Kremer, R.J., and Kitchen, N.R. (2017). Sensor data fusion for soil health assessment. *Geoderma* 305, 53–61. doi:10.1016/j.geoderma.2017.05.031

Webster, R., and Oliver, M.A. (1990). *Statistical methods in soil and land resource survey*. Oxford, UK: Oxford Univ. Press. p. 44–46.

Wegner, B.R., Kumar, S., Osborne, S., Schumacher, T.E., Vahyala, and Eynard, A. (2015). Soil response to corn residue removal and cover crops in eastern South Dakota. *Soil Sci. Soc. Am. J.* 79, 1179–1187. doi:10.2136/sssaj2014.10.0399

Wegner, B.R., Osborne, S.L., Lehman, R.M., and Kumar, S. (2018). Seven-year impact of cover crops on soil health when corn residue is removed. *BioEnergy Res.* 11, 239–248. doi:10.1007/s12155-017-9891-y

Wilson, G.W.T., Rice, C.W., Rillig, M.C., Springer, A., and Hartnett, D.C. (2009). Soil aggregation and carbon sequestration are tightly correlated with the abundance of arbuscular mycorrhizal fungi: Results from long-term field experiments. *Ecol. Lett.* 12, 452–461. doi:10.1111/j.1461-0248.2009.01303.x

Xiao, W., Chen, X., Jing, X., and Zhu, B. (2018). A meta-analysis of soil extracellular enzyme activities in response to global change. *Soil Biol. Biochem.* 123:21–32. doi:10.1016/j.soilbio.2018.05.001

Zak, D.R., Holmes, W.E., White, D.C., Aaron, D.P., and Tilman, D. (2003). Plant diversity, soil microbial communities, and ecosystem function: Are there any links? *Ecology* 84, 2042–2050. doi:10.1890/02-0433

Zhang, B., Li, Y., Ren, T., Tian, Z., Wang, G., He, X., and Tian, C. (2014). Short-term effect of tillage and crop rotation on microbial community structure and enzyme activities of a clay loam soil. *Biol. Fertil. Soils* 50, 1077–1085. doi:10.1007/s00374-014-0929-4

Zheng, F.-L., Merrill, S.D., Huang, C.-H., Tanaka, D.L., Darboux, F., Liebig, M.A., and Halvorson, A.D. (2004). Runoff, soil erosion, and erodibility of Conservation Reserve Program land under crop and hay production. *Soil Sci. Soc. Am. J.* 68, 1332–1341. doi:10.2136/sssaj2004.1332

Zornoza, R., Guerrero, C., Mataix-Solera, J., Arcenegui, V., García-Orenes, F., and Mataix-Beneyto, J. (2006). Assessing air-drying and rewetting pre-treatment effect on some soil enzyme activities under Mediterranean conditions. *Soil Biol. Biochem.* 38, 2125–2134. doi:10.1016/j.soilbio.2006.01.010

Zornoza, R., Landi, L., Nannipieri, P., and Renella, G. (2012). A protocol for the assay of arylesterase activity in soil. *Soil Biol. Biochem.* 41, 659–662. doi:10.1016/j.soilbio.2009.01.003

Zuber, S.M., and Villamil, M.B. (2016). Meta-analysis approach to assess effect of tillage on microbial biomass and enzyme activities. *Soil Biol. Biochem.* 97, 176–187. doi:10.1016/j.soilbio.2016.03.011

# 12

# PLFA and EL-FAME Indicators of Microbial Community Composition

*Kristen S. Veum, Veronica Acosta-Martinez, R. Michael Lehman, Chenhui Li, Amanda Cano, and Marcio R. Nunes*

## Abbreviations

| | |
|---|---|
| AMF | arbuscular mycorrhizal fungi |
| EL-FAME | ester-linked fatty acid methyl esters |
| FAME | fatty acid methyl ester |
| GC | gas chromatography |
| MS | mass spectrometry |
| NLFA | neutral lipid fatty acids |
| PLFA | phospholipid fatty acids |

## Introduction

Ecosystem level microbial interactions are complex. In the soil, microorganisms regulate biogeochemical transformations and improve soil structure by their role in aggregate formation. They are responsive to changes in the physical, chemical, and biological characteristics of the soil, serving as early indicators of soil health degradation (Nielsen & Winding, 2002; Schloter, Nannipieri, Sørensen, & van Elsas, 2018). Through biogeochemical cycling, microbes influence the sustainability of agronomic production systems and provide a wide range of ecosystem services. For example, soil microbial communities directly influence soil organic matter formation (Cotrufo, Wallenstein, Boot, Denef, & Paul, 2013; Kallenbach,

*Soil Health Series: Volume 2 Laboratory Methods for Soil Health Analysis*, First Edition.
Edited by Douglas L. Karlen, Diane E. Stott, and Maysoon M. Mikha.

Frey, & Grandy, 2016), which has been identified as the most important indicator of soil health (Lal, Kimble, Follett, & Stewart, 1998). Thus, diversity and function of the soil microbial community may be the most valuable biological components of any ecosystem, providing pathways for primary production, mineralization of organic matter, transformation of inorganic constituents, and production of metabolites (Lehman, Acosta-Martinez, et al., 2015, Lehman, Cambardella, et al., 2015).

Methods for quantifying and characterizing the microbial community have evolved considerably since early efforts to culture soil microorganisms and identify them via phenotyping (Perfil'ev & Gabe, 1969) or by selective enrichment methods designed to characterize functional diversity (Foster, 1962). These methods were highly selective and captured less than 1% of the existing biodiversity (Pinkart, Ringelberg, Piceno, Macnaughton, & White, 2002). More modern methods using culture-independent approaches have focused on fatty acid methyl ester (FAME) profiling including ester-linked fatty acid methyl esters (EL-FAME) or phospholipid fatty acids (PLFA) extracted from cell membranes (e.g., Frostegård & Bååth, 1996; Schutter & Dick, 2000; Zelles, 1999), or DNA-based extraction and identification techniques (Manter et al., 2017; Manter, Weir, & Vivanco, 2010).

Fatty acid profiling methods have been commonly used for assessing microbial community structure because they are cheaper, less time-consuming, and were available for routine analyses of microbial communities before DNA-based methods. The FAME approach uses biomarkers that are structural components of all microbial cell membranes and are linked with essential functions including storing energy and signaling to provide a profile representing the viable soil microbial community (Zelles et al., 1994). Both PLFA and EL-FAME methods extract fatty acids from soil samples and convert them into FAMEs using an alkaline reagent (Buyer & Sasser, 2012; Schutter & Dick, 2000) based on the early protocol of Bligh and Dyer (1959). Additional details on the development of these techniques can be found in (Quideau et al., 2016). The nomenclature of FAME generally follows that of Tunlid and White (1992), where individual fatty acids were designated according to the total number of carbon atoms, the number and position of double bonds, branching, and other structural and functional features. Two common nomenclature systems can be found in the literature (a) the omega system where carbons are counted from the methyl (omega) end, and (b) the delta system, where carbons are instead counted from the carboxylic acid end. This chapter will use the omega nomenclature system.

White (1983) pioneered the application of FAME analysis to microbial communities in natural environments, and efforts were made to assign FAME biomarkers to specific microbial groups (e.g., Zelles, 1999). Although this approach has been popular for over 40 years (Willers, Jansen van Rensburg, & Claassens, 2015), it has gained renewed attention as a soil health indicator, in part due to the development of a rapid, high-throughput microplate method (Buyer & Sasser, 2012) that identifies a large variety of PLFAs, including straight-chain saturated, mono- and polyunsaturated, 10-methyl, and cyclopropane fatty acids. These biomarkers have subsequently been assigned to microbial groups, including Gram positive and Gram negative bacteria, Actinobacteria, saprophytic fungi, arbuscular mycorrhizal fungi (AMF), and eukaryotes, based on the work of Zelles (1997, 1999) and others. It should be noted that microbial group assignments can be contentious, as many markers have been shown to be non-specific and more ubiquitous than originally assumed. For example, the 18:2ω6,9 and 18:1ω9 markers are generally assigned to fungi, but they are also found in eukaryotes and plants (Frostegård, Tunlid, & Bååth, 2011). As another example, the 10-methyl fatty acids used as biomarkers for the Actinobacteria group are made by multiple genera of Actinobacteria (e.g., Actinomycetes, Corynebacteria, Mycobacteria, among others), whereas other Actinobacteria genera (e.g., Bifidobacteria, Micrococcus, and Streptomyces among others) do not appear to produce them (personal communication, MIDI Corp).

Once FAME biomarkers have been identified, they are generally reported either on a mass or molar basis per gram of soil (e.g., ng/g soil or nmol/g soil). The molar conversion accounts for the molecular weight of each biomarker and results in a concentration value that can be compared across all biomarkers. In addition to estimating the biomass of individual microbial groups, total FAME content (the sum of all identified peaks; nmol/g soil) is commonly used to represent the total microbial biomass in a soil sample. Total FAME content has been shown to correlate with other microbial biomass methods providing an indication of microbial community size (Tunlid & White, 1992; Zelles, Bai, Rackwitz, Chadwick, & Beese, 1995). However, there are circumstances where the correlation of these microbial biomass measurements is decoupled depending on the type of soil system, interference from fatty acids of plant origin, and the fact that the composition of microbial membranes can change in response to environment conditions or physiological state (Tunlid & White, 1992; White, Stair, & Ringelberg, 1996).

General ecological roles related to soil processes, such as decomposition and nutrient cycling, can be attributed to the microbial groups identified by FAME biomarkers. For example, AMF contribute to ecosystem health by improving water and nutrient acquisition, particularly phosphorus uptake, and subsequently enhancing crop growth and productivity (Begum et al., 2019). Bacteria also play a major role in nutrient transformation and organic matter decomposition. Gram positive bacteria tend to utilize more recalcitrant carbon sources while Gram negative bacteria tend to use more plant-derived C sources. Therefore, the ratio of Gram positive to Gram negative bacteria has been used as an indicator of carbon availability (Fanin et al., 2019; Fierer, Schimel, & Holden, 2003; Kramer &

Gleixner, 2008). Saprophytic fungi and Actinobacteria, which are filamentous Gram positive bacteria, play a key role in decomposition, carbon cycling, soil aggregation, and the ability of plants to withstand water stress. In particular, Actinobacteria are often differentiated from other bacterial groups for their role in lignin decomposition in soil (Bailey, Smith, & Bolton, 2003; Moore-Kucera & Dick, 2008; Six, Frey, Thiet, & Batten, 2006).

Fatty acid profiling methods are thought to represent the "active" microbial biomass (Pinkart et al., 2002; White et al., 1996) and have been shown to reflect early changes in microbial community structure due to different land use and management practices (e.g., Bhandari, West, Acosta-Martinez, Cotton, & Cano, 2018; Bossio & Scow, 1998; Zhang et al., 2016) as well as natural climatic disturbances such as drought, hurricane, and extreme temperatures (Bérard, Bouchet, Sévenier, Pablo, & Gros, 2011; Cantrell et al., 2014; Pérez-Guzmán, Acosta-Martínez, Phillips, & Mauget, 2020). The sensitivity to management and climatic factors enhances the utility of FAMEs as a soil health indicator to elucidate the role of anthropogenic factors and climatic variability on soil function and microbial community structure.

## Available Methods

Multiple methods, including PLFA, EL-FAME, and DNA-based approaches, provide estimates of microbial biomass and information on community structure. However, advantages of FAME include sensitivity due to its rapid degradation in the soil, reasonable cost, and no requirement for bioinformatics skills. In contrast, when compared with DNA-based approaches, the information derived from FAME is broad, functional roles must be inferred, and detailed taxonomic information cannot be elicited.

The use of FAME analysis to estimate bacterial and fungal biomass in soil and to detect land use effects is well-founded (Frostegård & Bååth, 1996; Zelles, Bai, Beck, & Beese, 1992). Soil microbial biomass is widely accepted as an ecologically-significant quantity (proportionally related to biogeochemical cycling activities, etc.) and therefore it is a valuable response variable. Although the FAME method is quantitative, an exact conversion of FAME content to microbial biomass does not exist. Several studies have presented conversion factors for overall biomass, microbial groups, or specific microorganisms (Balkwill, Leach, Wilson, McNabb, & White, 1988; Franzmann, Patterson, Power, Nichols, & Davis, 1996; Green & Scow, 2000; Kieft, Ringelberg, & White, 1994), but most conversions were developed from cultures with organisms that were isolated from non-soil matrices. Further, large species-specific variation exists and there is a lack of knowledge regarding species composition and environmental/growth stage effects on biomarkers (Frostegård & Bååth, 1996; Frostegård et al., 2011; Willers et al., 2015). Therefore, caution must be employed when interpreting the data with respect to microbial group assignments and biomass estimation, especially when comparing different studies. On the other hand, relative comparisons of biomass and FAME profiles within a single study are generally robust.

The sensitivity of FAME biomarkers has been attributed to their response to short-term changes in the environment (Pinkart et al., 2002; White et al., 1996; Zelles, 1997). This is advantageous for monitoring and assessing the effects of seasonal changes and management practices on microbial community structure in agroecosystems; however, this requires adherence to strict FAME sample handling and storage protocols. Recommendations call for analysis of fresh soil, freezer storage until analysis, or lyophilization followed by freezer storage (Petersen & Klug, 1994; Schnecker, Wild, Fuchslueger, & Richter, 2012; White, Davis, Nickels, King, & Bobbie, 1979). Lack of proper sample handling, such as allowing a sample to air-dry or long-term moist storage, causes dramatic shifts in total FAME biomarkers that will not reflect the microbial community composition at time of sampling (Lee, Lorenz, Dick, & Dick, 2007; Schnecker et al., 2012; Veum, Lorenz, & Kremer, 2019). Realistically, it may not be practical for landowners to adhere to these handling and storage requirements, inducing artefacts that could be erroneously interpreted as effects related to land use or management. Overall, appropriate sampling and handling requirements can add significant cost to the analysis, and this should be carefully considered when planning for FAME analysis to produce reliable data.

## Selected Methods: PLFA and EL-FAME

For PLFA, the high throughput method of Buyer and Sasser (2012) has been selected due to low cost, ease of set-up, and pre-packaged peak assignments and interpretation using the Microbial IDentification Inc. (MIDI) Sherlock software (MIDI Corp, Newark, DE). This method relies on gas chromatography (GC) and does not require confirmation by mass spectrometry (MS), but for research purposes where confirmation of peak assignments is desired, downstream MS confirmation can be incorporated. Although the protocol presented here does not include the neutral lipid fraction, modifications to collect and identify this fraction will more reliably identify the biomarker for AMF (Sharma & Buyer, 2015). This protocol differs from that presented by Quideau et al. (2016) in that the Buyer and Sasser (2012) method uses a phosphate buffer instead of a citrate buffer and it has been scaled down and streamlined for high-throughput extraction. For EL-FAME, the economical and rapid method described in Schutter and Dick (2000) has been selected. The reduced cost and increased throughput compared to other method variants are primarily due to the first step that includes in-situ alkaline hydrolysis and methylation of the fatty acids in the soil sample without the phospholipid extraction steps. The primary differences in the extraction procedures for PLFA and EL-FAME are displayed in Figure 12.1. The EL-FAME protocol differs from its MIDI-FAME predecessor in that it uses a milder alkaline solution for hydrolysis, whereas MIDI-FAME uses harsher solvents at a higher pH. The high pH

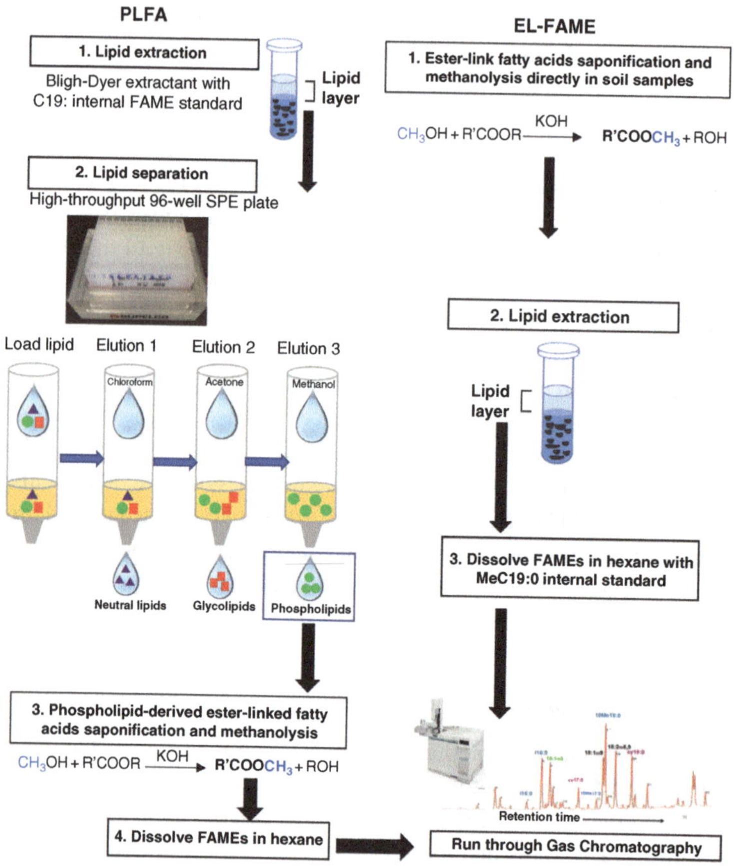


**Figure 12.1** The comparison of extraction procedures for phospholipid fatty acid (PLFA) and ester-linked fatty acid methyl ester (EL-FAME) as reported in Li, Cano, Acosta-Martinez, Veum, and Moore-Kucera (2020).

associated with the MIDI-FAME method is thought to cause loss of FAMEs during saponification and following incubation (Schutter & Dick, 2000).

## Advantages and Disadvantages of PLFA and EL-FAME Protocols

Various protocols are available for extraction of PLFA and EL-FAME from soil, as well as multiple methods for separation and identification by GC, with or without downstream MS confirmation. It should be noted that the methodology presented in this chapter relies solely on GC peak identification and does not include the recommended downstream peak confirmation by MS. The described PLFA protocol does not include the analysis of neutral lipid fatty acids (NLFA) for the most effective quantification of the AMF biomarker (Olsson, 1999; Sharma & Buyer, 2015). By using the EL-FAME method, the PLFA, glycolipids, and NLFA are collectively extracted, resulting in approximately double the yield of FAMEs relative to PLFA alone (Li et al., 2020). In contrast, during the PLFA extraction procedure, glycolipids and NLFA must be collected and identified separately. Similar to PLFA, EL-FAMEs can be identified using the MIDI-Sherlock software. Overall, each FAME profiling method is subject to potential interferences and tradeoffs with respect to biomarker extraction and purification (Li et al., 2020; Warren, 2019).

The Buyer and Sasser (2012) high-throughput PLFA method was modified from the Bligh and Dyer extraction (Bligh & Dyer, 1959; Frostegard, Tunlid, & Baath, 1993) and was scaled down to reduce sample size, consumable costs, time, and production of hazardous waste. The 2-d extraction procedure is followed by a rapid GC separation that allows for sample processing in approximately 15 min with automated peak identification and assignment by the MIDI-Sherlock software. This contracted GC run time results in a loss of peak resolution in comparison with longer run times (e.g., 45 min) and therefore, some peaks such as cis or trans isomers, are not resolved. Alternatively, if a modified GC run time is desired, peaks can be identified manually using standard FAMEs available from chemical suppliers. However, when time is valuable, the faster GC time for PLFA or EL-FAME increases throughput threefold and does not require expert knowledge for peak identification or microbial group assignment. Furthermore, use of the standardized and automated protocol allows for aggregation of data across laboratories and facilitates development of a national database. While sacrificing peak resolution, the selected high-throughput method is faster, cheaper, and better suited to service labs, despite the compromise on quality and quantity of data. As a result, the high-throughput method has been widely adopted and it has helped move FAME analysis from the research realm into service laboratories, where it is now available to producers and the public.

A comparison of PLFA and EL-FAME is provided in Table 12.1. Differences between the PLFA and EL-FAME methods include a longer soil residence time for the biomarkers identified with EL-FAME, leading to the conclusion that PLFA is more representative of the living and active microbial biomass than EL-FAME (Zelles, 1999). However, the turn-over (i.e., degradation) of FAMEs in the soil following cell death depends on several factors (Frostegård et al., 2011). Furthermore, due to the stronger, in-situ extraction conditions with EL-FAME, markers are more likely to originate from non-target lipids such as soil organic matter and plant residues, and thus the PLFA extraction is considered more selective for microbial biomarkers (Fernandes, Saxena, & Dick, 2013; Li et al., 2020; Zelles, 1999). From a more practical perspective, the PLFA extraction procedure is currently longer, more complicated, and more expensive on a per-sample basis than EL-FAME.

**Table 12.1** Comparison of PLFA and EL-FAME methods.

| | PLFA | EL-FAME |
|---|---|---|
| Source of fatty acids | Only phospholipids unless glycolipids and NLFA are collected and identified | Phospholipid, glycolipid, and NLFA |
| Cellular sources | Phospholipid membrane | Membrane and storage cells |
| Organism sources | Most likely living microbes | Living and dead microbes, soil organic matter, and plant residues |
| Consumable costs per sample (not including capital equipment costs or labor)[a] | High (can be > $5.01) | Low (can be < $2.48) |
| Labor and time for 96 samples | 2d extraction, 1d GC analysis | 1d extraction, 1d GC analysis |
| Extraction pH | 7.4 | 13.3 |
| Soil preparation | Freeze dried or fresh soil (never air-dried or from cold or room temperature storage) | |
| Results | EL-FAME has ~2X total FAME abundance compared with PLFA. Positive correlations between PLFA and EL-FAME for total FAME and microbial groups, including fungi (saprophytic fungi, AMF, and general fungi) and bacterial groups (Gram positive, Gram negative, and actinobacteria). | |
| Key biomarkers[a] | Microbial community composition analysis relies on a similar set of key biomarkers for PLFA and EL-FAME. The most abundant biomarkers based on mole percentage: *i*15:0, 16:0, 10Me16:0, 18:1ω7c; Key biomarkers for discriminating among samples: *i*15:0, 16:0, 10Me16:0, 10Me17:1ω7c, 18:1ω9c, 18:1ω7c, 18:0, cy19:0ω7c; Most sensitive to changes in soil properties: 16:0, 10Me17:1ω7c | |

| | | |
|---|---|---|
| Key biomarkers that are more important in one method than another | Most abundant | |
| | 16:1ω7c, cy19:0ω7c | 16:1ω5c, 18:1ω9c |
| | Key discriminators | |
| | 16:1ω7c | 16:1ω5c, 18:3ω6c |
| | Most sensitive to soil properties | |
| | *i*15:0, 18:1ω7c, cy19:0ω7c | *a*15:0, *i*16:0, 16:1ω5c, 10Me16:0, 10Me17:0, 18:0 |
| Sensitivity for interpretation | Similar trends but PLFA had higher correlations with soil properties such as pH, soil texture, soil organic matter, and active carbon | |

[a] Adapted from Li et al. (2020). EL-FAME = ester-linked fatty acid methyl esters; GC = gas chromatography; NLFA = neutral lipid fatty acids; PLFA = phospholipid fatty acids.

A recent comparison of PLFA and EL-FAME using soil samples from a wide geographic region in the U.S. found that PLFA was more sensitive to changes in soil characteristics than EL-FAME (Li et al., 2020). This result is primarily due to the greater extraction of non-target lipids with the EL-FAME protocol; these non-target lipids may reduce the signal-to-noise ratio for specific fatty acids. However, this same study also found that EL-FAME delivered more information for fungal biomarkers, particularly the AMF biomarker 16:1ω5 which is present in both the PLFA and NLFA fractions extracted with the EL-FAME procedure. Even though the PLFA biomarker 16:1ω5 is frequently assigned to AMF, it may appear in other taxa and has been shown to be an inferior marker for AMF compared with the same fatty acid present in the NLFA fraction (Olsson, 1999; Drijber, Doran, Parkhurst, & Lyon, 2000; Grigera et al., 2007; Lehman, Osborne, Taheri, Buyer, & Chim, 2019; Sharma & Buyer, 2015). The AMF biomarker 16:1ω5 in agricultural soils analyzed with the EL-FAME procedure (from both PLFA and NLFA) showed a stronger response to management than the same biomarker analyzed with the PLFA procedure (Drijber et al., 2000). In a greenhouse study, 16:1ω5 analyzed by both NLFA and EL-FAME were found to be superior biomarkers for AMF compared with the same fatty acid in the PFLA fraction (Sharma & Buyer, 2015). Our data from field trials also indicate that NLFA produces a more reliable AMF biomarker than PLFA (Lehman et al., 2019). The combined NLFA and PLFA biomarkers in EL-FAME may be considered advantageous for quantifying AMF in soils compared to PLFA, but also may dilute more reliable NLFA values by the co-extraction of the PLFA fraction. An alternative is to collect, save, and purify the NFLA fraction when performing the PLFA protocol and analyze it separately by GC to obtain the most informative fatty acid biomarker (16:1ω5) for AMF.

## PLFA Using the Buyer and Sasser (2012) High-throughput Extraction Method Paired with the MIDI-Sherlock System

### List of Supplies and Major Equipment for PLFA

This procedure and supply list was modified from personal communication with Jeff Buyer (USDA-ARS, Beltsville, MD) and details from MIDI Inc. (MIDI, 2014, 2015, 2018). No endorsement is implied by the use of Tradenames, part numbers may change, and functionally equivalent substitutes may be used where appropriate. Confirm all equipment with manufacturer before purchase, particularly capital equipment.

1) 7.5 ml, 13 × 100 mm glassware tubes (2 per sample)
    a) Fisher Scientific part #99447-13; Kimble part #45066A-13100
2) PTFE-lined screw caps for 13 mm tubes
    a) Fisher Scientific part #73802-13415

3) 50 mg SI-1 silica gel SPE 96-well plates
   a) Phenomenex part #8e-s012-dgb (currently no substitutes)
4) 1.5 ml Multi-Tier Micro Plates (Topas) with polypropylene 9 × 44 mm conical vials
   a) E&K Scientific part #EK-99238 (mfr 9915-812FBT)
5) Strata 96-well plate manifold with vacuum gauge
   a) Phenomenex part #AH0-8950 with spacer insert, polypropylene part #AH0-9214
   b) Waste collection plates with appropriate solvent resistance
6) PTFE/silicone molded liner cap mats for 96-well plates
   a) E&K Scientific part #EK-99249 (mfr 996050MR-96)
7) 0.5–5 ml BrandTech Dispensette S bottle-top dispenser with recirculation valve
   a) Brand Tech Scientific part #4600131; Fisher Scientific part #13-689-014
   b) Recommended: flexible discharge tube with recirculation valve part #708132
8) 0.5–5 ml BrandTech Dispensette S organic bottle-top dispenser with recirculation valve
   a) Brand Tech Scientific part #4630131; Fisher Scientific part #13-689-042
   b) Recommended: flexible discharge tube with recirculation valve part #708132
9) Finnpipette 4027/4500 Digital 40–200 µl pipette Millipore Sigma part #S7939 or functionally equivalent
   a) Millipore Sigma Solvent Safe pipette tips part #S7939
10) Repeating pipette BrandTech Handy Step touch (bundle) part #705210
   a) 250 µl–25 ml (25 ml capacity) BrandTech PD tips part #705746
   b) 500 µl–50 ml (50 ml capacity) BrandTech PD tips part #705748
   c) Extra adapters for 25 and 50 ml tips BrandTech part #702399
11) Labquake tube shakers/rotators (must rotate 360°; will need multiple)
   a) Thermo Scientific part #400110Q; Fisher Scientific part #13-687-10Q
12) Digital vortex mixer
   a) Fisher Scientific part #02-215-370
13) Poxygrid test tube racks (72 slots)
   a) Fisher Scientific part #14-793-14
14) Disposable borosilicate glass Pasteur pipettes
   a) Kimble Chase part #63A53; Fisher Scientific part #22-037-514
15) Manual or digital Pasteur pipette filler/controller; several vendors.
16) Lyophilizer (SP Scientific 25L or functionally equivalent).
   a) Fisher Scientific part #14-930-10A; SP Scientific VirTis Freezemobile Freeze Dryer part #25ES with 18-port stainless steel drum manifold
17) SpeedVac and Accessories (Labconco or functionally equivalent)
   a) Labconco CentriVap vacuum concentrator part #7810014 (USA); part #7810032 (Europe); part #7810036 (UK); part #7810040 (Australia/China)

b) Labconco CentriVap −84 °C cold trap; part #7460020 (USA); part #7460030 (Europe); part #7460035 (UK); part #7460037 (Australia/China).
c) Labconco CentriVap acid-resistant 12–13 mm rotor; part #7455101
d) Labconco microtiter plate rotor; part #7461900
e) Labconco rotary vane vacuum pump; part #1472100 (USA); part #7739400 (INTL) **OR** Labconco Vaccubrand hybrid pump; part #7584002 (USA); part #7584000 (INTL)
f) Freeze-dryer vacuum pressure gauge to monitor sample dryness; several vendors.

18) Ultrasonic cleaning bath (Branson or functionally equivalent).
    a) Branson ultrasonic cleaner with digital timer. Emerson Industrial; part #CPN-952-218 (USA); part #CPN-952-238 (INTL).
19) ThermoScientific multichannel pipettes and tips or functionally equivalent
    a) E1-ClipTip Equalizer 8 channel pipette (10–300 μl); part #4672080BT
        i) ClipTip 300 Ext (10–300 μl) racked pipet tips; part #94410610
    b) E1-ClipTip Equalizer 8 channel pipette (15–1,250 μl); part #4672100BT
        i) ClipTip 1,250 (15–1,250 μl) racked pipet tips; part #94410810
20) a) Properly equipped GC with a flame ionization detector: Agilent 6850, 6890, 7820, or 7890 series with Agilent ChemStation software
        - Column 50 for 6850 GCs
        - Column 90 for 6890, 7820, and 7890 GCs
        - Appropriate injection port liners (1221 or 1221-F)

    **OR**

    b) Shimadzu GC-2010/2030 with LabSolutions software
        - Column J&W Ultra 2, 25 m × 0.2 mm × 0.33 μm film thickness
        - Split liner for focusing; Shimadzu part #220-94766-00
21) 2 ml autosampler GC vials
    a) Thomas Scientific part #2702-A01; Thermo part #C4000-1; Fisher part #03-377A
22) 250 μl limited volume inserts for GC vials
    a) Agilent part #5183-2085; Thermo part #C4010630; Fisher part #03-375-3A
23) GC vial screw caps with PTFE/silicone/PTFE septa
    a) Thomas Scientific part #2702-A68; Restek part #24498; Fisher part #06-718-916
24) Appropriate GC gases and CGA components or generators: hydrogen carrier gas (99.999% purity or generator); nitrogen make-up gas (99.999% purity or generator); industrial grade air or zero air generator.

## Chemicals

Potassium phosphate dibasic ($K_2HPO_4$), ACS reagent grade
Potassium hydroxide pellets (KOH), ACS reagent grade
6 N hydrochloric acid (HCl), ACS reagent grade
Glacial acetic acid, HPLC grade
Methanol, HPLC grade
Chloroform, HPLC grade
Acetone, HPLC grade
Toluene, HPLC grade
Hexanes, HPLC grade
Internal standard: 1,2-dinonadecanoyl-sn-glycero-3-phosphocholine. Avanti Polar Lipids Catalog # 850367P (white powder)

## Reagents

### Internal Standard Solution

Dissolve 40.9 mg 1,2-dinonadecanoyl-*sn*-glycero-3-phosphocholine in 20 ml of 1:1 chloroform:methanol (2.5 mM solution). Also used to assess overall recovery.

Store at −20 °C in aliquots of 1 ml to prevent multiple freeze–thaw cycles. Seal with Teflon tape and secondary containment to prevent evaporation.

### Phosphate Buffer (50 mM)

Dissolve 8.7 g $K_2HPO_4$ dibasic per liter deionized water. Adjust pH with 6 N HCl to pH 7.4.

### Bligh-Dyer Extractant (enough for ~2 × 96 well plates)

Mix 200 ml of phosphate buffer, 500 ml methanol, 250 ml chloroform, and 475 µl of Internal Standard Solution (at room temperature). Mix well.

Add the internal standard solution to the Bligh-Dyer extractant at a rate of 0.5 µl $ml^{-1}$ of extractant and mix. This is equivalent to adding 10 nmoles of 19:0.

Mix fresh daily, or at least weekly if many runs are anticipated. Excess may be needed to account for solution remaining in the bottle. Store in dedicated organic dispensette.

### Transesterification Reagent

Dissolve 0.561 g KOH in 75 ml methanol. Pellets will dissolve slowly. Use a deep-walled, flat-bottomed dish with a wide diameter to accommodate the eight-channel pipette. Add 25 ml toluene after KOH has completely dissolved. Prepare weekly.

**5:5:1 methanol/chloroform/$H_2O$**

Need 0.5 ml per sample; make fresh daily.

Example: 50 ml methanol, 50 ml chloroform, and 10 ml ultrapure water.

**0.075 M acetic acid**

For 500 ml, add 2.147 ml glacial acetic acid to 400 ml deionized water, and QS to 500 ml in a volumetric flask.

## Standard Operating Procedure for PLFA Extraction

### Freeze dry soil

1) Weigh pre-labeled 13 × 100 mm screw-cap glass test tubes without caps on and record tare weight (+/−0.01 g). Note that some inks are easily removed from the test tubes by the solvents in the extraction, and adhesive labels will interfere with fitting the tubes in the centrifuge rotor.
2) Add 1.5–2.0 g moist soil to each test tube, ensuring no soil is left in threads at top of tube—this will cause tube to leak. Weigh at the high end for low organic matter soils and at the low end for rich, high organic matter soils.
3) Freeze dry soil
   a) If using a lyophilizer (preferred), cap and freeze test tubes first (−20 °C) then loosen caps and freeze dry. Some lyophilizers have a built-in pre-freeze step. Lyophilization is preferred as tubes may occasionally break in SpeedVac during this step.
   b) If using a SpeedVac (Labconco CentriVap acid-resistant 12–13 mm rotor; part #7455101), run overnight at room temperature (no caps).
4) Weigh tubes + dry soil (no cap) to calculate the dry weight of soil in each sample.
5) Samples can be stored after lyophilization and before extraction, by capping each tube with Teflon-lined screw cap and storing at −20 °C. If working with previously lyophilized soils, simply weigh 1–2 g of dry soil into labeled test tubes. If moving directly to extraction step, no need to cap tubes yet.

### Extraction

1) For each 96-well plate, assign a minimum of one blank (reagents, no soil) and create a plate map with sample IDs to match test tube IDs (numbered 1–96).
2) Add 4 ml of Bligh-Dyer Extractant containing internal standard to each test tube using the dedicated organic dispensette, then cap each tube and invert.
3) Sonicate the rack(s) of tubes for 10 min in an ultrasonic cleaning bath at room temperature. Spread samples out in racks so not crowd the sonicator and make sure the water level is just above the sample level.
4) Incubate at room temperature with slow, end-over-end shaking for 2 hours (Labquake tube shaker/rotator; Figure 12.2). Check that soil is properly mixing

Figure 12.2 Soil samples on end-over-end shaker.

with extractant and that tubes are not leaking. Alternatively, you can vortex each tube individually, sonicate for 10 min, and repeat vortexing and sonication step 2 times.

5) Check tubes to see if soil is stuck at the top, and gently shake soil down before centrifugation. Centrifuge with caps for 10 min in SpeedVac without vacuum (Labconco CentriVap acid-resistant 12–13 mm rotor part #7455101). Only half of the samples will fit in the rotor with caps on, so complete in two sets.
6) Transfer liquid phase to new, labeled 13 × 100 test tube by carefully decanting/pouring (Fisher Scientific part #14-930-10A). Minimize soil transfer, but some transfer is OK. Dispose of original test tubes with soil or set tubes aside for washing if intended for reuse. Keep tubes uncapped for next step.

**Separation**

1) Add 1 ml each of chloroform to each test tube using dedicated organic dispensette.
2) Add 1 ml of deionized water to each test tube using Handy Step dispenser or dispensette.
3) Cap with new PTFE-lined screw cap and vortex 10 seconds then centrifuge 10 min in SpeedVac with caps and without vacuum (Labconco CentriVap acid-resistant 12–13 mm rotor part #7455101). Only half of the samples will fit in the rotor with caps on, so complete in two sets (Figure 12.3).

Aspirate top (aqueous) phase using disposable borosilicate glass Pasteur pipettes attached to a vacuum system with a collection vessel (Figure 12.4). If the pipette tip crosses the interface, the pipette must be replaced.

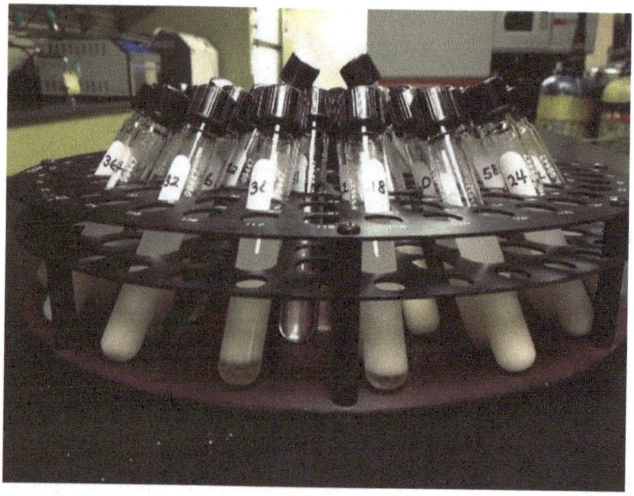

**Figure 12.3** Capped samples in SpeedVac rotor using one half of the spaces.

**Figure 12.4** Aspirating aqueous layer from top of each sample.

4) SpeedVac to dryness (~500 μm; personal communication, Jeffrey Buyer, 2014) in test tubes with vacuum and without caps at 30 °C (~1 hr). Set caps aside in order so that they can be returned to the proper tube and not cause cross-contamination. If necessary, you can stop here and save samples for the next day (capped and stored at −20 °C). Note that freezer storage of samples containing flammable solvents is hazardous, so store samples only after solvents have been evaporated.
5) Add 1 ml chloroform from a dedicated organic dispensette to each test tube to dissolve the sample extract for chromatography.

**Chromatography**

1) Mount a 50 mg silica gel SPE 96-well plate (Phenomenex part #8E-S012-DGB; currently no substitutes) on the vacuum manifold with appropriate solvent-resistant waste collection tray underneath. Follow your plate map to ensure samples end up in the proper well.
2) Wash each well of the SPE plate three times with 1 ml methanol, delivered from dedicated dispensette (organic or standard) and using the manifold and a gentle vacuum after the final wash. Make sure the collection tray (with appropriate solvent resistance) is emptied as needed in this and the following steps. Allow methanol to dry completely before moving to next step.
3) Wash each well of the SPE plate three times with 1 ml chloroform delivered from dedicated organic dispensette, using the manifold and a gentle vacuum after the final wash. Make sure collection tray is emptied as needed. This conditions the SPE plate.
4) Transfer sample extract to the appropriate well by pouring from the test tube carefully into the well. This takes some practice to avoid spilling the sample. Some soil may transfer.
5) Let samples drain into SPE columns without vacuum.
6) Repeat transfer by rinsing the tube with 1 ml chloroform from the dedicated organic dispensette (swirl then pour into the proper well), allowing sample to drain into column followed by a gentle vacuum. This is to optimize sample transfer.
7) Wash each well in the SPE plate with 1 ml chloroform from dedicated organic dispensette and with gentle vacuum (Figure 12.5).
8) Wash each well in the SPE plate with 1 ml acetone from dedicated dispensette (organic or standard) and with gentle vacuum. Empty collection trough.
9) Place clean 1.5 ml multi-tier plate with glass flat bottom vials (E&K Scientific part #EK-99238) in bottom of 96-well plate manifold (Phenomenex part #AH0-8950) to collect lipids. Situate SPE plate on top.
10) Elute phospholipids with 0.5 ml of 5:5:1 methanol/chloroform/$H_2O$ into glass vials using the multichannel pipette.

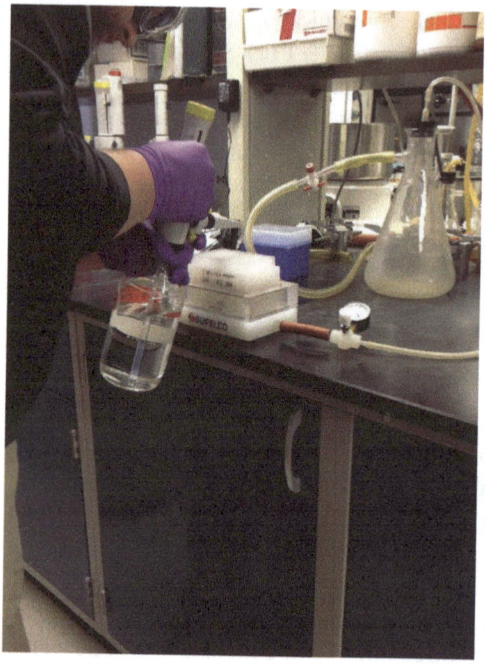

**Figure 12.5** Washing wells with solvents using dedicated dispensette.

11) Balance the plates for centrifugation. SpeedVac the plates at 70 °C for 30 min, then 37 °C until dry (500 μm), ~1.5–2 hr total. The two-stage temperature ramp is necessary to eliminate the water (Labconco microtiter plate rotor part #7461900).

**Transesterification and Transfer to GC vials**

1) Add 0.2 ml transesterification reagent to each vial using the multichannel pipette and cover each plate rack tightly with a cap mat (E & K Scientific part #EK-99249) and mix/shake by hand. Mark cap mats so orientation is preserved and cross-contamination is avoided when mats are used in following steps.
2) Incubate racks with cap mats (E & K Scientific part #EK-99249) on at 37 °C for 15 min in SpeedVac (without vacuum or centrifugation) or in an incubator.
3) Add 0.4 ml of 0.075 M acetic acid to each vial using the repeating pipette (BrandTech Handy Step Touch part #705210 and 25 ml PD tips; part #705746 with adapter) or a dedicated dispensette (standard or organic).
4) Add 0.4 ml chloroform to each vial using dedicated organic dispensette.
5) Reseal each plate rack with designated cap mat (E & K Scientific part #EK-99249) and very carefully shake vigorously by hand, then let stand a few minutes while phases separate.

6) Transfer the **bottom** 300 μl of sample to a 1 ml multi-tier plate (E & K Scientific part #EK-99234) using the multichannel pipettor and the 300 μl extended (Ext) ClipTips. Fresh tips must be used for each sample. The surface tension of the sample will result in the pipette tips dripping, so transfer quickly without cross-contaminating wells. The tall/extended tips are necessary to reach the bottom of the wells without causing the sample to overflow the vial.
7) Repeat the shaking step by adding another 0.4 ml chloroform, capping tightly, and shaking by hand. Allow samples to settle and let the phases separate for a few minutes.
8) Transfer the **bottom** 200 μl of sample twice, for a total transfer of 400 μl in this step, using the multichannel pipette and the 300 μl extended (Ext) ClipTips, into the same 1 ml multi-tier plate. Use fresh tips for each sample.
9) If any aqueous phase was accidentally transferred (seen on top of the transferred chloroform), remove with clean disposable Pasteur pipets.
10) Evaporate the chloroform in SpeedVac under vacuum at room temperature and remove as soon as dry (500 μm), ~30–45 min. (Labconco microtiter plate rotor; part #7461900.)
11) Dissolve extract by adding 75 μl hexane to each vial, one at a time using the 200 μl pipette with Solvent Safe (1–200 μl) tips.
12) Quantitatively transfer each sample with a fresh, disposable Pasteur pipette using a pipette controller, into the limited-volume insert (with conical bottom) inside a labeled GC vial, and screw caps with PTFE/silicone/PTFE septa (250 μl limited volume insert; Agilent part #5183–2085; 2 ml autosampler GC vial; Thomas Scientific part #2702-A01; Screw cap with PTFE/silicone/PTFE septa; Thomas Scientific part #2702-A68) and analyze on the GC following MIDI-Sherlock PLFA protocols.

## EL-FAME Extraction Paired with the MIDI-Sherlock System

### List of Supplies and Major equipment for EL-FAME method

1) 9 ml, 13 × 100 mm Reusable glass tubes with plain end (1 per sample)
   a) Fisher Scientific part #14-957C
2) PTFE-lined screw caps for 13 mm tubes
   a) Fisher Scientific, part #14-933A
3) 25 ml, 20 × 150 mm Reusable glassware tubes with rubber-lined phenolic caps (1 per sample)
   a) Fisher Scientific part #14-932C
4) 5.75 in., Disposable glass, non-sterile, unplugged Pasteur pipets (3 per sample)
   a) Corning, part #7095B-5X

5) Vortex mixer
6) Techne sample concentrator
   a) Cole Parmer part #UX-36620-40
7) VWR Water bath
8) Centrifuge
   a) Fisherbrand
9) Test tube racks
   a) For 13 mm glass tubes, Fisher Scientific part #14-793-14
   b) For 20 mm glass tubes, Fisher Scientific part #14-809-26
10) BrandTech Dispensette bottletop dispensers
   a) Fisher Scientific part #18901-128
11) 2 ml Autosampler GC vials
   a) Fisher Scientific part #50-635-160
12) 250 μl Limited volume inserts for GC vials
   a) Fisher Scientific part #C4010-627L
13) GC Vial screw caps with PTFE/silicone/PTFE septa
   a) Fisher Scientific part #03-376-480
14) GC Screw top vial racks for 50 vials
   a) Thermo Scientific part #03-375-9
15) Properly equipped GC with a flame ionization detector
   a) Agilent 6850 or single channel 7890 Series GC with Agilent ChemStation software
      i) Column 50 for 6850 GCs
      ii) Column 90 for 6890, 7820, and 7890 GCs
      iii) Appropriate injection port liners

   OR

   b) Shimadzu GC-2010/2030 with LabSolutions software
      i) Column J&W Ultra 2, 25 m × 0.2 mm × 0.33 μm film thickness
      ii) Split liner for focusing; Shimadzu part #220-94766-00

## Reagents

Methylation reagent
0.2 M KOH in Methanol

- While stirring 500 ml methanol (HPLC grade), add 5.61 g KOH pellets (certified ACS), stir until dissolved.
- Make fresh for each extraction

Neutralization reagent
1.0 M Acetic acid

- Fill a 1-L volumetric flask with ~500 ml DI water and add 57.5 ml of glacial acetic acid (certified ACS).
- Bring volume to 1 L with DI water, stir to mix.

Extraction solvent

- Enough HPLC grade hexane for 3 ml per sample

Standard redissolve solution

- Hexane/MTBE (1:1 HPLC grade hexanes mixed with HPLC grade Methyl-tert-butyl ether)
- Internal standard 0.64 mM C19:0 (Methyl nonadecanoic acid)
  - Dissolve 0.20 μg C19:0 in 1 μl Hexane/MTBE
  - To make a working stock solution use 1 mg C 19:0 in 5 ml Hexane/MTBE and store at −20 °C.

## Standard Operating Procedure for EL-FAME Extraction

### Methylation

1) Add 3 g · field moist equivalent soil to an appropriately-sized (usually 20 × 150 mm) teflon lined screw-cap test tube.
2) Add 15 ml of methylation reagent and vortex 10 s.
3) Heat in a 37 °C water bath for 60 min. Total, vortexing 5 s every 15 min (4 times total).
4) Cool at ambient temp (~5 min).

### Neutralization

1) Add 3 ml of neutralization reagent and vortex 5 s.
2) Let samples sit at ambient temp for reaction to fully occur (~5 min).

### Extraction

1) Add 3 ml of extraction solvent.
2) Invert tubes by hand five times.
3) Centrifuge at 2200 rpm for 10 min.
4) Carefully transfer the top organic phase to a 13 × 100 mm teflon lined screw-cap test tube.

### FAME Concentration in N Evaporation Chamber

1) Place the tubes with the organic phase under nitrogen flow (20 psi) until complete dryness
2) Add 100 ml of standard redissolve solution by pipette to each tube, cap, and vortex to allow all concentrate to go into solution.

3) Using Pasteur pipette (new for each sample), transfer all of each sample to a new GC vial containing a micro-insert, then immediately cap each vial (make sure to tightly cap each vial to prevent sample loss due to evaporation).
4) Samples can now be analyzed by GC.

## Tips and Tricks for FAME Extraction and Analysis

- Other general laboratory requirements include climate control for proper functioning of sensitive instrumentation, appropriate personal protective equipment for personnel, standard laboratory glassware for analytical chemistry, including volumetric flasks of various sizes, an analytical balance, a source of deionized water (ideally double deionized water, 18 MΩ-cm, or better), a pH meter, a house vacuum system, sufficient power supply, freezer storage for samples, fume hoods, proper storage for solvents/flammables and corrosives/acids, waste containers, and disposal protocols for safe handling of wastes.
- Sample handling prior to lyophilization for FAMEs is very important. Ideally, keep soil cold using blue ice in a cooler during field transport, followed by immediate lyophilization or freezer/cryogenic storage (−20 to −80 °C freezer) to preserve sample integrity. Samples should not be air-dried or stored at room temperature or in refrigerators.
- Lipids can spread across the surface of wet glass, so gloves are necessary to avoid contamination from your skin, but latex gloves should not be used due to potential contamination issues. Unpowdered nitrile gloves are suitable.
- Do not use sealing tape to cover microplates. The adhesives can cause contamination.
- Cleaning reusable labware: All glassware should be scrubbed carefully with detergent and thoroughly rinsed with deionized water (3×) while wearing gloves. An ultrasonic cleaning bath is helpful. If possible, bake non-volumetric glassware at 400–500 °C for 2 h in a dedicated muffle furnace (volumetric glassware can be damaged at these temperatures). Cap mats should be gently scrubbed with soap and water, then rinsed sequentially with DI water, methanol, and chloroform, then dried in a fume hood. If reusing screw caps, they should be shaken with hexane in a large test tube. Make sure the PTFE liner is still in place before reusing the caps.
- Evaporation under nitrogen gas is recommended for procedures where larger volumes of solvent are evaporated to prevent oxidation of the sample. The high-throughput PLFA protocol does not require evaporation under nitrogen due to small sample size and rapid evaporation, but this can be incorporated into the procedure if desired.

- Analysis of high organic matter soils or soil with amendments, such as composts or biochars, present challenges with absorption of extraction solvents. You may have to adjust your sample mass via trial and error to address this during the extraction step.
- Problems can occur at both low and high FAME concentrations. With low concentration samples, there will be limit of detection issues (minimum peak size identified by GC). This is a common problem with deep soil profile samples where organic matter and microbial biomass are very low; this problem can be compounded by interference from root lipids mistaken for microbial markers. For high concentration samples, the GC may experience peak overload, swamping the detector cell in the GC and requiring sample dilution with additional hexanes and reanalysis. Soil sample mass can be adjusted to address these problems, but some samples are not amenable to this procedure.
- Quality assurance and quality control (QA/QC) are essential components of any laboratory process, and each laboratory must decide how to implement a QA/QC program depending on labor, cost, and expectations. In this procedure, each sample has a known amount of added internal standard added prior to PLFA extraction to account for the multiple, non-quantitative transfer steps in the procedure. Some protocols may call for a second internal standard to be prepared and added to each sample just prior to GC analysis to assess GC recovery. Protocols may also include calibration mixes of fatty acids for additional QA/QC information. Furthermore, it is wise to allocate one or more wells in each batch as sample blanks (reagents but no soil) and hexane blanks to monitor extraction and GC performance. In addition, sample spikes or check standards (see MIDI standards) can be utilized. Depending on the QA/QC program of the laboratory, sample duplicates can be considered or a reference soil can be homogenized and preserved at −80 °C for use as a check standard.

## Calculations and Interpretation of FAME Data

The Sherlock PLFA Analysis System was developed in conjunction with the Buyer and Sasser (2012) high-throughput PLFA or the EL-FAME extraction method to automatically identify each biomarker peak and assign each peak to its respective microbial or fat type groups. Multiple GC and GC–MS methods and instrument types can be used in addition to the systems described above; however, if the prescribed GC settings and automated software system are not used, an experienced biochemist must carefully identify biomarker peaks based on retention times developed from known standards before peak assignments can be confirmed.

Results for each individual peak, each microbial group, and the sum of all peaks are typically reported in nanomoles per gram of dry soil (nmol/g) following corrections based on the internal standard (19:0 FAME) as well as molecular weight conversions. Results are also reported as molar percentages (mole %) for individual peaks and microbial groups as a proportion of the total sample. Common calculations include total fungi represented by the sum of saprophytic fungi and AMF. Total bacteria is calculated as the sum of markers assigned to Gram negative bacteria, Gram positive bacteria, and Actinobacteria. Microbial group peak assignments can be found in Table 12.2.

Ratios of microbial groups or specific biomarkers are commonly used to evaluate shifts in community structure. These ratios include the fungi to bacteria ratio (Bardgett & McAlister, 1999), which has been used to describe the flow of energy and nutrients through the microbial community (Bardgett & McAlister, 1999; Frostegård & Bååth, 1996; Parekh & Bardgett, 2002), where lower fungal biomass has been associated with increased soil disturbance from tillage (Drijber et al., 2000; Frey, Elliott, & Paustian, 1999; Stevenson, Hunter, & Rhodes, 2014) and decreased soil C:N ratios (Bossuyt et al., 2001). Overall, an increase in the ratio of fungi to bacteria suggests improved soil health based on studies of soil restoration practices such as reduced soil disturbance or perennialization such as conservation reserve program systems (e.g., Acosta-Martínez, Bell, Morris, Zak, & Allen, 2010; Ananyeva, Castaldi, Stolnikova, Kudeyarov, & Valentini, 2015; Bardgett & McAlister, 1999; Li, Fultz, Moore-Kucera, Acosta-Martínez, & Kakarla, 2018). The Gram negative to Gram positive ratio (Guckert, Hood, & White, 1986; Keynan & Sandler, 1983; Pennanen, Frostegard, Fritze, & Baath, 1996) may indicate stressful environmental conditions (e.g., low oxygen, suboptimal pH, water stress, or low nutrient supply). These shifts have been attributed to enhanced robustness of Gram positive bacteria due to an ability to form spores, or to the enhanced resilience of Gram negative bacteria due to cyclo-fatty acids that allow them to withstand harsh environmental conditions (Guckert et al., 1986; Keynan & Sandler, 1983; Pennanen et al., 1996). Metabolic status and "stress ratios" are also potentially useful ways to interpret shifts in PLFA profiles. During active metabolism, high levels of monounsaturated fatty acids (e.g., 16:1ω7 and 18:1ω7) are produced by Gram negative bacteria, but these are converted to cyclopropane fatty acids (e.g., 17:0 cyclopropane and 19:0 cyclopropane) when metabolism slows down. This shift has been attributed to a range of environmental stressors, where higher proportions of the cyclopropanes suggest greater stress (Kaur, Chaudhary, Kaur, Choudhary, & Kaushik, 2005; Moore-Kucera & Dick, 2008; Petersen & Klug, 1994; Villanueva, Navarette, Urmeneta, White, & Guerrero, 2004). The stress ratio is calculated as (cy17:0 + cy19:0)/(16:1ω7c + 18:1ω7c) following Petersen and Klug (1994). Overall, interpretation of these ratios is largely based on correlations with other variables or on observed trends. It is generally not possible

**Table 12.2** Key PLFA microbial group assignments used by the MIDI Sherlock System. Markers for anaerobes, methanotrophs, and general assignment are not presented here.

| Microbial group | Markers | | | |
|---|---|---|---|---|
| AM fungi | 16:1 ω5c | | | |
| Fungi | 18:2 ω6c | | | |
| Gram negative | 10:0 2OH | 10:0 3OH | 12:1 ω8c | 12:1 ω5c |
| | 13:1 ω5c | 13:1 ω4c | 13:1 ω3c | 12:0 2OH |
| | 14:1 ω9c | 14:1 ω8C | 14:1 ω7c | 14:1 ω5c |
| | 15:1 ω9c | 15:1 ω8c | 15:1 ω7c | 15:1 ω6c |
| | 15:1 ω5c | 14:0 2OH | 16:1 ω9c | 16:1 ω7c |
| | 16:1 ω6c | 16:1 ω4c | 16:1 ω3c | 17:1 ω9c |
| | 17:1 ω7c | 17:1 ω6c | 17:0 cyclo ω7c | 17:1 ω5c |
| | 17:1 ω4c | 17:1 ω3c | 16:0 2OH | 17:1 ω8c |
| | 18:1 ω8c | 18:1 ω7c | 18:1 ω6c | 18:1 ω5c |
| | 18:1 ω3c | 19:1 ω9c | 19:1 ω8c | 19:1 ω7c |
| | 19:1 ω6c | 19:0 cyclo ω9c | 19:0 cyclo ω7c | 19:0 cyclo ω6c |
| | 20:1 ω9c | 20:1 ω8c | 20:1 ω6c | 20:1 ω4c |
| | 21:1 ω9c | 20:0 cyclo ω6c | 21:1 ω9c | 21:1 ω8c |
| | 21:1 ω6c | 21:1 ω5c | 21:1 ω4c | 21:1 ω3c |
| | 22:1 ω9c | 22:1 ω8c | 22:1 ω6c | 22:1 ω5c |
| | 22:1 ω3c | 22:0 cyclo ω6c | 24:1 ω9c | 24:1 ω7c |
| Eukaryote | 15:4 ω3c | 15:3 ω3c | 16:4 ω3c | 16:3 ω6c |
| | 18:3 ω6c | 19:4 ω6c | 19:3 ω6c | 19:3 ω3c |
| | 20:4 ω6c | 20:5 ω3c | 20:3 ω6c | 20:2 ω6c |
| | 21:3 ω6c | 21:3 ω3c | 22:5 ω6c | 22:6 ω3c |
| | 22:4 ω6c | 22:5 ω3c | 22:2 ω6c | 23:4 ω6c |
| | 23:3 ω6c | 23:3 ω3c | 23:1 ω5c | 23:1 ω4c |
| | 24:4 ω6c | 24:3 ω6c | 24:3 ω3c | 24:1 ω3c |
| Gram positive | 11:0 iso | 11:0 anteiso | 12:0 iso | 12:0 anteiso |
| | 13:0 iso | 13:0 anteiso | 14:1 iso ω7c | 14:0 iso |
| | 14:0 anteiso | 15:1 iso ω9c | 15:1 iso ω6c | 15:1 anteiso ω9c |
| | 15:0 iso | 15:0 anteiso | 16:0 iso | 16:0 anteiso |
| | 17:1 iso ω9c | 17:0 iso | 17:0 anteiso | 18:0 iso |

*(Continued)*

**Table 12.2** (Continued)

| Microbial group | Markers | | | |
|---|---|---|---|---|
| | 19:0 iso | 19:0 anteiso | 20:0 iso | 22:0 iso |
| Actinobacteria | 16:0 10-methyl | 17:1 ω7c 10-methyl | 17:0 10-methyl | 20:0 10-methyl |
| (Actinomycetes) | 18:1 ω7c 10-methyl | 18:0 10-methyl | 19:1 ω7c 10-methyl | |

to distinguish between changes in FAME profiles due to shifts in species composition resulting from changes in cell membranes or physiological status induced by environmental conditions. However, FAME ratios are still useful indicators of changes in microbial community structure (Frostegård et al., 2011).

The Sherlock software is customizable, and microbial community calculations can be modified if the desired calculation is not already provided. In addition, dendrogram plots, neighbor-joining trees, principal component analysis (PCA), and histograms are some of the options available within the software program. Furthermore, the data can be output into multiple standard formats and analyzed with any appropriate statistical approach based on the experimental design of the study, and limited only by the creativity and skillset of the analyst.

## Management Implications

PLFA and EL-FAME methods provide an estimate of the "active" or living microbial biomass in the soil (Pinkart et al., 2002; White et al., 1996) and reflect broad land use changes as well as the effects of specific management practices within general agricultural production systems. Management practices that increase availability and diversity of microbial substrates and improve microbial habitat, such as improved organic matter content, soil structure, and pore space (Lehman, Acosta-Martinez, et al., 2015, Lehman, Cambardella et al., 2015), are expected to result in greater FAME content. For cropping systems, this includes increased soil cover via cover crops or perennial systems, residue retention, extended crop rotations, reduced soil disturbance, use of soil amendments, or management that reduces soil erosion. In contrast, long-term intensive tillage practices (e.g., plowing, disking, and weeding) are well known to decrease soil organic matter and impact biological soil properties and the soil microbial community (Acosta-Martínez, Klose, & Zobeck, 2003; Doran, Fraser, Culik, & Liebhardt, 1987; Lal, 2005; Nunes, Karlen, Veum, Moorman, & Cambardella, 2020; Veum et al., 2015; Veum, Goyne, Kremer, Miles, & Sudduth, 2014).

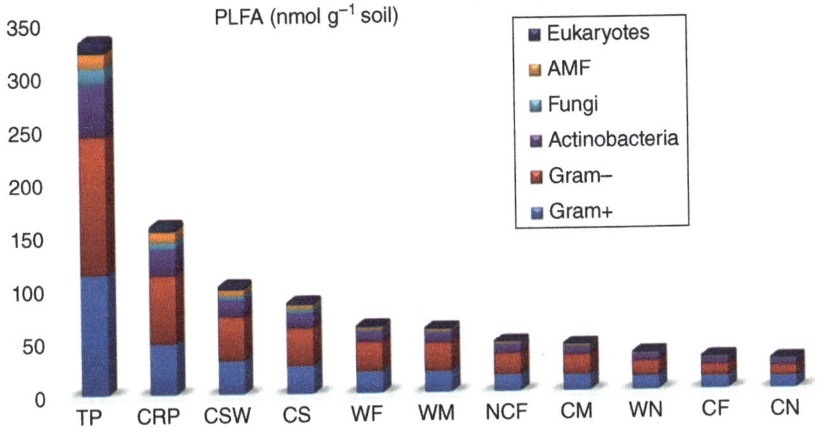


**Figure 12.6** Total phospholipid fatty acids and by microbial group for native prairie (TP); Conservation Reserve Program (CRP); diversified corn-soybean-wheat system with cover crops (CSW); mulch-till, corn-soybean rotation (CS); continuous wheat with inorganic fertilizer (WF) or manure (WM); no-till continuous corn with inorganic fertilizer (NCF); continuous corn with manure (CM); continuous wheat with no amendments (WN); continuous corn with inorganic fertilizer (CF); continuous corn with no amendments (CN). Data from Veum et al. (2018).

For example, a 2008 study at the USDA Agricultural Research Service's Long-Term Agroecosystem Research site in Columbia, MO, compared PLFA profiles from the surface soil layer (0–15 cm) from long-term cropping systems and an uncultivated, remnant tallgrass prairie (Veum, Lorenz, & Kremer, 2018). This comparison illustrated the long-term, positive impact of perennialization over annual row cropping systems, reduced tillage over conventional inversion tillage, inclusion of cover crops, extended crop rotation diversity, and the addition of organic soil amendments (e.g., manure) on total PLFA microbial biomass. Similar biomass trends across management systems were observed for all microbial groups (Figure 12.6). Although total microbial biomass declined dramatically as a result of conversion from native prairie and as a result of increased disturbance and reduced diversity and inputs within annual cropping systems, interpretation of the PLFA ratios with respect to management was not as straightforward (Figure 12.7). Most notably, the native prairie exhibited a lower fungi to bacteria ratio, higher Gram positive to Gram negative ratio, and a higher stress ratio relative to CRP and the diversified cropping systems. This may be the result of lower nutrient status in native systems relative to managed agricultural systems with high inputs. Within the long-term Sanborn field plots, the results followed the expected trends of reduced fungi to bacteria ratios, increased Gram positive to

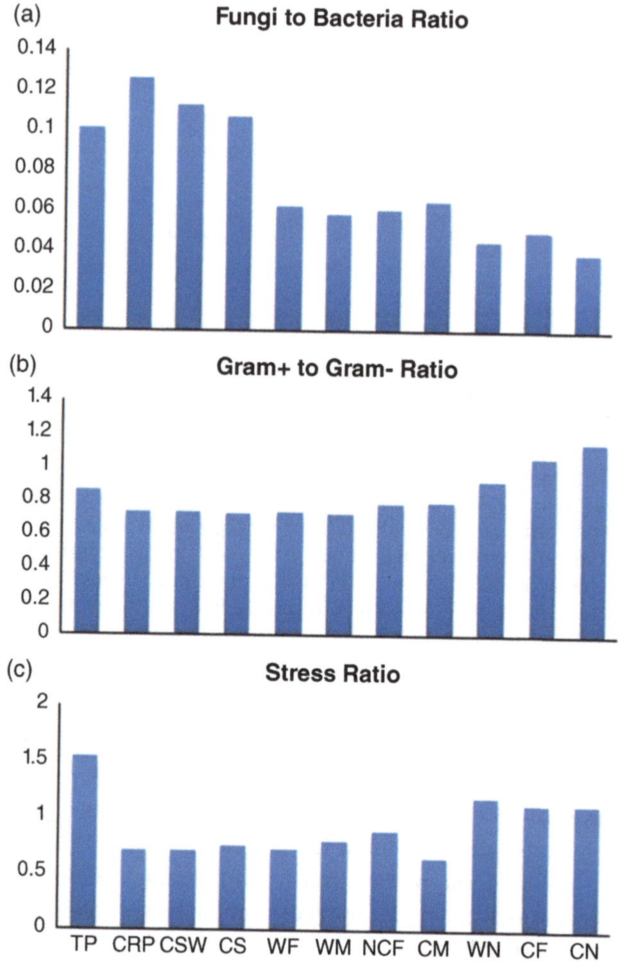


**Figure 12.7** (a) Fungi to bacteria ratio; (b) Gram positive to Gram negative ratio; and (c) stress ratio (cy17:0 + cy19:0)/(16:1ω7c + 18:1ω7c) for systems in Missouri, USA. Total phospholipid fatty acids and by microbial group for native prairie (TP); Conservation Reserve Program (CRP); diversified corn-soybean-wheat system with cover crops (CSW); mulch-till, corn-soybean rotation (CS); continuous wheat with inorganic fertilizer (WF) or manure (WM); no-till continuous corn with inorganic fertilizer (NCF); continuous corn with manure (CM); continuous wheat with no amendments (WN); continuous corn with inorganic fertilizer (CF); continuous corn with no amendments (CN). Data from Veum et al. (2018).

Gram negative ratios, and increased stress ratios in the systems with low crop diversification and no amendments.

Another study conducted in Texas (Cotton & Acosta-Martínez, 2018) examined shifts in microbial community composition in the 0–10 cm soil layer via EL-FAME in a commercial field that had perennial grasses for over 10 year followed by intensive tillage to convert it to row cropping. The initial planting of cotton failed due to a lack of precipitation and the site was replanted to corn. In total, deep tillage occurred four times throughout the growing season (8 months) following conversion. The initial total FAMEs before tillage in the converted site resembled the grassland control site and within a month of conversion, the FAME profiles resembled the nearby row crop control site. This study showed that total FAMEs were reduced by about 60% within the first month of disturbance, representing reductions of 40–60% for saprophytic fungi, bacteria, and AMF microbial groups. This example demonstrates how rapidly the soil microbial community can be impacted by tillage of a long-term perennial grass system (Figure 12.8).

Across climates and soils, FAME profiles have demonstrated sensitivity to agricultural management practices in the U.S. (Acosta-Martínez et al., 2003; Lehman et al., 2019; Rankoth et al., 2019; Sekaran et al., 2020; Unger, Goyne, Kremer, & Kennedy, 2013; Xu, Silveira, Inglett, Sollenberger, & Gerber, 2017), Canada (Lupwayi, Larney, Blackshaw, Kanashiro, & Pearson, 2017; Mann, Lynch, Fillmore, & Mills, 2019), Indonesia (Miura et al., 2016), Argentina (Chavarría et al., 2016; Vargas Gil et al., 2011), China (Li, Wu, et al., 2018; Wang et al., 2020; Zhang et al., 2014; Zhong, Zeng, & Jin, 2016), Brazil (Lopes & Fernandes, 2020), and Europe (D'Hose et al., 2018). Combined, these studies illustrate the strong relationship between the soil microbial community and management practices across a wide range of climate and edaphic conditions (Hartmann et al., 2012; Lewandowski et al., 2015; Malik et al., 2016). Ultimately, shifts in total FAME biomarkers tell us that the soil ecosystem is supporting more or less microbial biomass of any specific group of microorganisms or overall, and one can subsequently infer that factors such as (a) soil characteristics, (b) climate/weather conditions, or (c) management practices are the cause of the observed shift. Therefore, FAME data may be most useful for site-specific comparisons across management practices or to understand the role of climate and edaphic factors in microbial community structure and function at the regional or continental scale. In non-replicated situations (e.g., on-farm assessments), a nearby reference soil can account for climate and edaphic conditions or serve as a baseline for comparative purposes. This could be a nearby remnant prairie, a fence row, or simply represent a different land use or management scenario. Another defensible approach would be repeated measures of soil samples collected at georeferenced locations within fields over multiple years to generate a temporal comparison of FAME profiles from baseline conditions with those following changes in management practices.

FAME analysis is a popular and reliable method for estimating microbial biomass and assessing soil microbial community structure at a coarse level. While there is strong interest in the use of FAME content or microbial group ratios as benchmarks for soil health assessment and interpretation, threshold values of biomarker levels for soil health interpretation do not currently exist. Further

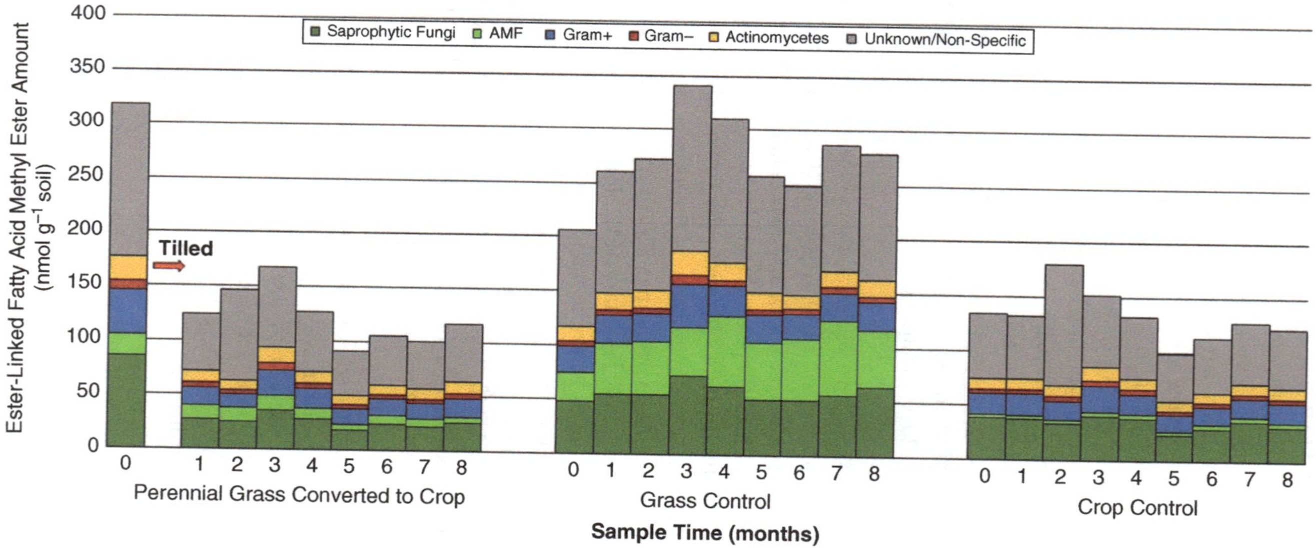


**Figure 12.8** EL-FAME profiles before and after grassland conversion tillage compared with the grass control site and the crop control site. *Source:* Cotton and Acosta-Martínez (2018).

research is required to interpret specific FAME biomarker values or ratios that support establishment of robust threshold levels for FAME biomarkers in the assessment and prescription of management practices promoting soil health. As with other soil biological measurements, there are several factors that should be carefully considered before selecting FAME as a soil health indicator and when interpreting FAME results. Soil depth, soil moisture and temperature, surrounding vegetation and proximity to the rhizosphere, time or seasonality of sampling, soil chemical and physical attributes, as well as current and historical land management information should be acquired to provide context for interpretation of FAME data. When considered in the proper context, FAME profiles can serve as useful soil health indicators by providing a sensitive measure of total soil microbial biomass, which is critical to organic matter formation, and a coarse assessment of community structure, which provides insight into the response of the soil food web to management practices.

## References

Acosta-Martínez, V., C.W. Bell, B.E.L. Morris, J. Zak and V.G. Allen. 2010. Long-term soil microbial community and enzyme activity responses to an integrated cropping-livestock system in a semi-arid region. *Agriculture, Ecosystems and Environment* 137: 231–240.

Acosta-Martínez, V., S. Klose and T.M. Zobeck. 2003. Enzyme activities in semiarid soils under conservation reserve program, native rangeland, and cropland. *Journal of Plant Nutrition and Soil Science* 166: 699–707.

Ananyeva, N.D., S. Castaldi, E.V. Stolnikova, V.N. Kudeyarov and R. Valentini. 2015. Fungi-to-bacteria ratio in soils of European Russia. *Archives of Agronomy and Soil Science* 61: 427–446.

Bailey, V.L., J.L. Smith and H. Bolton. 2003. Novel antibiotics as inhibitors for the selective respiratory inhibition method of measuring fungal: Bacterial ratios in soil. *Biology and Fertility of Soils* 38: 154–160.

Balkwill, D.L., F.R. Leach, J.T. Wilson, J.F. McNabb and D.C. White. 1988. Equivalence of microbial biomass measures based on membrane lipid and cell wall components, adenosine triphosphate, and direct counts in subsurface aquifer sediments. *Microbial Ecology* 16: 73–84.

Bardgett, R.D. and E. McAlister. 1999. The measurement of soil fungal: Bacterial biomass ratios as an indicator of ecosystem self-regulation in temperate meadow grasslands. *Biology and Fertility of Soils* 29: 282–290.

Begum, N., C. Qin, M.A. Ahanger, S. Raza, M.I. Khan, M. Ashraf, . . ., L. Zhang. 2019. Role of arbuscular mycorrhizal fungi in plant growth regulation: implications in abiotic stress tolerance. *Frontiers in Plant Science* 10: 1068.

Bérard, A., T. Bouchet, G. Sévenier, A.L. Pablo and R. Gros. 2011. Resilience of soil microbial communities impacted by severe drought and high temperature in the context of Mediterranean heat waves. *European Journal of Soil Biology* 47: 333–342.

Bhandari, K.B., C.P. West, V. Acosta-Martinez, J. Cotton and A. Cano. 2018. Soil health indicators as affected by diverse forage species and mixtures in semi-arid pastures. *Applied Soil Ecology* 132: 179–186.

Bligh, E.G. and W.J. Dyer. 1959. A rapid method of total lipid extraction and purification. *Canadian Journal of Biochemistry and Physiology* 37: 911–917.

Bossio, D.A. and K.M. Scow. 1998. Impacts of carbon and flooding on soil microbial communities: Phospholipid fatty acid profiles and substrate utilization patterns. *Microbial Ecology* 35: 265–278.

Bossuyt, H., K. Denef, J. Six, S.D. Frey, R. Merckx and K. Paustian. 2001. Influence of microbial populations and residue. *Applied Soil Ecology* 16: 195–208.

Buyer, J.S. and M. Sasser. 2012. High throughput phospholipid fatty acid analysis of soils. *Applied Soil Ecology* 61: 127–130.

Cantrell, S.A., M. Molina, D.J. Lodge, F.J. Rivera-Figueroa, M.L. Ortiz-Hernández, A.A. Marchetti, . . ., J.R. Pérez-Jiménez. 2014. Effects of a simulated hurricane disturbance on forest floor microbial communities. *Forest Ecology and Management* 332: 22–31.

Chavarría, D.N., R.A. Verdenelli, D.L. Serri, S.B. Restovich, A.E. Andriulo, J.M. Meriles, . . ., S. Vargas-Gil. 2016. Effect of cover crops on microbial community structure and related enzyme activities and macronutrient availability. *European Journal of Soil Biology* 76: 74–82.

Cotrufo, M.F., M.D. Wallenstein, C.M. Boot, K. Denef and E. Paul. 2013. The Microbial Efficiency-Matrix Stabilization (MEMS) framework integrates plant litter decomposition with soil organic matter stabilization: Do labile plant inputs form stable soil organic matter? *Global Change Biology* 19: 988–995.

Cotton, J. and V. Acosta-Martínez. 2018. Intensive tillage converting grassland to cropland immediately reduces soil microbial community size and organic carbon. *Agricultural & Environmental Letters* 3: 1–4.

D'Hose, T., L. Molendijk, L. Van Vooren, W. van den Berg, H. Hoek, W. Runia, . . ., Greet Ruysschaert 2018. Responses of soil biota to non-inversion tillage and organic amendments: An analysis on European multiyear field experiments. *Pedobiologia* 66: 18–28.

Doran, J.W., D.G. Fraser, M.N. Culik and W.C. Liebhardt. 1987. Influence of alternative and conventional agricultural management on soil microbial processes and nitrogen availability. *American Journal of Alternative Agriculture* 2: 99–106.

Drijber, R.A., J.W. Doran, A.M. Parkhurst and D.J. Lyon. 2000. Changes in soil microbial community structure with tillage under long-term wheat-fallow management. *Soil Biology and Biochemistry* 32: 1419–1430.

Fanin, N., P. Kardol, M. Farrell, M.-C. Nilsson, M.J. Gundale and D.A. Wardle. 2019. The ratio of Gram-positive to Gram-negative bacterial PLFA markers as an indicator of carbon availability in organic soils. *Soil Biology and Biochemistry* 128: 111–114.

Fernandes, M.F., J. Saxena and R.P. Dick. 2013. Comparison of whole-cell fatty acid (MIDI) or Phospholipid Fatty Acid (PLFA) extractants as biomarkers to profile soil microbial communities. *Microbial Ecology* 66: 145–157.

Fierer, N., J.P. Schimel and P.A. Holden. 2003. Variations in microbial community composition through two soil depth profiles. *Soil Biology and Biochemistry* 35: 167–176.

Foster, J.W. 1962. Hydrocarbons as substrates for microorganisms. *Antonie van Leewenhoek Journal of Microbiology and Serology* 28: 241–274.

Franzmann, P.D., B.M. Patterson, T.R. Power, P.D. Nichols and G.B. Davis. 1996. Microbial biomass in a shallow, urban aquifer contaminated with aromatic hydrocarbons: Analysis by phospholipid fatty acid content and composition. *The Journal of Applied Bacteriology* 80: 617–625.

Frey, S.D., E.T. Elliott and K. Paustian. 1999. Bacterial and fungal abundance and biomass in conventional and no-tillage agroecosystems along two climatic gradients. *Soil Biology and Biochemistry* 31: 573–585.

Frostegård, A. and E. Bååth. 1996. The use of phospholipid fatty acid analysis to estimate bacterial and fungal biomass in soil. *Biology and Fertility of Soils* 22: 59–65.

Frostegard, A., A. Tunlid and E. Baath. 1993. Phospholipid fatty acid composition, biomass, and activity of microbial communities from two soil types experimentally exposed to different heavy metals. *Applied and Environmental Microbiology* 59: 3605–3617.

Frostegård, Å., A. Tunlid and E. Bååth. 2011. Use and misuse of PLFA measurements in soils. *Soil Biology and Biochemistry* 43: 1621–1625.

Green, C.T. and K.M. Scow. 2000. Analysis of phospholipid fatty acids (PLFA) to characterize microbial communities in aquifers. *Hydrogeology Journal* 8: 126–141.

Grigera, M. S., Drijber, R. A., Shores-Morrow, R. H., & Wienhold, B. J. (2007). Distribution of the arbuscular mycorrhizal biomarker C16:1cis11 among neutral, glyco and phospholipids extracted from soil during the reproductive growth of corn. *Soil Biology &. Biochemistry*, 39, 1589–1596.

Guckert, J.B., M.A. Hood and D.C. White. 1986. Phospholipid ester-linked fatty acid profile changes during nutrient deprivation of Vibrio cholerae: Increases in the trans/cis ratio and proportions of cyclopropyl fatty acids. *Applied and Environmental Microbiology* 52: 794–801.

Hartmann, M., C.G. Howes, D. VanInsberghe, H. Yu, D. Bachar, R. Christen, . . ., William W Mohn 2012. Significant and persistent impact of timber harvesting on soil microbial communities in Northern coniferous forests. *The ISME Journal* 6: 2199–2218.

Kallenbach, C.M., S.D. Frey and A.S. Grandy. 2016. Direct evidence for microbial-derived soil organic matter formation and its ecophysiological controls. *Nature Communications* 7: 13630.

Kaur, A., A. Chaudhary, A. Kaur, R. Choudhary and R. Kaushik. 2005. Phospholipid fatty acid—A bioindicator of environment monitoring and assessment in soil ecosystem. *Current Science* 89: 1103–1112.

Keynan, A. and N. Sandler. 1983. Spore research in historical perspective. In: A. Hurst and G. W. Gould, editors, *The bacterial spore*, Vol. 2. Academic Press, New York, NY. p. 1–48.

Kieft, T.L., D.B. Ringelberg and D.C. White. 1994. Changes in ester-linked phospholipid fatty-acid profiles of subsurface bacteria during starvation and dissication in a porous medium. *Applied and Environmental Microbiology* 60: 3292–3299.

Kramer, C. and G. Gleixner. 2008. Soil organic matter in soil depth profiles: Distinct carbon preferences of microbial groups during carbon transformation. *Soil Biology and Biochemistry* 40: 425–433.

Lal, R. 2005. Soil erosion and carbon dynamics *Soil and Tillage Research* 81: 137–142.

Lal, R., J.M. Kimble, R.F. Follett and B.A. Stewart. 1998. *Soil processes and the carbon cycle*.CRC Press, Boca Raton, FL.

Lee, Y.B., N. Lorenz, L.K. Dick and R.P. Dick. 2007. Cold storage and pretreatment incubation effects on soil microbial properties *Soil Science Society of America Journal* 71: 1299–1305.

Lehman, R.M., V. Acosta-Martinez, J.S. Buyer, C.A. Cambardella, H.P. Collins, T.F. Ducey, . . ., D.E. Stott 2015. Soil biology for resilient, healthy soil. *Journal of Soil and Water Conservation* 70: 12A–18A.

Lehman, R.M., C. Cambardella, D. Stott, V. Acosta-Martinez, D.K. Manter, J.S. Buyer, . . ., D.L. Karlen 2015. Understanding and enhancing soil biological health: The solution for reversing soil degradation. *Sustainability* 7: 988.

Lehman, R.M., S.L. Osborne, W.I. Taheri, J.S. Buyer and B.K. Chim. 2019. Comparative measurements of arbuscular mycorrhizal fungal responses to agricultural management practices. *Mycorrhiza* 29: 227–235.

Lewandowski, T.E., J.A. Forrester, D.J. Mladenoff, J.L. Stoffel, S.T. Gower, A.W. D'Amato, . . ., T.C. Balser. 2015. Soil microbial community response and recovery following group selection harvest: Temporal patterns from an experimental harvest in a US northern hardwood forest. *Forest Ecology and Management* 340: 82–94.

Li, C., A. Cano, V. Acosta-Martinez, K.S. Veum and J. Moore-Kucera. 2020. A comparison between fatty acid methyl ester profiling methods (PLFA and EL-FAME) as soil health indicators. *Soil Science Society of America Journal* 2020: 1–17.

Li, C., L.M. Fultz, J. Moore-Kucera, V. Acosta-Martínez and M. Kakarla. 2018. Soil microbial community restoration in Conservation Reserve Program semi-arid grasslands. *Soil Biology & Biochemistry* 118: 166–177.

Li, J., X. Wu, M.T. Gebremikael, H. Wu, D. Cai, B. Wang, . . ., J. Xi 2018. Response of soil organic carbon fractions, microbial community composition and carbon mineralization to high-input fertilizer practices under an intensive agricultural system. *PLoS One* 13: e0195144.

Lopes, L.D. and M.F. Fernandes. 2020. Changes in microbial community structure and physiological profile in a kaolinitic tropical soil under different conservation agricultural practices. *Applied Soil Ecology* 152: 103545.

Lupwayi, N.Z., F.J. Larney, R.E. Blackshaw, D.A. Kanashiro and D.C. Pearson. 2017. Phospholipid fatty acid biomarkers show positive soil microbial community responses to conservation soil management of irrigated crop rotations. *Soil and Tillage Research* 168: 1–10.

Malik, A.A., S. Chowdhury, V. Schlager, A. Oliver, J. Puissant, P.G.M. Vazquez, . . ., G. Gleixner 2016. Soil fungal:bacterial ratios are linked to altered carbon cycling. *Frontiers in Microbiology* 7.

Mann, C., D. Lynch, S. Fillmore and A. Mills. 2019. Relationships between field management, soil health, and microbial community composition. *Applied Soil Ecology* 144: 12–21.

Manter, D.K., J.A. Delgado, H.D. Blackburn, D. Harmel, A.A. Pérez de León and C.W. Honeycutt. 2017. Opinion: Why we need a National Living Soil Repository. *Proceedings of the National Academy of Sciences of the United States of America* 114: 13587–13590.

Manter, D.K., T.L. Weir and J.M. Vivanco. 2010. Negative effects of sample pooling on PCR-based estimates of soil microbial richness and community structure. *Applied and Environmental Microbiology* 76: 2086–2090.

MIDI. 2014. *Sherlock PLFA tools users' guide version 1.2*. MIDI, Inc., Newark, Deleware. p. 14.

MIDI. 2015. *Sherlock equipment and consumables list: Phospholipid fatty acid (PLFA) extraction method*. MIDI, Inc., Newark, Delaware. p. 5.

MIDI. 2018. *Automated Phospholipid Fatty Acid (PLFA) analysis using the Sherlock PLFA software package on a Shimadzu GC-2010/2030. Application note: Soil microbial community analysis*. MIDI, Newark, Deleware. p. 5.

Miura, T., A. Niswati, I.G. Swibawa, S. Haryani, H. Gunito, M. Arai, . . ., K. Fujie 2016. Shifts in the composition and potential functions of soil microbial communities responding to a no-tillage practice and bagasse mulching on a sugarcane plantation. *Biology and Fertility of Soils* 52: 307–322.

Moore-Kucera, J. and R.P. Dick. 2008. PLFA profiling of microbial community structure and seasonal shifts in soils of a Douglas-fir chronosequence. *Microbial Ecology* 55: 500–511.

Nielsen, M.N. and A. Winding. 2002. Microorganisms as Indicators of Soil Health. NERI Technical Report No. 388. National Environmental Research Institute, Denmark. p. 84.

Nunes, M.R., D.L. Karlen, K.S. Veum, T.B. Moorman and C.A. Cambardella. 2020. Biological soil health indicators respond to tillage intensity: A US meta-analysis. *Geoderma* 369: 114335.

Olsson, P.A. 1999. Signature fatty acids provide tools for determination of the distribution and interactions of mycorrhizal fungi in soil. *FEMS Microbiology Ecology* 29: 303–310.

Parekh, N.R. and R.D. Bardgett. 2002. The characterisation of microbial communities in environmental samples. *Radioactivity in the Environment* 2: 37–60.

Pennanen, T., A. Frostegard, H. Fritze and E. Baath. 1996. Phospholipid fatty acid composition and heavy metal tolerance of soil microbial communities along two heavy metal-polluted gradients in coniferous forests. *Applied and Environmental Microbiology* 62: 420–428.

Pérez-Guzmán, L., V. Acosta-Martínez, L.A. Phillips and S.A. Mauget. 2020. Resilience of the microbial communities and their enzymatic activities in semi-arid agricultural soils during natural climatic variability events. *Applied Soil Ecology* 149.

Perfil'ev, B.V. and D.R. Gabe. 1969. *Capillary methods of investigating micro-organisms.* Oliver and Boyd, Edinburgh.

Petersen, S.O. and M.J. Klug. 1994. Effects of sieving, storage, and incubation temperature on the phospholipid fatty acid profile of a soil microbial community. *Applied and Environmental Microbiology* 60: 2421–2430.

Pinkart, H.C., D.B. Ringelberg, Y.M. Piceno, S.J. Macnaughton and D.C. White. 2002. Biochemical approaches to biomass measurements and community structure analysis. In: C.J. Hurst, R.L. Crawford, G.R. Knudsen, M.J. McInerney and L.D. Stetzenbach, editors, *Manual of environmental microbiology*, 2. ASM Press, Washington DC. p. 101–113.

Quideau, S.A., A.C.S. McIntosh, C.E. Norris, E. Lloret, M.J.B. Swallow and K. Hannam. 2016. Extraction and analysis of microbial phospholipid fatty acids in soils. *Journal of Visualized Experiments* 114: 54360.

Rankoth, L.M., R.J. Udawatta, C.J. Gantzer, S. Jose, K.S. Veum and H.A. Dewanto. 2019. Cover crops on temporal and spatial variations in soil microbial communities by PLFA profiling. *Agronomy Journal* 111: 1–11.

Schloter, M., P. Nannipieri, S.J. Sørensen and J.D. van Elsas. 2018. Microbial indicators for soil quality. *Biology and Fertility of Soils* 54: 1–10.

Schnecker, J., B. Wild, L. Fuchslueger and A. Richter. 2012. A field method to store samples from temperate mountain grassland soils for analysis of phospholipid fatty acids. *Soil Biology and Biochemistry* 51: 81–83.

Schutter, M.E. and R.P. Dick. 2000. Comparison of fatty acid methyl ester (FAME) methods for characterizing microbial communities. *Soil Science Society of America Journal* 64: 1659–1668.

Sekaran, U., K.L. Sagar, L.G.D.O. Denardin, J. Singh, N. Singh, G.O. Abagandura, . . . A.P. Martins 2020. Responses of soil biochemical properties and microbial community

structure to short and long-term no-till systems. *European Journal of Soil Science* 2020: 1–16.
Sharma, M.P. and J.S. Buyer. 2015. Comparison of biochemical and microscopic methods for quantification of arbuscular mycorrhizal fungi in soil and roots. *Applied Soil Ecology* 95: 86–89.
Six, J., S.D. Frey, R.K. Thiet and K.M. Batten. 2006. Bacterial and fungal contributions to carbon sequestration in agroecosystems. *Soil Science Society of America Journal* 70: 555–569.
Stevenson, B.A., D.W.F. Hunter and P.L. Rhodes. 2014. Temporal and seasonal change in microbial community structure of an undisturbed, disturbed, and carbon-amended pasture soil. *Soil Biology and Biochemistry* 75: 175–185.
Tunlid, A. and D.C. White. 1992. Biochemical analysis of biomass, community structure, nutritional status, and metabolic activity of microbial communities in soil. In: G. Stotzky and J.M. Bollag, eds., *Soil biochemistry*. Vol. 7. Marcel Dekker, Inc., New York. p. 229–262.
Unger, I., K. Goyne, R. Kremer and A. Kennedy. 2013. Microbial community diversity in agroforestry and grass vegetative filter strips. *Agroforestry Systems* 87: 395–402.
Vargas Gil, S., J. Meriles, C. Conforto, M. Basanta, V. Radl, A. Hagn, . . ., G.J. March 2011. Response of soil microbial communities to different management practices in surface soils of a soybean agroecosystem in Argentina. *European Journal of Soil Biology* 47: 55–60.
Veum, K.S., K.W. Goyne, R.J. Kremer, R.J. Miles and K.A. Sudduth. 2014. Biological indicators of soil quality and soil organic matter characteristics in an agricultural management continuum. *Biogeochemistry* 117: 81–99.
Veum, K.S., R.J. Kremer, K.A. Sudduth, N.R. Kitchen, R.N. Lerch, C. Baffaut, . . . E.J. Sadler 2015. Conservation effects on soil quality indicators in the Missouri Salt River basin. *Journal of Soil and Water Conservation* 70: 232–246.
Veum, K.S., T.L. Lorenz and R.J. Kremer. 2018. Microbial community structure in Missouri prairie soils. *Missouri Prairie Journal* 39: 18–20.
Veum, K.S., T.L. Lorenz and R.J. Kremer. 2019. Phospholipid fatty acid profiles of soils under variable handling and storage conditions. *Agronomy Journal* 111: 1090–1096.
Villanueva, L., A. Navarette, J. Urmeneta, D.C. White and R. Guerrero. 2004. Combined phospholipid biomarker-16S rRNA gene denaturing gradient gel electrophoresis analysis of bacterial diversity and physiological status in an intertidal microbial mat. *Applied and Environmental Microbiology* 70: 6920–6926.
Wang, Q.-q., L.-l. Liu, Y. Li, S. Qin, C.-j. Wang, A.-d. Cai, . . ., Wen-ju Zhang, 2020. Long-term fertilization leads to specific PLFA finger-prints in Chinese Hapludults soil. *Journal of Integrative Agriculture* 19: 1354–1362.
Warren, C.R. 2019. Does silica solid-phase extraction of soil lipids isolate a pure phospholipid fraction? *Soil Biology and Biochemistry* 128: 175–178.

White, D.C. 1983. Analysis of microorganisms in terms of quantity and activity in natural environments. Microbes in their natural environments *Society for General Microbiology*. p. 37–66.

White, D.C., W.M. Davis, J.S. Nickels, J.D. King and R.J. Bobbie. 1979. Determination of the sedimentary microbial biomass by extractible lipid phosphate. *Oecologia* 40: 51–62.

White, D.C., J.O. Stair and D.B. Ringelberg. 1996. Quantitative comparisons of in situ microbial biodiversity by signature biomarker analysis. *Journal of Industrial Microbiology* 17: 185–196.

Willers, C., P.J. Jansen van Rensburg and S. Claassens. 2015. Phospholipid fatty acid profiling of microbial communities—a review of interpretations and recent applications. *Journal of Applied Microbiology* 119: 1207–1218.

Xu, S., M.L. Silveira, K.S. Inglett, L.E. Sollenberger and S. Gerber. 2017. Soil microbial community responses to long-term land use intensification in subtropical grazing lands. *Geoderma* 293: 73–81.

Zelles, L. 1997. Phospholipid fatty acid profiles in selected members of soil microbial communities. *Chemosphere* 35: 275–294.

Zelles, L. 1999. Fatty acid patterns of phospholipids and lipopolysaccharides in the characterisation of microbial communities in soil: A review. *Biology and Fertility of Soils* 29: 111–129.

Zelles, L., Q.Y. Bai, T. Beck and F. Beese. 1992. Signature fatty acids in phospholipids and lipopolysaccharides as indicators of microbial biomass and community structure in agricultural soils. *Soil Biology and Biochemistry* 24: 317–323.

Zelles, L., Q.Y. Bai, R.X. Ma, R. Rackwitz, K. Winter and F. Beese. 1994. Microbial biomass, metabolic activity and nutritional status determined from fatty acid patterns and poly-hydroxybutyrate in agriculturally-managed soils. *Soil Biology and Biochemistry* 26: 439–446.

Zelles, L., Q.Y. Bai, R. Rackwitz, D. Chadwick and F. Beese. 1995. Determination of phospholipid- and lipopolysaccharide-derived fatty acids as an estimate of microbial biomass and community structures in soils. *Biology and Fertility of Soils* 19: 115–123.

Zhang, B., Y. Li, T. Ren, Z. Tian, G. Wang, X. He, . . ., C. Tian 2014. Short-term effect of tillage and crop rotation on microbial community structure and enzyme activities of a clay loam soil. *Biology and Fertility of Soils* 50: 1077–1085.

Zhang, Q., J. Wu, F. Yang, Y. Lei, Q. Zhang and X. Cheng. 2016. Alterations in soil microbial community composition and biomass following agricultural land use change. *Scientific Reports* 6: 36587.

Zhong, S., H.-C. Zeng and Z.-Q. Jin. 2016. Response of soil nematode community composition and diversity to different crop rotations and tillage in the tropics. *Applied Soil Ecology* 107: 134–143.

# 13

# Microbial Community Composition, Diversity, and Function

*Daniel K. Manter, J. Michael Moore, R. M. Lehman, and Alison K. Hamm*

Soil health is dependent on abundant, diverse, and active biological communities. Several soil enzymes, phospholipid fatty acid (PLFA), ester-linked fatty acid methyl esters (EL-FAME), and active carbon (POXC) methods have been discussed in this book as indirect methods for quantifying those communities and their contribution to soil health. This chapter presents a single molecular assay of microbial community characteristics that can directly assess the abundance of key taxa associated with various soil functions important in soil health.

## Introduction

The soil microbial community provides multiple ecosystem services by directly and indirectly influencing key functions to support a healthy soil. These functions that support crop production include the formation and cycling of soil organic matter (SOM) and soil organic carbon (SOC); enhancing nutrient cycling and availability; improving soil structure to store water, nutrients, and resist erosion; controlling plant diseases; and reducing pressures from insects and weeds (Chaparro et al., 2012; Grichko & Glick, 2001; Six et al., 2006; Tilak et al., 2005; Van Bruggen & Semenov, 2000). Soil microbes have recently been recognized as critical drivers for increasing resilience and resistance in response to environmental

*Soil Health Series: Volume 2 Laboratory Methods for Soil Health Analysis*, First Edition.
Edited by Douglas L. Karlen, Diane E. Stott, and Maysoon M. Mikha.

disturbances including droughts (Naylor & Coleman-Derr, 2018), floods (Grichko & Glick, 2001), fires (Docherty et al., 2012), and temperature extremes (Verma et al., 2017; Yadav et al., 2018).

Microbial biomass estimates are often considered a general indicator of soil biological activity, but they do not capture community structure (i.e., diversity) or functional capacity. Individually microbes often influence unique ecosystem functions (e.g., pathogen, nutrient cycling, etc.), but many argue that due to vast microbial diversity and presumed ubiquity of many taxa, maintenance of soil microbial biodiversity is of little concern (Finlay et al., 1997) and that functional capacity is independent of community composition. However, others argue that community composition is a critical driver of soil function (Reed & Martiny, 2007) and that it is influenced by environmental (Pagaling et al., 2014) and management (Doran & Zeiss, 2000) histories. Our position is that the perceived importance of taxonomic structure on soil function(s) is dependent on the functional redundancy of the relevant genes within the soil microbial community. For example, taxonomic-specific abundance may be less important for ubiquitous functions such as those coded for by 6-phospho-β-glucosidase (*bglB*), than for more taxonomic-specific genes, such as nitrogen fixation (i.e., *nifH*). Furthermore, a suite of common genes associated with the positive effects of root-colonizing bacteria has recently emerged (Table 13.1); and while many PGPR (plant growth promoting rhizobacteria) strains may contain more than one gene, none are common to a single isolate (Bruto et al., 2014). The redundancy of genes across divergent taxa, and unique gene combinations within taxa, suggests that PGPRs work in cooperation; and that a diversity of taxonomic groups contribute to overall soil health. As a result, methods that assess the total functional contribution of all soil organisms (e.g., gene abundance across all relevant taxa) may be critical to understanding microbial contributions to soil health.

Soil microbes have a wide-ranging impact on plant health through a variety of mechanisms, as discussed in several reviews on this subject (e.g., Babalola, 2010; Esitken, 2011; Lugtenberg & Kamilova, 2009). Most knowledge regarding plant-microbial interaction mechanisms have come from studies involving individual microbial strains; but the mechanistic basis (i.e., key genes and their functions) continues to be elucidated (e.g., Saharan & Nehra, 2011). Interrelationships between soil microbes and plant roots are complex and not easily isolated. Multiple lines of evidence document that soil microbes, as measured by total abundance, specific taxonomic groups, or their functional genes, have been associated with enhanced plant productivity and quality. We identified six groups of important soil health processes: 1) SOM and C formation and cycling; 2) physical stability (i.e., aggregation); 3) nutrient cycling; 4) disease suppression; 5) plant stress resiliency, and 6) soil stress resiliency (Table 13.1). For each group, taxonomic and

**Table 13.1** Taxonomic and functional indicators of various soil functions that collectively contribute to soil health.

| Soil Process | Taxonomic Indicator | Functional Indicator[1] |
|---|---|---|
| SOM/C Formation & Cycling | Total microbial biomass | C-mineralizing genes (*bgIB*) |
| Physical stability (aggregation) | Actinobacteria<br>Firmicutes and other extra-cellular polysaccharide producers<br>Mycorrhizae (Glomeromycota)<br>Saprophytic fungi | Unknown |
| Nutrient availability | Nitrogen fixers (*Rhizobium*, *Azospirillum*)<br>P solubilizers (*Bacillus*, *Pseudomonas*, *Rhizobium*) | N-fixing genes (*nifH*)<br>N-mineralizing genes<br>P-solubilizing genes (*pqqC*)<br>P-mineralizing genes (*phoD*)<br>S-mineralizing genes (*aslA*) |
| Disease suppression | *Pseudomonas*<br>*Bacillus*<br>*Trichoderma* | Antibiotic-producing genes (*phzE*)<br>Siderophore-producing genes (*entA*) |
| Plant stress resiliency | Hormone producing microbes (*Pseudomonas*, *Bacillus*)<br>Ethylene pathway disruptors | Hormone-producing genes (*ipdC*)<br>Ethylene pathway disruptor genes (*acdS*) |
| Soil stress resiliency | General taxonomic diversity (evenness and richness)<br>Gram+/Gram- ratio | Functional diversity of beneficial (PGPR) genes |

[1] Multiple genes exist for each functional indicator; a single candidate is shown in parenthesis.

functional molecular indicators are identified. The first three soil processes are currently measured indirectly as part of many health assessments (e.g., Chapters 3 to 7 in this volume). In contrast, the method proposed in this chapter allows for a single molecular assay of microbial community characteristics that can assess the abundance of key taxa associated with various soil processes (e.g., Actinobacteria, Glomeromycota, *Bacillus*, *Pseudomonas*, etc.) and functional (i.e., gene) abundances through phylogenetic reconstruction (Langille et al., 2013). The combination of a single assay and its ability to identify multiple taxonomic and functional soil microbial groups thus provides a robust, cost-effective indicator of the collective microbial processes that constitute soil health.

## Methods for Identifying and Quantifying Microbial Communities

### Traditional Techniques

Traditional techniques for obtaining, quantifying, and analyzing microorganisms largely rely on organismal growth on various solid or liquid media. Although functional and/or physiological diversity inherent in such selection processes may be important, the strong selection pressure and artificial growth conditions can severely limit the microbial diversity observed.

#### Plate Counts

The classic measure for identifying microorganisms is through serial dilution plating and counting of CFUs (colony forming units). Use of selective indicator media enables identification of microorganisms that influence specific functions (e.g., chrome azurol S; Louden et al., 2011) may be used to indicate the presence of siderophores). Those different media target specific taxonomic groups by varying nutrient composition and pH. However, it has been estimated that <1% of the soil bacteria population can be cultured by standard laboratory practices (Torsvik et al., 1990). Temperature, light, and potential difficulties in dislodging bacteria or spores from soil particles (Sørheim et al., 1989; Tabacchioni et al., 2000; Vieira & Nahas, 2005) are among the challenges that influence the diversity of cultivable organisms and thus affect plate counts. Despite those potential taxonomic and/or functional selectivity challenges associated with plate counts; this technique is still widely used as a coarse measure of microbial enumeration but is perhaps best used to generate pure cultures for further studies.

#### Community Level Physiological Profiles (CLPP)

Community level Physiological Profiles (CLPP) analyze physiological diversity by measuring microbial growth on a variety of sole source carbon substrates, e.g., BIOLOG plates (Garland & Mills, 1991). In theory, microbial functional diversity of a mixed sample is based on the number of carbon sources capable of being utilized, although statistical approaches vary and have not been standardized (Broughton & Gross, 2000). One major drawback of BIOLOG plates is that the dye(s) enabling detection can bias microbial growth. For example, tetrazolium, a redox active dye that is used as alternative electron acceptor in place of oxygen, cannot be metabolized by some bacteria (Winding & Hendriksen, 1997) and fungi (Dobranic & Zak, 1999), but selection of alternative dyes may help alleviate this problem (Preston-Mafham et al., 2002). Overall, CLPP is essentially a culture-based method, affected by the same potential limitations as for plate counts described above. However, it remains popular because it provides an inexpensive and quick assessment tool to monitor microbial growth in the presence of several carbon sources.

## Biochemical Techniques

Many different biochemical techniques have been used to measure microbial biomass. This includes total biomass based on adenosine triphosphate (ATP) concentration (Sorokin & Kadota, 1972) or respiration (Anderson & Domsch, 1978) as well as fungal biomass measurements based on ergosterol concentrations (Osswald et al., 1986). The biggest drawback of these methods is that they do not provide an assessment of community composition. The use of fatty acid profiles is one of the most used biochemical approaches and provides some estimate of composition, but it lacks the resolution necessary to identify key functional and taxonomic responses.

### Fatty Acids

Fatty acids have been used extensively to identify individual microorganisms (Banowetz et al., 2006) and the composition of fatty acid profiles (phospholipid fatty acids, PLFA; ester-linked fatty acid methyl esters, EL-FAME) is thought to represent different compositions of communities of microorganisms (Tunlid & White, 1992). Indeed, many studies have shown that fatty acid profiles are responsive to a wide range of environmental conditions, soil types, and management practices (Broughton & Gross, 2000; Ibekwe et al., 2002; Zelles et al., 1992). Fatty acid profiling does not require cultivation of microorganisms, is relatively inexpensive, quantitative, and can have high reproducibility. However, individual fatty acid profiles have been shown to change with growth conditions (Scherer et al., 2003) and do not have the specificity to target individual members (i.e., genus/species levels) within an environmental sample, such as soil. Despite these limitations, identification of functional groups using potential fatty acid indicators has been used in many ecosystems to track shifts due to management or environmental drivers (Bååth & Anderson, 2003; Findlay 1996; Moore-Kucera & Dick, 2008; Olsson, 1999). However, caution must be exercised when interpreting results as many of the proposed PLFA indicators are present in multiple functional groups and overinterpretation or misinterpretation has been documented (Frostegård et al., 2011). For more detailed discussions of PLFAs please see Chapter 12 in this book.

## Molecular-Based Techniques

It is widely accepted that more than 99% of the microorganisms in any environment cannot be cultured (Hugenholtz, 2002) and often a large portion of potentially cultivable organisms are in a viable but unculturable state (Oliver, 2005). Furthermore, many candidate phyla (BRC1, OP10, OP11, SC3, TM7, WS2, WS3) have no cultured representatives and are defined only by molecular (e.g., 16S rRNA) sequences (Schloss & Handelsman, 2004). For those and numerous other reasons, estimates of community composition are arguably better quantified through molecular techniques. Several molecular techniques are available and have been grouped into whole community (DNA-DNA reassociation, G+C fractionation, whole genome sequencing, metagenomics, and transcriptomics) or partial community protocols (T-RFLP, DGGE, RISA, qPCR, FISH, microarray, and amplicon sequencing) whereby the latter typically rely on nucleic acid amplification and target selection using polymerase chain reaction (PCR) technologies (Rastogi & Sani, 2011). For brevity, only the techniques we consider most promising for microbial community studies are presented herein, since other excellent reviews on the subject can be found elsewhere (Fakruddin & Mannan, 2013; Franzosa et al., 2015; Rastogi & Sani, 2011).

### Partial Community Analysis

***Quantitative Polymerase Chain Reaction (qPCR)*** Quantitative PCR (qPCR) uses fluorescent dyes to measure accumulated PCR amplicons in real-time, and can therefore be used to determine initial template concentrations based on either absolute or relative quantitation (Bustin et al., 2009). Like all partial community approaches, primer selection is critical for selectivity and greatly influences abundance estimates and potential diversity of amplicons. However, qPCR has been successfully used in a variety of environmental samples and primers are available for both taxonomic (e.g., 16S, 18S, ITS [Fierer et al., 2005]) and function/gene (e.g., *nifH*, *nosZ* [Wallenstein & Vilgalys, 2005]) selectivity. Due to its popularity, numerous factors influencing the efficiency, success, and interpretation of qPCR have been identified. This has resulted in development of guidelines for reporting qPCR assays, also known as The Minimum Information for Publication of Quantitative Real-Time PCR Experiments (MIQE) (Bustin et al., 2009). Although, qPCR may be the 'gold-standard' for estimating taxonomic/gene abundances, it alone cannot provide information on taxonomic or functional composition (i.e., amplicon diversity) unless it is coupled with additional downstream analyses, such as melt-curve analysis, RFLP, or DNA sequencing. As such, we propose that qPCR is a critical component of any molecular study for community composition and therefore discuss it in more detail below.

***Amplicon Sequencing*** Use of single gene DNA sequence variation can perhaps be traced back to Woese and colleagues (Woese & Fox, 1977) who used the small subunit of the ribosomal RNA gene to infer phylogenetic relationships of prokaryotes (16S) and eukaryotes (18S). Use of rRNA genes for phylogenetic analyses is particularly attractive, since it: (i) is a component of all self-replicating living organisms; (ii) is highly conserved and serves as a reliable molecular chronometer; (iii) is seldom transferred horizontally; (iv) possesses both conserved and variable regions; and (v) is amenable in size to many molecular techniques (Hugerth & Andersson, 2017). Furthermore, due to phenotypic and/or biochemical difficulties in species definition, it is common practice to use rRNA sequence analysis to define new bacterial species (Stackebrandt & Groebel, 1994). Early phylogenetic studies with 16S rRNA relied on sequencing bulk cellular rRNA preparations (Lane et al., 1985) and eventually progressed to various cloning strategies (Pace et al., 1986). With the rise of next-generation sequencing in combination with PCR amplification, the analytical capacity to explore DNA sequence variants from natural environments exploded. For example, the first known study using Roche 454 pyrosequencing obtained a total approximately 118,000 16S rRNA gene sequences (Sogin et al., 2006). With the further refinement including development of sample-specific barcodes (Hamady et al., 2008) and advances in sequencing technology (next-generation sequencing; Illumina platform), it is now possible to simultaneously obtain upward of $10^8$ sequences from a variety of samples (Caporaso et al., 2012). As these are a partial-community technique and typically reliant on PCR amplification, it is amenable to any gene target with suitable primers such as rRNA (Kozich et al., 2013) or *nifH* (Gaby et al., 2018).

**Whole Community Analysis**

***Metagenomics*** Metagenomics differs from amplicon sequencing by relying on direct sequencing of all genetic material present in a sample or "shotgun sequencing" following DNA fractionation. Despite its great potential and ability to analyze 100% of the genetic information in a sample, we suggest metagenomics is currently too expensive and faces too many difficulties in highly diverse samples such as soil, to be applicable as a soil health assay for soil microbial community analyses (Quince et al., 2017). Key questions such as which type of analytical pipeline may be most appropriate (i.e., metagenome assembly or assembly-free) remain unanswered. For metagenome assembly, reference genomes for guiding assembly software are lacking. Indeed, genomes of some uncultured bacterial strains exist (Albertsen et al., 2013), but most candidate divisions are not represented and the success of assembly is still largely driven by genome coverage. Assembly-free methods may overcome some of those difficulties, but questions remain regarding appropriate binning strategies of short DNA sequences. For instance, at what level of similarity should a sequence be assigned to a known gene or protein family?

***Transcriptomics*** Transcriptomics is the sequencing or analysis of complete RNA transcripts, usually mRNA, which is a direct measure of gene expression. This approach been used successfully in simplified (e.g., mesocosm) soil studies

(Perazzolli et al., 2016) and dual cultures of plant pathogens and biocontrol (Mela et al., 2011); however, its use in native soils is still in its infancy. Transcriptomics should represent a greater understanding of soil microbial functions as this method measures functioning genes and their expression levels at the time of sampling. However, many of the same complexity and bioinformatic issues associated with metagenomics also exist with transcriptomics. In addition, transcriptomics relies on successful preservation and isolation of RNA from soils, which is notoriously unstable. Furthermore, gene expression adds a time-course (e.g., diurnal changes in gene expression [Church et al., 2005]) component not seen in genomic data that is all but unknown in soil systems. For example, gene expression profiles may change in response to environmental stimuli in minutes, increasing or decreasing many fold and are not persistent even when environmental stimuli do not change. Different methods of sample handling and storage are additional examples of environmental stimuli that can confound transcriptomics interpretation. Transcriptomic, like metagenomics, offers great potential and with time and further technological advancements, likely will become more cost effective and be more applicable to highly diverse environmental samples.

## Criteria for Method Selection

We were tasked in this chapter with identifying a suitable method for analysis of soil microbial community composition, diversity, and function, addressing the six important soil processes outlined in Table 1. Arguably each of those processes may require a unique indicator and/or molecular technique to completely characterize any one component considering the complexity found in soil microbial communities and the functions they perform. It was beyond the scope of this chapter to propose methods to assess soil fauna. The importance of the entire biological community cannot be understated, however, to help limit this discussion, we focus on the microbial community. Our goal is to provide a cost-effective method that best predicts taxonomic composition and diversity of soil microbial communities but may be flexible enough to provide additional information on the functions.

Molecular tools are currently the most advanced method for measuring microbial taxonomy and function in complex samples. For example, as our concept of microbial species has evolved and rRNA amplicon sequencing has become a major determinant of new species descriptions using relatively robust databases, rRNA sequence analysis is an obvious place to start. One caveat is that rRNA copy number differs between individuals making it difficult to accurately quantify individuals within the community. Whole genome sequencing may overcome such problems, but this comes with an increased price and computational complexity that may be inappropriate for samples containing 1000's of distinct genomes (Singh et al., 2009).

Given cost considerations, flexibility, and ability to target both taxonomic and functional genes, we suggest a combined qPCR, amplicon sequencing, and phylogenetic reconstruction approach will provide the best current assessment of microbial abundance, diversity, and function. A variety of analytical pipelines exist for handling and processing sequencing data (e.g., QIIME [Caporaso et al., 2010], Mothur [Schloss et al., 2009]), but myPhyloDB (Manter et al., 2016) is specifically designed for agricultural soils and can be used for all aspects of data handling (e.g., processing, storage, and analysis). Although not measured directly, 16S rRNA amplicon sequencing can be used to address soil functions (Pantigoso et al., 2018; Zhu et al., 2016) through a phylogenetic reconstruction approach (e.g., PICRUSt [Langille et al., 2013]). Our proposed method is: (1) practical for molecular biology labs of all sizes, (2) relatively fast (~182 samples processed and analyzed per week), and (3) cost-effective (~$50 per sample not including labor). The combined flexibility of those two techniques allows for targeted and untargeted approaches at user-defined taxonomic or functional specificity levels.

## Soil Health Indicator Effectiveness and Interpretation

Soil biota are responsive to both short-term and long-term environmental changes (Fierer, 2017) and previous research has shown shifts in microbial community composition as a result of irrigation (Mavrodi et al., 2018), tillage (Degrune et al., 2017), and inputs of nutrients (Zhu et al., 2016), herbicides (Borowik et al., 2017) and pesticides (Regar et al., 2019). Our proposed qPCR, amplicon sequencing, and phylogenetic reconstruction can quantify taxonomic and functional abundances within the soil microbial community that directly influence the soil processes outlined in Table 13.1. This information along with agricultural management histories, key soil characteristics, and climatic variables can then be used to identify thresholds and microbial assemblages of healthy soils across multiple environments and systems. This information will enable the interpretative power to properly assess a single sample in time using indicator curves or other relative comparisons to healthy and productive cropping systems. In the short-term, as this database grows, it is recommended that sample collection include a 'side-by-side' management comparison or samples that track management changes over time.

The list of potential taxa that are responsive to management and correlated with soil health continues to grow as our knowledge of soil microbial communities develops. For example, bacteria of the phyla Verrucomicrobia have been shown to be associated with minimally disturbed tallgrass prairie soils (Bergmann et al., 2011; Fierer et al., 2013). This phyla tends to be more abundant under low nutrient conditions (da Rocha et al., 2010) and are found less in crop soils that had been fertilized (Ramirez et al., 2012). Thus, higher abundance is likely more a reflection of nutritional status rather than disturbance. Bacteroidetes, in contrast, are more common under nutrient rich conditions (Schattenhofer et al., 2009). Other studies have identified key bacterial and archaeal taxa correlated with soil fertility management (Wessén et al., 2010) could be used as indicators of nutritional stress or abundance. Similarly, different groups of Acidobacteria have been shown to vary with land use and soil properties (Eichorst et al., 2011) and should be explored as potential indicators. For example, Acidobacteria along with Firmicutes, and Actinobacteria are involved in degradation of complex lignin components of plant cell walls (Saini et al., 2015). The phylum Rubrobacteria are resistant to high temperatures (Davinic et al., 2012), gamma radiation (Ferreira et al., 1999) and desiccation (Singleton et al., 2003), and thus could be used as indicators of drought or temperature stress.

The abundance and variety of functional genes affecting C, N, P, and S transformations have been linked to land management and may be useful for assessing soil health. Genes related to the N cycle are the most quantified, particularly those influencing N fixation and nitrification which are thought to be especially sensitive to both environmental and management (Levy-Booth et al., 2014). The abundance of functional genes for N transformation (denitrification, ammonification) was closely linked with independent measures of soil N pools and fluxes in a study comparing conventional versus conservation agricultural practices (Xue et al., 2013). The abundance of *nifH* genes quantified by qPCR in agricultural soils has also been correlated to crop residue removal, crop rotation, and N fertility (Reardon et al., 2014; Wakelin et al., 2007). Ammonia oxidizing (*amoA*) gene abundances have been correlated with the amount and form of N present (Banerjee et al., 2018; Banning et al., 2015). Genes involved in denitrification have been a qPCR target in soil studies connecting gene distribution and abundance to processing of C and N in different terrestrial environments (Kong et al., 2010; Morales et al., 2010; Wallenstein et al., 2006). Other potential structural gene targets (e.g., *nifH*, *bglB*, *acdS*; Table 13.1) reflecting soil health include both well-defined, economically important plant pathogens and poorly defined "plant growth promoting bacteria" or soil bacteria associated with weed-suppression. Use of qPCR to quantify antibiotic resistance (AR) genes is another common environmental application, further supporting quantification of AR gene types and abundance as potential indicators of biological soil health.

The number of 16S or 18S rRNA gene copies for a given taxon can be considered a proxy for the relative amount of bacterial and fungal biomass. A common soil health index is the fungal to bacterial (F:B) ratio, which has been correlated with biome type, vegetation successional stage, land management, soil properties, soil trophic interactions, and biogeochemical cycles (Crowther et al., 2015; de Vries et al., 2012; Fierer et al., 2009; Grandy et al., 2009; Harris 2009; Six et al., 2006). Elevated F:B ratios have been positively correlated with ecosystem services and nutrient retention within several terrestrial settings (de Vries et al., 2013), although there are contrary observations (Strickland & Rousk, 2010; Thiet et al., 2006). Although typically estimated using fatty acid or cell counting techniques, F:B ratios can also be calculated by coupling amplicon sequencing with qPCR.

Among the fungal groups, arbuscular mycorrhizal (AM) fungi have been used as a soil health indicator for assessing positive (Dai et al., 2014; de Vries et al., 2013; Wagg et al., 2014) and negative (Jansa et al., 2006) effects of soil and crop management practices. Several traditional methods to quantify AM fungi in soils or plants exist (Sylvia, 2018), but none seem to be satisfactory for routine soil health evaluation. The PLFA biomarker (C16:1*cis*11) is one example often discussed as an indicator of AM fungi. Unfortunately, it is not reliable in the phospholipid fraction and although it is reliable in the neutral lipid fraction, that material is seldom analyzed (Drijber et al., 2000; Sharma & Buyer, 2015). To date, we are not aware of any suitable AM fungi-specific primer sets that can be used in a qPCR only assay to estimate AM biomass; however, as AM fungi are limited to the phylum Glomeromycota, a combined qPCR/amplicon sequencing approach can be used by limiting biomass estimates to the Glomeromycota phylum.

As highlighted above, the major advantage of the qPCR/amplicon sequencing approach is that it involves enumeration of DNA extracts that can be used for multiple purposes (e.g., taxonomy, diversity, and function). Once extracted DNA can be stored for long periods of time and re-evaluated as new primers or platforms become available. Furthermore, given the vast diversity of microbes in the soil, apparent geographic specificity of microbes, and functional redundancy of microbial genes across taxa, we suggest that the functional indicators of soil health will capture the true status of soil health in agricultural systems.

## Production Readiness

The qPCR and library preparation steps can be done in a typical molecular biology lab manually or with advanced automation. We recommend use of Qiagen's QIAcube instrument for automated DNA extraction, and QIAgility for PCR-prep liquid handling. One sequencing run consists of two, 96-well PCR plates. With extraction and PCR-prep automation, multiple libraries can be prepared in two to 3 d at a staggered pace. All equipment and reagents for this protocol are commercially available and require no significant modification. Recommended software for analysis and interpretation are open source and easily run on a simple desktop computer.

## Measurement Repeatability

The method described herein should be detailed enough for accurate replication in a variety of commercial or research molecular laboratories. While the combination of qPCR and amplicon sequencing is not currently standard practice, each step is common, and can be easily combined. The procedure is based on amplification of DNA extracted from soils, which is very stable if handled and stored properly, thus allowing for further analyses years or even decades into the future. The raw data obtained by next-generation sequencing platforms (e.g., MiSeq sequencing (fastq files) can safely be stored in myPhyloDB for future re-analysis; both features are attractive as technology and databases are always rapidly evolving. There are inherent biases in molecular techniques that are discussed below; however, by following the standardized protocol, these should not impact repeatability or precision of the method.

Quantification of gene (rRNA) copies by qPCR is robust, using a standard curve and multiple controls that include a mock community of known bacterial concentrations, rRNA copy numbers, a calibration soil sample, and negative control.

## Other Considerations

DNA extraction and PCR have inherent biases that cannot be avoided. During extraction, the homogenization step introduces variability in lysis of the diverse cell types (e.g., bacteria vs. fungi, Gram positive vs. Gram negative). When comparing samples extracted at different times, it is important the kits and homogenization methods are identical. During DNA extraction, polyphenols, humic acids, and other PCR inhibitors are also extracted with genomic DNA and carried over to qPCR affecting accurate quantification. Ensuring all samples are diluted and treated in a similar manner is paramount. The PCR polymerase chosen is also a source of variability as their fidelity and proofreading capacities differ. The amplicon DNA sequences themselves present bias, as higher GC content is less efficient at annealing. Bacteria also possess various gene copy numbers, such that one amplicon sequenced does not equal presence of one bacterial/fungal cell. It is therefore possible for some microbial DNA sequences to be under or over amplified during PCR. Protocols must be carefully designed and meticulously followed to eliminate such biases and maximize PCR efficiency. Although the protocol described below (see Section 3) is designed to provide a standardized method of soil microbial communities using 16S and 18S rRNA primers, other primers targeting a gene of interest (e.g., *nifH*, *acdS*, or other variable regions of the rRNA genes) can be easily incorporated into the protocol with the addition of Illumina-specific overhang adaptor sequences and appropriate modification of the PCR conditions.

## Selected Method Protocol

A two-step library preparation for quantification of microbial biomass (e.g., rRNA copies per gram soil) and amplicon sequencing with the Illumina MiSeq sequencing platform is used to quantify abundance and identify community structure. Primers presented here amplify the V3–4 variable regions of the bacterial 16S rRNA and the V4–5 variable regions of the eukaryote 18S rRNA genes with Illumina overhang adaptor sequences. We recommend using approximately 150 samples in a pooled library to ensure at least 10,000 reads per sample.

### List of Specialized Equipment

Required

- Illumina MiSeq System
- qPCR thermal cycler
- Centrifuges (tube and 96-well plate)
- Shell blender
- Electrophoresis rig
- Gel imaging system
- Pipettes
- Qubit Fluorometer
- Vortex
- Freezer (−20 °C)
- Freezer (−80 °C)
- PCR/biological hood with UV light
- Analytical balance
- Oven

Optional

- Qiagen QIAcube
- Qiagen QIAgility
- Tissue lyzer homogenizer

### Soil Handling

All soils should be kept cool during sampling and transport (i.e., placed in a cooler with blue ice) avoiding excessive delays and temperature extremes. Upon receipt in the laboratory, soils should be passed through a 4-mm sieve, homogenized by hand, and stored at −20 °C until further analysis.

## Soil Moisture Determination

Weigh out approximately 5 g of field most soil into a pre-weighed aluminum weigh boat. Record both tin and tin + field-moist soil weights to four decimal places using an analytical balance. Oven dry sample to a constant weight (>24 h) at 105 °C and record tin + dry soil weight. Calculate the dry weight (DW) to fresh weight (FW) ratio for each sample, which will be used to determine microbial biomass on a soil DW basis.

## DNA Extraction

Genomic DNA (gDNA) is extracted from samples by hand or using Qiagen's QIAube platform. Samples should be extracted in triplicate. It is recommended to also extract a positive control of a calibration soil and a negative control of molecular grade water and to process all through complete sequencing. We are unaware of any commercially available calibration soils for molecular analyses; therefore, we propose that labs include a subsample of a well-homogenized soil sample (~10 kg soil from a single field collection, sieved to 4 mm, and mixed thoroughly (e.g., V type blender), aliquoted into 1-g subsamples, and stored at −80 °C) in all qPCR runs to monitor qPCR efficiency and consistency.

### Reagents and Consumables

- Qiagen PowerLyzer PowerSoil Pro kit (cat# 12855-100)
- 1000 μl and 100 μl filtered tips
- 1.5 ml tubes

### Procedure

1) Weigh out 0.25 ± 0.01 g of soil onto pre-tared weighing paper and record to four decimal places. Transfer soil to the DNA extraction bead tube.
2) Follow DNA extraction per manufacturer's instructions. Elute into 100 μl elution buffer.

### Quality Check (optional, but Recommended)

Using Qubit fluorometer, quantify 1 μl of gDNA according to manufacturer's recommendations. If the gDNA eluted from your calibration soil (>10% lower than expected) or soil samples (<10 ng/μl) is low you want to double-check your DNA extraction kits and re-extract samples as appropriate.

## Quantitative Polymerase Chain Reaction (qPCR)

The first round of qPCR amplifies the target gene region and incorporates Illumina-specific overhang adapters for attachment of barcodes/indices during the second round of PCR. Amplicons are quantified during this first round of qPCR using a standard curve of known concentrations. Using the exact starting weight of soil extracted, the number of gene copies per gram of soil (dry weight) can be calculated to quantify total bacterial and/or fungal abundance (see 'Analysis Section: qPCR', p. 308).

### Reagents and Consumables

- Extracted soil gDNA; 10–20 ng/µl

  Note: The starting concentration of input DNA template may need to be optimized for your project and sample. In general, starting with 10 ng/µl of soil gDNA will result in cycle threshold (Ct) values within the standard curve and no PCR inhibition using the specified dilutions.

- Bacterial gDNA standard: *Pseudomonas putida* KT2440 (cat# ATCC 47054D-5)
- Fungal gDNA standard: *Aspergillus niger* 4247 (cat# ATCC 6275D-1)
- Mock community gDNA, 1 ng/µl (ZymoBIOMICS, cat# D6305)
- Molecular/PCR grade water
- Maxima SYBR Green Master Mix (ThermoFisher, cat# K0242)
- Qubit fluorometer dsDNA HS Assay Kit (Invitrogen, cat# Q32854)
- Qubit 500 µl assay tubes (Invitrogen, cat# Q32856)
- 96-well PCR plate and clear microseal film
- 1.5–2.0 ml lo-bind micro centrifuge tube
- 1000 µl, 200 µl, and 10 µl filtered tips

### Primer Selection

Primers for this protocol are not available for purchase through Illumina and must be custom ordered. The bacteria 16S primers targeting the V3–4 rRNA region are 341F and 785R (Klindworth et al., 2013). With the overhang, adaptor and index sequences, the total amplicon length is approximately 601 base pairs as measured on a TapeStation bioanalyzer during QC prior to sequencing. The full-length primer sequences with overhang adaptor (underlined) are:

16S Forward: 5' TCGTCGGCAGCGTCAGATGTGTATAAGAGACAGCCTACGGGNGGCWGCAG 3'

16S Reverse: 5' GTCTCGTGGGCTCGGAGATGTGTATAAGAGACAGGGACTACHVGGGTATCTAATCC 3'

The eukaryotic V4–5 18S rRNA primers are nu-SSU-0817–5' forward and nu-SSU-1196–3' reverse (Borneman & Hartin, 2000). The amplicon length with overhang adaptor and indices is approximately 516 base pairs as measured by TapeStation. The full-length primer sequences with overhang adaptor (underlined) are:

18S Forward: 5' TCGTCGGCAGCGTCAGATGTGTATAAGAGACAGGGAAACTCACCAGGTCCAGA 3'

18S Reverse: 5' GTCTCGTGGGCTCGGAGATGTGTATAAGAGACAGATTGCAATGCYCTATCCCCA 3'

## Procedure

1) Optional: For new laboratories or new soil types it is recommended that samples are tested for the presence of PCR inhibitors.
   a) Run an independent qPCR analyses on selected test gDNA extracts serially diluted with molecular grade $H_2O$ and check for increasing target concentrations with dilution. In our experience, the dilutions in step 5 below are adequate for most soil types.
2) Set up the PCR reaction master mix with SYBR, water, forward and reverse primers (see PCR conditions below).
3) Create gDNA standards by diluting the stock gDNA standards (e.g., *Pseudomonas putida* or *Aspergillus niger*) with molecular-grade $H_2O$ to 1000, 100, 10, 1, and 0.1 pg/µl.
4) Pipette 18 µl of master mix into each PCR plate well.
5) Pipette 2 µl diluted gDNA from each soil sample, positive controls (i.e., Zymo mock community, calibration soil), negative controls (i.e., molecular-grade $H_2O$) into each well.
   a) For 16S PCR, dilute gDNA 20-fold to avoid PCR inhibitors (see also step 1), or to a starting concentration of approximately 0.5–1 ng/µl.
   b) For 18S PCR, dilute gDNA 10-fold (see also step 1), or to a starting concentration of 1–2 ng/µl.
   c) Each sample should be run in triplicates. Controls and standards should be run in duplicates or triplicates.
6) Cover, and spin down plate.
7) Run plate in thermal cycler under conditions specific to the target amplicon (see thermal cycler conditions below).
8) When PCR is finished, immediately remove plate from thermal cycler, spin down briefly, and proceed to PCR bead cleanup step.

**PCR Conditions: 16S**

- 2 µl DNA template (20X dilution)
- 10 µl 2X Maxima SYBR
- 4 µl PCR grade water
- 2 µl Forward primer (10 µM)
- 2 µl Reverse primer (10 µM)
  - Total: 20 µl

**Thermal Cycler Conditions**

1) 95 °C for 5 min
2) 30 cycles
   a) 95 °C for 40 s
   b) 55 °C for 120 s
   c) 72 °C for 60 s
3) 72 °C for 7 min

**PCR Conditions: 18S**

- 2 µl DNA template (20X dilution)
- 10 µl 2X Maxima SYBR
- 4 µl PCR grade water
- 2 µl Forward primer (10 µM)
- 2 µl Reverse primer (10 µM)
  - Total: 20 µl

**Thermal Cycler Conditions**

1) 95 °C for 5 min
2) 35 cycles
   a) 94 °C for 15 s
   b) 55 °C for 30 s
   c) 72 °C for 60 s
3) 72 °C for 5 min

## Quality Check

Visualize 5 µl pooled PCR product on a 0.8% agarose gel with 1 kb ladder or other appropriate DNA standard. Verify target size, amplification of sample wells, and lack of amplification in negative controls. If no bands are visible, verify there are no inhibitors (i.e., dilute gDNA an additional 10X and re-run PCR) or increase cycle number for low abundance targets.

## Amplicon Bead Clean-up

Next, the PCR product is cleaned using magnetic solid-phase reversible immobilization (SPRI) beads to isolate the target amplicon DNA, and to remove unwanted contaminants such as proteins and non-target oligomers. Use of a handheld magnetic bead extractor is recommended to reduce pipetting and time. A 96-well magnetic separation block may also be used.

### Reagents and Consumables

1) 96-well magnetic bead extractor (V&P Scientific, cat# VP407AM-N) and 96-well plate co\ver (cat# VP 407AM-N-PCR)
2) Solid phase reversible immobilization (SPRI) magnetic beads such as AmPureXP beads (cat# A63880) or self-made SPRI beads (Rohland & Reich, 2012)
3) Molecular-grade ethanol (EtOH), 200 proof (100% EtOH)
4) Molecular-grade water
5) 96-well plate, >0.3 ml well volume (Corning/Costar 3797, cat# 07200105)

**Procedure**

1) Before starting:
   a) Bring beads and PCR samples to room temperature.
   b) Make fresh 75% EtOH by diluting with molecular-grade $H_2O$ enough for 200 µl per sample.
   c) Use a new magnetic cover plate or thoroughly cleaned with a 10% Chlorox solution (Prince & Andrus, 1992) and rinsed with molecular-grade $H_2O$. Fit onto Magnetic Bead Extractor so there are no gaps between magnetic pins and end of the cover plate's pointed wells.
   d) Prepare four 96-well plates: One for mixing beads and PCR product, two for EtOH washing, and one for DNA elution into water. Plates used for bead mixing and EtOH washing can be reused, if cleaned and sterilized with 10% bleach and UV irradiation.
2) Add 1.0X of SPRI beads to each well of the first plate (20 µl beads + 20 µl PCR product). Note: Ratio may vary. To target longer amplicons, use a lower ratio of beads such as 0.8X. To target maximum DNA recovery regardless of size, use a higher ratio of beads such as 1.2X.
3) Using a multichannel pipette, add PCR reaction to beads. Pipette up and down slowly 10 times to mix.
4) Incubate at room temperature for 5 min.
5) Insert magnetic pins with cover plate attached and allow beads to bind for 1 min. Try to concentrate beads at the tip (bottom) of the plate as much as possible for maximum elution yield.
6) Transfer magnet with beads to second plate with 100 µl of 75% EtOH for 30 s. Do not remove magnet.
7) Transfer magnet with beads to third plate with 100 µl of 75% EtOH for 30 s. Do not remove magnet.
8) Remove from third plate and allow to air-dry for up to 5 min. Do not over dry to the point of cracking, which will reduce DNA elution and recovery.
9) Release plate cover with attached beads from the magnet into final fourth elution plate with ≥40 µl water. Swirl to release beads and incubate 2 min. Note: For better efficiency, elution water can be heated to 50 °C.
10) Add magnet back to cover to capture beads, remove magnet and plate, and discard.
11) Check amplicon size on 1.2% agarose gel.
12) Proceed to next amplification step, or seal plate and store at 4 °C overnight or −20 °C longer term.

## Indexing PCR

A second round of PCR is performed to attach an index/barcode sequence to the Illumina-specific overhang adapters. Each sample has a unique set of forward and reverse indices for multiplexing up to 182 samples. Illumina Nextera index sequences can be found in Illumina document #1000000002694 v12.

### Reagents and Consumables

1) Illumina Nextera XT DNA Library Preparation Kit 96 samples (Illumina, cat# FC-131-1024)
2) PCR grade water
3) 2X Maxima SYBR Green polymerase (Thermofisher, cat# K0242)
4) 96-well PCR plate and clear microseal film
5) 200 µl, and 10 µl filtered tips

### Procedure

1) Into each well, add SYBR and water (see PCR conditions below).
2) Add the appropriate forward and reverse index into each well. Use extreme care not to cross contaminate index tubes or wells.
3) Cover, and spin down plate.
4) Run plate in thermal cycler (see thermal cycler conditions below).
5) When PCR is finished, immediately remove plate from thermal cycler, spin down briefly, and proceed to PCR bead cleanup step.

### Indexing PCR Conditions

- 5 µl DNA template (no dilution)
- 25 µl 2X Maxima SYBR
- 10 µl PCR grade water
- 5 µl Forward i7 index (5 µM
- 5 µl Reverse i5 index (5 µM
  - Total: 50 µl

### Thermal Cycler Conditions

4) 95 °C for 3 min
5) 8 cycles
   a) 95 °C for 30 s
   b) 55 °C for 30 s
   c) 72 °C for 30 s
6) 72 °C for 5 min

### Indexed Amplicon Bead Clean-up

See ‘Amplicon Bead Clean-up’, p. 305

### Normalization and Pooling of Final Library

The indexing and bead clean-up steps should result in all samples being of equal concentration, typically 5–15 ng/µl. However, if concentrations are variable, it may be necessary to quantify and dilute each sample prior to pooling. For a typical MiSeq run, 3 mM or approximately 1 ng/µl is a sufficient starting concentration.

**Materials and Consumables**

1) Qubit fluorometer dsDNA HS Assay Kit (Invitrogen, cat# Q32854)
2) Qubit 500 μl assay tubes (Invitrogen, cat# Q32856)
3) 1.5–2 ml lo-bind Eppendorf tubes
4) 4. 200 μl, 10ul filtered tips

**Procedure**

1) Using a Qubit fluorometer, test at least 10 samples to ensure they are of similar concentration (2–4 ng/μl). If not, quantify and dilute each sample to 2 ng/μl with molecular grade $H_2O$.
2) Pool 10 μl of each sample into a 1.5 ml tube.
3) Quantify final pool using Qubit.
4) Aliquot >20 μl into 1.5 ml tube for sequencing.
5) Store remaining library in −20 °C as backup.

## Library Amplicon Size Determination

Prior to sequencing, it is important to check amplicon quality by running a gel to ensure there are no unindexed amplicons or other oligomers in the final library. The absolute amplicon size must be determined using a bioanalyzer to dilute the final library to the desired nanomolar concentration.

**Materials and Consumables**

1) Agarose gel with ethidium bromide or equivalent
2) Reagents and consumables for capillary electrophoresis (bioanalyzer) system (e.g., Agilent TapeStation)

**Procedure**

1) Run library on 1.2% agarose gel with ladder to confirm amplicon size and purity.
2) Run bioanalyzer and record average amplicon size in base pairs.

## Library Quantification and Dilution

The precise concentration of the library must be determined by qPCR to calculate absolute nanomolar concentration. Using the resulting ng/ul concentration and amplicon size determined in step 4.9, the library is then diluted to 4 nM in preparation for denaturation.

**Materials and Consumables**

3) KAPA Library Quantification Kit (cat#KK4824)
4) Sequencing library with known concentration as positive control

5) PCR grade water
6) 200 µl, 10 µl filtered tips
7) 96-well PCR plate and clear microseal film
8) Dilution buffer: Tris-Cl 10 mM, pH 8.5 with 0.1% Tween 20

**Procedure**

1) Use the KAPA Library Quantification Kit per manufacturer's instructions.
   - Run in triplicates of two dilutions: 10,000X and 20,000X.
   - Also run a previous library with known concentration as a control.
2) Calculate absolute nanomolar concentration using resulting ng/ul and amplicon length measured by bioanalyzer.
3) Dilute library to 4 nM using dilution buffer.

## MiSeq Library Denaturation and Dilution

The final library is denatured using NaOH and diluted to a final volume of 600 µl before loading onto the MiSeq. A detailed protocol for denaturation and dilution can be found on the Illumina website (document #15039740).

**Materials and Consumables**

1) 4 nM library
2) Illumina MiSeq Reagent Kit v3 (600 cycle, cat# MS-102-3003)
3) Illumina PhiX Control (cat #FC-110-3001)
4) 1.0 N NaOH molecular biology grade
5) Dilution buffer: Tris-Cl 10mM, pH 8.5 with 0.1% Tween 20
6) Tris-HCl, pH 7.0
7) Molecular grade water
8) 1.5–2 ml Eppendorf tubes
9) Ice bucket

**Procedure**

1) Remove sequencing cartridge from −20 °C, and thaw to room temperature.
2) Remove HT1 (Hybridization Buffer) from −20 °C, and thaw on ice.
3) Prepare fresh dilution of 0.2 N NaOH: 800 µl water + 200 µl stock 1.0 N NaOH. Use within 12 h.
4) Denature library:
   a) In a 1.5ml tube, combine 5 µl of 4 nM diluted library + 5 µl of 0.2 N NaOH
   b) Vortex briefly, spin down for 1 min, and incubate at room temperature for 5 min.

5) Add 990 µl prechilled HT1. The result is 1 ml of a 20 pM denatured library.
6) Dilute library:
   a) Dilute to desired concentration. The final concentration depends on your amplicon length and complexity of sample with 15 pM (450 µl library and 150 µl prechilled HT1 buffer) a good starting point.

| Concentration | 6 pM | 8 pM | 10 pM | 12 pM | 15 pM | 20 pM |
|---|---|---|---|---|---|---|
| 20 pM library | 180 ul | 240 ul | 300 ul | 360 ul | 450 ul | 600 ul |
| Prechilled HT1 | 420 ul | 360 ul | 300 ul | 240 ul | 150 ul | 0 ul |

7) Add 20 nM PhiX control at 15% volume of the library. PhiX will then show up in your run as 15% of your indexed reads. The amount of PhiX used depends on the complexity and diversity of the samples: The lower the diversity, the more PhiX should be used. Performance of previous sequence runs can indicate if PhiX needs to be adjusted to a slightly higher or lower concentrations.

### MiSeq Loading, Sequencing, and Data Transfer

Beginning a sequencing run consists of uploading an Excel meta sequencing sheet including run parameters, sample names, and index sequences to de-multiplex. A template can be found on the Illumina website or provided by the MiSeq operator. The library is then loaded onto the cartridge and into the instrument. For detailed instructions, a complete MiSeq System User Guide (document# 15027617) can be found on the Illumina website.

#### Procedure

1) Upload Excel meta file to MiSeq instrument using the USB port
2) Load entire library onto reagent cartridge in position 17, labeled 'load samples'
3) Insert cartridge into MiSeq instrument
4) Insert flow cell into MiSeq instrument flow cell compartment
5) Begin run. Run time including demultiplexing is 48–72 h
6) When run is complete, download the fastq files from Illumina BaseSpace Sequence Hub

## Analysis Section

### qPCR

A typical qPCR standard curve is shown in Figure 13.1a. The typical sigmoidal shape of DNA fluorescence during PCR amplification can be seen for all standards (solid lines) in the graph. Once fluorescence increases beyond the

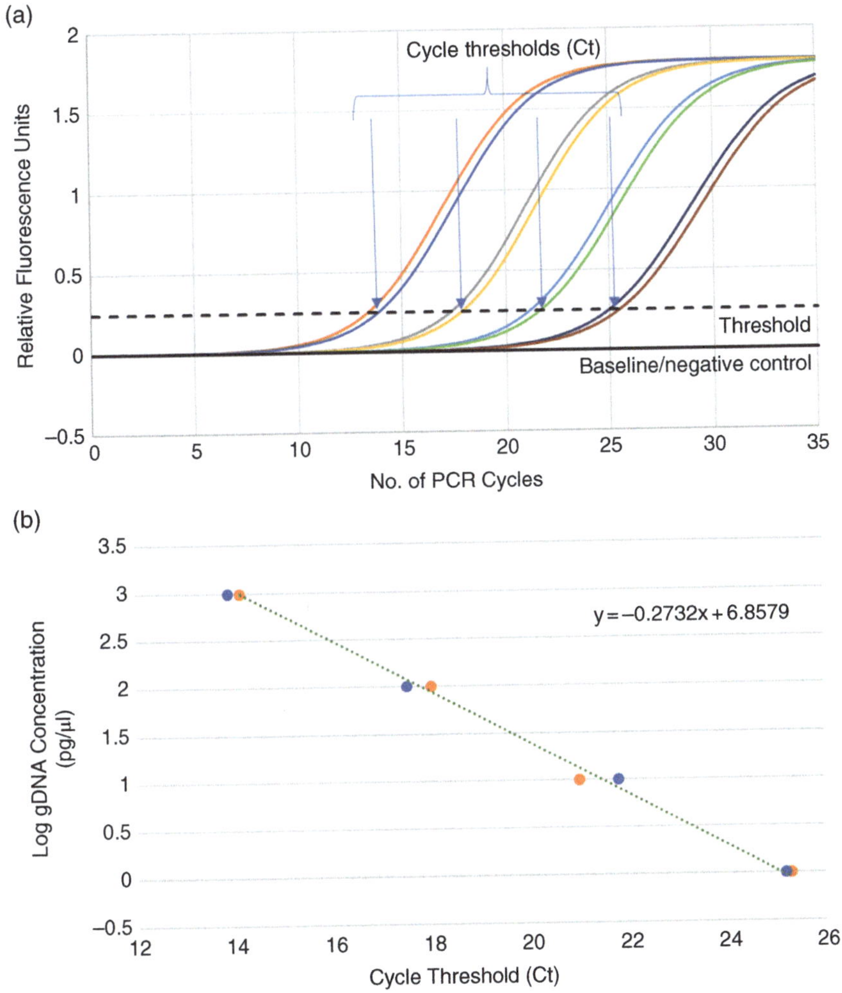


**Figure 13.1** Typical qPCR standard curves using two replicates per concentration. *A*) Amplification curve using four gDNA standards (e.g., 1000, 100, 10, 1 pg/µl) and one negative control (i.e., no template control). *B*) Regression line relating gDNA starting concentration and cycle threshold (Ct).

background (i.e., below detection) and above an instrument software derived threshold, a cycle threshold (Ct) value is calculated for each sample (i.e., number of PCR cycles required to reach the threshold). These Ct values and starting genomic DNA (gDNA) concentrations can be used to develop a standard curve as shown in Figure 13.1b. This (hypothetical) regression equation can then be used to determine the starting concentration (pg/g soil dry weight) of target gDNA in a soil sample ($A_i$) as follows:

$$A_i = 10^{\left(-0.2732 \times C_i + 6.8579\right)} \times D_i \div V_i \times W_i$$

where $C_i$ is the observed sample qPCR Ct value, $D_i$ is the dilution factor for the given sample, $V_i$ is the extracted soil fresh weight (FW), and $W_i$ is the FW/DW ratio for the soil sample. If desired, this value can be further converted to rRNA copies per g soil DW using the known rRNA copy number for DNA standards that have sequenced genomes. For example, the *Pseudomonas putida* KT2440 (ATCC 47054D-5) gDNA standard has seven 16S rRNA copies in a single genome or 1120 copies/pg gDNA, and the *Aspergillus niger* 4247 (ATCC 6275D-1) gDNA standard has 100 18S rRNA copies in a single genome or 2660 copies/pg gDNA. If you run the Zymo mock community, use its calculated concentration to correct for any differences between PCR runs.

## Amplicon Sequence Processing and Analysis

There are a wide variety of analytical pipelines and software available for handling amplicon sequence data; however, we routinely use the default DADA2 pipeline (Callahan et al., 2016) that is contained within myPhyloDB v1.2 (Manter et al., 2016). Sequence reads clustered into operational taxonomic units (OTUs) at a defined genetic identity (i.e., 97%) can result in spurious OTUs. De-noising packages, such as DADA2, greatly reduce these OTUs and result in α-diversity metrics that better represent mock communities (Nearing et al., 2018). Briefly, the standard myPhyloDB pipeline removes PCR primers using the open source Python program Cutadapt (Martin, 2011), infers amplicon sequence variants (ASV), and removes Chimeras using the default DADA2 pipeline. It then assigns each ASV to the closest reference sequence (99% match) contained within the Green Genes reference database (16S rRNA; DeSantis et al., 2006) or SILVA reference database (18S rRNA; Quast et al., 2012) using VSEARCH (Rognes et al., 2016). For 16S data, taxonomic relative abundances are converted to gene relative abundances by phylogenetic reconstruction using PICRUSt (Langille et al., 2013), which results in a community-level gene relative abundance (i.e., abundances are not corrected for potential differences in copy number). For instance, if 50% of the community contains a gene of interest then the gene abundance will be 50%. Both taxonomic and gene relative abundance can be converted to total abundance (copies/g soil DW) by multiplying relative abundances by the appropriate qPCR values. Example calculations are as follows:

$$p_i = s_i /$$

where $p_i$ is the relative abundance (proportion) of the taxon of interest, $s_i$ is the number of sequences assigned to the taxon of interest, and $S$ is the total number of sequences for that sample. Converting to total abundance of the specified taxon ($N_i$) in a soil sample is calculated as:

$$N_i = p_i \cdot \mathrm{A}_i$$

where $p_i$ is the relative abundance (proportion) of the taxon of interest, and $A_i$ is the appropriate starting concentration (pg/g soil DW) of target gDNA in a soil sample. Any taxonomic level contained in the supplied reference database or gene with a known KEGG orthology (Kanehisa et al., 2016) can be targeted using the analytical capabilities of myPhyloDB.

## References

Albertsen, M., Hugenholtz, P., Skarshewski, A., Nielsen, K. L., Tyson, G. W., & Nielsen, P. H. (2013). Genome sequences of rare, uncultured bacteria obtained by differential coverage binning of multiple metagenomes. *Nature Biotechnology*, 31, 533. doi:10.1038/nbt.2579

Anderson, J. P. E., & Domsch, K. H. (1978). A physiological method for the quantitative measurement of microbial biomass in soils. *Soil Biology & Biochemistry*, 10(3), 215–221. doi:10.1016/0038-0717(78)90099-8

Bååth, E., & Anderson, T. H. (2003). Comparison of soil fungal/bacterial ratios in a pH gradient using physiological and PLFA-based techniques. *Soil Biology & Biochemistry*, 35(7), 955–963. doi:10.1016/S0038-0717(03)00154-8

Babalola, O. O. (2010). Beneficial bacteria of agricultural importance. *Biotechnology Letters*, 32(11), 1559–1570. doi:10.1007/s10529-010-0347-0

Banerjee, S., Thrall, P. H., Bissett, A., van der Heijden, M. G., & Richardson, A. E. (2018). Linking microbial co-occurrences to soil ecological processes across a woodland-grassland ecotone. *Ecology Evolution*, 8(16), 8217–8230. doi:10.1002/ece3.4346

Banning, N. C., Maccarone, L. D., Fisk, L. M., & Murphy, D. V. (2015). Ammonia-oxidising bacteria not archaea dominate nitrification activity in semi-arid agricultural soil. *Scientific Reports*, 5(1), 11146. doi:10.1038/srep11146

Banowetz, G. M., Whitaker, G. W., Kierksen, K. P., Azevedo, M. D., Kennedy, A. C., Giffith, S. M., & Steiner, J. J. (2006). Fatty acid methl ester analysis to identify sources of soil in surface water. *Journal of Environmental Quality*, 35, 133–140. doi:10.2134/jeq2005.0048

Bergmann, G. T., Bates, S. T., Eilers, K. G., Lauber C. L., Caporaso, J. G., Walters, W. A., Knight, R., & Fierer, N. (2011). The under-recognized dominance of

Verrucomicrobia in soil bacterial communities. *Soil Biology & Biochemistry*, 43(7), 1450–1455. doi:10.1016/j.soilbio.2011.03.012

Borneman, J., & Hartin, R. J. (2000). PCR primers that amplify fungal rRNA genes from environmental samples. *Applied & Environmental Microbiology*, 66(10), 4356–4360. doi:10.1128/AEM.66.10.4356-4360.2000

Borowik, A., Wyszkowska, J., Kucharski, J., Baćmaga, M., & Tomkiel, M. (2017). Response of microorganisms and enzymes to soil contamination with a mixture of terbuthylazine, mesotrione, and S-metolachlor. *Environmental Science & Pollution Research*, 24(2), 1910–1925. doi:10.1007/s11356-016-7919-z

Broughton, L. C., & Gross, K. L. (2000). Patterns of diversity in plant and soil microbial communities along a productivity gradient in a Michigan old-field. *Oecologia*, 125, 420–427. doi:10.1007/s004420000456

Bruto, M., Prigent-Combaret, C., Muller, D., & Moenne-Locoz, Y. (2014). Analysis of genes contributing to plant-beneficial functions in plant-growth-promoting rhizobacteria and related Proteobacteria. *Scientific Reports*, 4, 6261. doi:10.1038/srep06261

Bustin, S. A., Benes, V., Garson, J. A., Hellemans, J., Huggett, J., Kubista, M., Mueller, R., Nolan, T., Pfaffl, M. W., Shipley, G. L., Vandesompele, J., & Wittwer, C. T. (2009). The MIQE Guidelines: Minimum Information for publication of quantitative real-time PCR experiments. *Clinical Chemistry*, 55(4), 611–622. doi:10.1373/clinchem.2008.112797

Callahan, B. J., McMurdie, P. J., Rosen, M. J., Han, A. W., Johnson, A. A., & Holmes, S.P. (2016). DADA2: High-resolution sample inference from Illumina amplicon data. *Nature Methods*, 13(7), 581–583. doi:10.1038/nmeth.3869

Caporaso, J. G., Kuczynski, J., Stombaugh, J., Bittinger, K., Bushman, F. D., Costello, E. K., Fierer, N., Peña, A. G., Goodrich, J. K., Gordon, J. I., Huttley, G. A., Kelley, S. T., Knights, D., Koenig, J. E., Ley, R. E., Lozupone, C. A., McDonald, D., Muegge, B. D., Pirrung, M., Reeder, J., Sevinsky, J. R., Turnbaugh, P. J., Walters, W. A., Widmann, J., Yatsunenko, T., Zaneveld, J., & Knight, R. (2010). QIIME allows analysis of high-throughput community sequencing data. *Nature Methods*, 7(5), 335–336. doi:10.1038/nmeth.f.303

Caporaso, J. G., Lauber, C. L., Walters, W. A., Berg-Lyons, D., Huntley, J., Fierer, N., Owens, S. M., Betley, J., Fraser, L., Bauer, M., Gormley, N., Gilbert, J., Smith, G., & Knight, R. (2012). Ultra-high-throughput microbial community analysis on the Illumina HiSeq and MiSeq platforms. *ISME Journal*, 6, 1621–1624. doi:10.1038/ismej.2012.8

Chaparro, J. M., Sheflin, A. M., Manter, D. K., & Vivanco, J. M. (2012). Manipulating the soil microbiome to increase soil health and plant fertility. *Biology & Fertility of Soils*, 48, 489–499. doi:10.1007/s00374-012-0691-4

Church, M. J., Short, C. M., Jenkins, B. D., Karl, D. M., & Zehr, J. P. (2005). Temporal patterns of nitrogenase gene (nifH) expression in the oligotrophic North Pacific

Ocean. *Applied & Environmental Microbiology*, 71(9), 5362–5370. doi:10.1128/AEM.71.9.5362-5370.2005

Crowther, T. W., Thomas, S. M., Maynard, D. S., Baldrian, P., Covey, K., Frey, S. D., van Diepen, L. T. A., & Bradford, M. A. (2015). Biotic interactions mediate soil microbial feedbacks to climate change. *Proceedings of the National Academy of Science USA*, 112(22), 7033–7038. doi:10.1073/pnas.1502956112

Rocha, U. N., Andreote, F. D., de Azevedo, J. L., van Elsas, J. K., & van Overbeek, L. S. (2010). Cultivation of hitherto-uncultured bacteria belonging to the Verrucomicrobia subdivision 1 from the potato (*Solanum tuberosum* L.) rhizosphere. *Journal of Soils & Sediments*, 10(2), 326–339. doi:10.1007/s11368-009-0160-3

Dai, M., Hamel, C., Bainard, L. D., Arnaud, M. S., Grant, C. A., Lupwayi, N. Z., Malhi, S. S., & Lemke, R. (2014). Negative and positive contributions of arbuscular mycorrhizal fungal taxa to wheat production and nutrient uptake efficiency in organic and conventional systems in the Canadian prairie. *Soil Biology & Biochemistry*, 74(0), 156–166. doi:10.1016/j.soilbio.2014.03.016

Davinic, M., Fultz, L. M., Acosta-Martinez, V., Calderon, J. F., Cox, S. B., Dowd, S. E., Allen, V. G., Zak, J. C., & Moore-Kucera, J. (2012). Pyrosequencing and mid-infrared spectroscopy reveal distinct aggregate stratification of soil bacterial communities and organic matter composition. *Soil Biology & Biochemistry*, 46, 63–72. doi:10.1016/j.soilbio.2011.11.012

Vries, F. T., Bloem, J., Quirk, H., Stevens, C. J., Bol, R., & Bardgett, R. D. (2012). Extensive management promotes plant and microbial nitrogen retention in temperate grassland. *PLoS One*, 7(12), e51201.

Vries, F. T., Thébault, E., Liiri, M., Birkhofer, K., Tsiafouli, M. A., Bjørnlund, L., Jørgensen, H. B., Brady, M. V., Christensen, S., & de Ruiter, P. C. (2013). Soil food web properties explain ecosystem services across European land use systems. *Proceedings of the National Academy of Sciences USA*, 110(35), 14296–14301.

Degrune, F., Theodorakopoulos, N., Colinet, G., Hiel, M. P., Bodson, B., Taminiau, B., Daube, G., Vandenbol, M., & Hartmann, M. (2017). Temporal dynamics of soil microbial communities below the seedbed under two contrasting tillage regimes. *Frontiers in Microbiology*, 8, 1127. doi:10.3389/fmicb.2017.01127

DeSantis, T. Z., Hugenholtz, P., Larsen, N., Rojas, M., Brodie, E. L., Keller, K., Huber, T., Dalevi, D., Hu, P., & Andersen, G. L. (2006). Greengenes, a chimera-checked 16S rRNA gene database and workbench compatible with ARB. *Applied & Environmental Microbiology*, 72(7), 5069–5072. doi:10.1128/AEM.03006-05

Dobranic, J. K., & Zak, J.C. (1999). A microtiter plate procedure for evaluating fungal functional diversity. *Mycologia*, 91(5), 756–765. doi:10.1080/00275514.1999.12061081

Docherty, K. M., Balser, T. C., Bohanna, B. J., & Gutknecht, J. L. (2012). Soil microbial responses to fire and interacting global change factors in a California annual grassland. *Biogeochemistry*, 109, 63–83. doi:10.1007/s10533-011-9654-3

Doran, J. W., & Zeiss, M. R. (2000). Soil health and sustainability: Managing the biotic component of soil quality. *Applied Soil Ecology*, 15(1), 3–11. doi:10.1016/S0929-1393(00)00067-6

Drijber, R. A., Doran, W., Parkhurst, A. M., & Lyon, D. J. (2000). Changes in soil microbial community structure with tillage under long-term wheat-fallow management. *Soil Biology & Biochemistry*, 32, 1419–1430. doi:10.1016/S0038-0717(00)00060-2

Eichorst, S. A., Kuske, C. R., & Schmidt, T. M. (2011). Influence of plant polymers on the distribution and cultivation of bacteria in the phylum Acidobacteria. *Applied & Environmental Microbiology*, 77(2), 586–596. doi:10.1128/AEM.01080-10

Esitken, A. (2011). Use of plant growth promoting Rhizobacteria in horticultural crops. In D. K. Maheshwari (Ed.) *Bacteria in agrobiology: Crop ecosystems* (pp. 189–235). Berlin: Springer.

Fakruddin, M., & Mannan, K. S. B. (2013). Methods for analyzing diversity of microbial communities in natural environments. *Ceylon Journal of Science Biological Sciences*, 42(1), 19–33. doi:10.4038/cjsbs.v42i1.5896

Ferreira, A. C., Nobre, M. F., Moore, E., Rainey, F. A., Batista, J. R., & da Costa, M. S. (1999). Characterization and radiation resistance of new isolates of Rubrobacter radiotolerans and Rubrobacter xylanophilus. *Extremophiles*, 3(4), 235–238. doi:10.1007/s007920050121

Fierer, N. (2017). Embracing the unknown: Disentangling the complexities of the soil microbiome. *Nature Reviews Microbiology* 15, 579–590. doi:10.1038/nrmicro.2017.87

Fierer, N., Ladau, J., Clemente, J. C., Leff, J. W., Owens, S. M., Pollard, K. S., Knight, R., Gilbert, J. A., & McCulley, R. L. (2005). Assessment of soil microbial community structure by use of taxon-specific quantitative PCR assays. *Applied Environmental Microbiology*, 71, 4117–4120. doi:10.1128/AEM.71.7.4117-4120.2005

Fierer, N., Strickland, M. S., Liptzin, D., Bradford, M. A., & Cleveland, C. C. (2009). Global patterns in belowground communities. *Ecology Letters*, 12(11), 1238–1249. doi:10.1111/j.1461-0248.2009.01360.x

Fierer, N., Ladau, J., Clemente, J. C., Leff, J. W., Owens, S. M., Pollard, K. S., Knight, R., Gilbert, J. A., & McCulley, R. L. (2013). Reconstructing the microbial diversity and function of pre-agricultural tallgrass prairie soils in the United States. *Science*, 342(6158), 621–624. doi:10.1126/science.1243768

Findlay, R. H. (1996). The use of phospholipid fatty acids to determine microbial community structure. In A. D. L. Akkermans, J. D. Van Elsas, & F. J. DeBruijn (Eds.), *Molecular microbial ecology manual* (pp. 77–93). Dordrecht: Springer. doi:10.1007/978-94-009-0215-2_7

Finlay, B. J., Maberly, S. C., & Cooper, J. I. (1997). Microbial diversity and ecosystem function. *Oikos*, 80(2), 209–213. doi:10.2307/3546587

Franzosa, E. A., Hsu, T., Sirota-Madi, A., Shafquat, A., Abu-Ali, G., Morgan, X. C., & Huttenhower, C. (2015). Sequencing and beyond: Integrating molecular 'omics' for microbial community profiling. *Nature Reviews Microbiology*, 13, 360–372. doi:10.1038/nrmicro3451

Frostegård, Å., Tunlid, A., & Bååth, E. (2011). Use and misuse of PLFA measurements in soils. *Soil Biology & Biochemistry*, 43(8), 1621–1625. doi:10.1016/j.soilbio.2010.11.021

Gaby, J. C., Rishishwar, L., Valderrama-Aguirre, L. C., Green, S. J., Valderrama-Aguirre, A., Jordan, I. K., & Kostka, J. E. (2018). Diazotroph community characterization via a high-throughput *nifH* amplicon sequencing and analysis pipeline. *Applied Environmental Microbiology*, 84, 1512–1517.

Garland, J. L., & Mills, A. L. (1991). Classification and characterization of heterotrophic microbial communities on the basis of patterns of community-level sole-carbon-source utilization. *Applied Environmental Microbiology*, 57(8), 2351–2359. doi:10.1128/AEM.57.8.2351-2359.1991

Grandy, A. S., Strickland, M. S., Lauber, C. L., Bradford, M. A., & Fierer, N. (2009). The influence of microbial communities, management, and soil texture on soil organic matter chemistry. *Geoderma*, 150, 278–286. doi:10.1016/j.geoderma.2009.02.007

Grichko, V. P., & Glick, B.R. (2001). Amelioration of flooding stress by ACC deaminase-containing plant growth-promoting bacteria. *Plant Physiology & Biochemistry*, 39(1), 11–17. doi:10.1016/S0981-9428(00)01212-2

Hamady, M., Walker, J. J., Harris, J. K., Gold, N. J., & Knight, R. (2008). Error-correcting barcoded primers for pyrosequencing hundreds of samples in multiplex. *Nature Methods*, 5(3), 235–237. doi:10.1038/nmeth.1184

Harris, J. (2009). Soil microbial communities and restoration ecology: Facilitators or followers? *Science*, 325(5940), 573–574. doi:10.1126/science.1172975

Hugenholtz, P. (2002). Exploring prokaryotic diversity in the genomic era. *Genome Biology*, 3.

Hugerth, L. W., & Andersson, A. F. (2017). Analyzing microbial community composition through amplicon sequencing: From sampling to hypothesis testing. *Frontiers in Microbiology*, 8, 1561. doi:10.3389/fmicb.2017.01561

Ibekwe, A. M., Kennedy, A. C., Frohne, P. S., Papiernik, S. K., Yang, C., & Crowley, D. E. (2002). Microbial diversity along a transect of agronomic zones. *FEMS Microbiology Ecology*, 39, 183–191. doi:10.1111/j.1574-6941.2002.tb00921.x

Jansa, J., Wiemken, A., & Frossard, E. (2006). The effects of agricultural practices on arbuscular mycorrhizal fungi. *Geological Society London Special Publications*, 266(1), 89–115. doi:10.1144/GSL.SP.2006.266.01.08

Kanehisa, M., & Goto, S. (2016). KEGG as a reference resource for gene and protein annotation. *Nucleic Acids Research*, 44, D457–D462. doi:10.1093/nar/gkv1070

Klindworth, A., Pruesse, E., Schweer, T., Peplies, J., Quast, C., Horn, M., & Glöckner, F. O. (2013). Evaluation of general 16S ribosomal RNA gene PCR primers for classical and next-generation sequencing-based diversity studies. *Nucleic Acids Research*, 41, e1. doi:10.1093/nar/gks808

Kong, A. Y. Y., Hristova, K., Scow, K. M., & Six, J. (2010). Impacts of different N management regimes on nitrifier and denitrifier communities and N cycling in soil microenvironments. *Soil Biology & Biochemistry*, 42, 1523–33.

Kozich, J. J., Westcott, S. L., Baxter, N. T., Highlander, S. K., & Schloss, P. D. (2013). Development of a dual-index sequencing strategy and curation pipeline for analyzing amplicon sequence data on the MiSeq Illumina sequencing platform. *Applied Environmental Microbiology*, 79, 5112–5120.

Lane, D. J., Pace. B., Olssen, G. J., Stahl, D. A., Sogin, M. L., & Pace, N. R. (1985). Rapid determination of 16S ribosomal RNA sequences for phylogenetic analyses. *Proceedings of the National Academy of Sciences USA* 82, 6955–6959. doi:10.1073/pnas.82.20.6955

Langille, M. G. I., Zaneveld, J., Caporaso, J. G., McDonald, D., Knights, D., Reyes, J. A., Clemente, J. C., Burkepile, D. E., Vega Thurber, R. L., Knight, R., Beiko, R. G., & Huttenhower, C. (2013). Predictive functional profiling of microbial communities using 16S rRNA marker gene sequences. *Nature Biotechnology*, 31(9), 814–821. doi:10.1038/nbt.2676

Levy-Booth, D. J., Prescott, C. E., & Grayston, S. J. (2014). Microbial functional genes involved in nitrogen fixation, nitrification and denitrification in forest ecosystems. *Soil Biology & Biochemistry*, 75, 11–25. doi:10.1016/j.soilbio.2014.03.021

Louden, B. C., Haarmann, D., & Lynne, A. M. (2011). Use of blue agar CAS assay for siderophore detection. *Journal of Microbiology & Biology Education*, 12(1):51–53. doi:10.1128/jmbe.v12i1.249

Lugtenberg, B., & Kamilova, F. (2009). Plant-growth-promoting Rhizobacteria. *Annual Reviews Microbiology*, 63, 541–556. doi:10.1146/annurev.micro.62.081307.162918

Manter, D. K., Korsa, M., Tebbe, C., & Delgado, J. A. (2016). myPhyloDB: A local web server for the storage and analysis of metagenomic data. *Database*, *2016*, baw037.

Martin, M. (2011). Cutadapt removes adapter sequences from high-throughput sequencing reads. *EMBnet Journal*, 17, 10–12. doi:10.14806/ej.17.1.200

Mavrodi, D. V., Mavrodi, O. V., Elbourne, L. D. H., Tetu, S., Bonsall, R. F., Parejko, J., Yang, M., Paulsen, I. T., Weller, D. M., & Thomashow, L. S. (2018). Long-term irrigation affects the dynamics and activity of the wheat Rhizosphere microbiome. *Frontiers in Plant Science*, 9, 345.

Mela, F., Fritsche, K., de Boer, W., van Veen, J. A., de Graaff, L. H., van den Berg, M., & Leveau, J. H. J. (2011). Dual transcriptional profiling of a bacterial/fungal confrontation: Collimonas fungivorans versus Aspergillus niger. *ISME Journal*, 5(9), 1494–1504. doi:10.1038/ismej.2011.29

Moore-Kucera, J., & Dick, R. P. (2008). PLFA profiling of microbial community structure and seasonal shifts in soils of a Douglas-fir chronosequence. *Microbial Ecology*, 55(3), 500–511. doi:10.1007/s00248-007-9295-1

Morales, S. E., Cosart, T., & Holben, W. E. (2010). Bacterial gene abundances as indicators of greenhouse gas emission in soils. *ISME Journal*, 4(6), 799. doi:10.1038/ismej.2010.8

Naylor, D., & Coleman-Derr, D. (2018). Drought stress and root-associated bacterial communities. *Frontiers in Plant Science*, 8, 2223. doi:10.3389/fpls.2017.02223

Nearing, J. T., Douglas, G. M., Comeau, A. M., & Langille, M. G. I. (2018). Denoising the denoisers: An independent evaluation of microbiome sequence error-correction approaches. *PeerJ*, 6, e5364. doi:10.7717/peerj.5364

Oliver, J. D. (2005). The viable but nonculturable state in bacteria. *Journal of Microbiology*, 43, 93–100.

Olsson, P. A. (1999). Signature fatty acids provide tools for determination of the distribution and interactions of mycorrhizal fungi in soil. *FEMS Microbiology Ecology*, 29(4), 303–310. doi:10.1111/j.1574-6941.1999.tb00621.x

Osswald, W. F., Holl, W., & Elstner, E. F. (1986). Ergosterol as a biochemical indicator of fungal infection in spruce and fir needles from different sources. *Zeitschrift fuer Naturforschung Section C*, 41, 542–546. doi:10.1515/znc-1986-5-609

Pace, N. R., Stahl, D. A., Lane, D. J., & Olsen G. J. (1986). The analysis of natural microbial populations by ribosomal RNA sequences. In K. C. Marshall (Ed.), *Advances in microbial ecology* (pp. 1–55). Boston, MA: Springer. doi:10.1007/978-1-4757-0611-6_1

Pagaling, E., Strathdee, F., Spears, B. M., Cates, M. E., Allen, R. J., & Free, A. (2014). Community history affects the predictability of microbial ecosystem development. *ISME Journal*, 8(1), 19–30. doi:10.1038/ismej.2013.150

Pantigoso, H. A., Manter, D. K., & Vivanco, J. M. (2018). Phosphorus addition shifts the microbial community in the rhizosphere of blueberry (*Vaccinium corymbosum* L.). *Rhizosphere*, 7, 1–7. doi:10.1016/j.rhisph.2018.06.008

Perazzolli, M., Herrero, N., Sterck, L., Lenzi, L., Pellegrini, A., Puopolo, G., Van de Peer, Y., & Pertot, I. (2016). Transcriptomic responses of a simplified soil microcosm to a plant pathogen and its biocontrol agent reveal a complex reaction to harsh habitat. *BMC Genomics*, 17(1), 838.

Preston-Mafham, J., Boddy, L., & Randerson, P. F. (2002). Analysis of microbial community functional diversity using sole-carbon-source utilisation profiles- a critique. *FEMS Microbiology Ecology*, 42, 1–14.

Prince, A. M., & Andrus, L. (1992). PCR: How to kill unwanted DNA. *Biotechniques*, 12(3), 358–360.

Quast, C., Pruesse, E., Yilmaz, P., Gerken, J., Schweer, T., Yarza, P., Peplies, J., & Glockner, F. O. (2012). The SILVA ribosomal RNA gene database project: Improved data processing and web-based tools. *Nucleic Acids Research*, 41(D1), D590–D596. doi:10.1093/nar/gks1219

Quince, C., Walker, A. W., Simpson, J. T., Loman, N. J., & Segata, N. (2017). Shotgun metagenomics, from sampling to analysis. *Nature Biotechnology*, 35, 833–844. doi:10.1038/nbt.3935

Ramirez, K. S., Craine, J. M., & Fierer, N. (2012). Consistent effects of nitrogen amendments on soil microbial communities and processes across biomes. *Global Change Biology*, 18(6), 1918–1927. doi:10.1111/j.1365-2486.2012.02639.x

Rastogi, G., & Sani, R. K. (2011). Molecular techniques to assess microbial community structure, function, and dynamics in the environment. In I. Ahmad, F. Ahmad, & J. Pichtel (Eds.), *Microbes and microbial technology* (pp. 29–57). New York: Springer-Verlag. doi:10.1007/978-1-4419-7931-5_2

Reardon, C. L., Gollany, H. T., & Wuest, S. B. (2014). Diazotroph community structure and abundance in wheat–fallow and wheat–pea crop rotations. *Soil Biology & Biochemistry*, 69, 406–412. doi:10.1016/j.soilbio.2013.10.038

Reed, H. E., & Martiny, J. B. H. (2007). Testing the functional significance of microbial composition in natural communities. *FEMS Microbiology Ecology*, 62(2), 161–170. doi:10.1111/j.1574-6941.2007.00386.x

Regar, R. K., Gaur, V. K., Bajaj, A., Tambat, S., & Manickam, N. (2019). Comparative microbiome analysis of two different long-term pesticide contaminated soils revealed the anthropogenic influence on functional potential of microbial communities. *Science of the Total Environment*, 681, 413–423. doi:10.1016/j.scitotenv.2019.05.090

Rognes, T., Flouri, T., Nichold, B., Quince, C., & Mahe, F. (2016). VSEARCH: A versatile open source tool for metagenomics. *PeerJ*, 4, e2584. doi:10.7717/peerj.2584

Rohland, N., & Reich, D. (2012). Cost-effective, high-throughput DNA sequencing libraries for multiplexed target capture. *Genome Research*, 22, 939–946. doi:10.1101/gr.128124.111

Saharan, B. S., & Nehra, V. (2011). Plant growth promoting rhizobacteria: A critical review. *Life Science Medical Research*, LSMR-21.

Saini, A., Aggarwal, N. K., Sharma, A., & Yadav, A. (2015). Actinomycetes: A source of lignocellulolytic enzymes. *Enzyme Research*, 2015, 279381. doi:10.1155/2015/279381

Schattenhofer, M., Fuchs, B. M., Amann, R., Zubkov, M. V., Tarran, G. A., & Pernthaler, J. (2009). Latitudinal distribution of prokaryotic picoplankton populations in the Atlantic Ocean. *Environmental Microbiology*, 11(8), 2078–2093.

Scherer, C., Müller, K.-D., Rath, P.-M., & Ansorg, R. A. M. (2003). Influence of culture conditions on the fatty acid profiles of laboratory-adapted and freshly isolated strains of Helicobacter pylori. *Journal of Clinical Microbiology*, 41(3), 1114–1117. doi:10.1128/JCM.41.3.1114-1117.2003

Schloss, P. D., Westcott, S. L., Ryabin, T., Hall, J. R., Hartmann, M., Hollister, E. B., Lesniewski, R. A., Oakley, B. B., Parks, D. H., Robinson, C. J., Sahl, J. W., Stres, B., Thallinger, G. G., Van Horn, D. J., & Weber, C. F. (2009). Introducing mothur: Open source, platform-independent, community-supported software for describing and comparing microbial communities. *Applied Environmental Microbiology*, 75(23), 7537–7541. doi:10.1128/AEM.01541-09

Schloss, P. D., & Handelsman, J. (2004). Status of the microbial census. *Microbiology Molecular Biology Reviews* 68(4), 686–691.

Sharma, M. P., & Buyer, J. S. (2015). Comparison of biochemical and microscopic methods for quantification of arbuscular mycorrhizal fungi in soil and roots. *Applied Soil Ecology*, 95, 86–89.

Singh, B. K., Campbell, C. D., Sorenson, S. J., & Zhou, J. (2009). Soil genomics. *Nature Reviews Microbiology*, 7, 756. doi:10.1038/nrmicro2119-c1

Singleton, D. R., Furlong, M. A., Peacock, A. D., White, D. C., Coleman, D. C., & Whitman, W. B. (2003). Solirubrobacter pauli gen. nov., sp. nov., a mesophilic bacterium within the Rubrobacteridae related to common soil clones. *International Journal of Systematic & Evolutionary Microbiology*, 53(2), 485–490. doi:10.1099/ijs.0.02438-0

Six, J., Frey, S., Thiet, R., & Batten, K. (2006). Bacterial and fungal contributions to carbon sequestration in agroecosystems. *Soil Science Society of America Journal*, 70(2), 555–569. doi:10.2136/sssaj2004.0347

Sogin, M. L., Morrison, H. G., Huber, J. A., Welch, D. M., Huse, S. M., Neal, P. R., Arrieta, J. M., & Herndl, G. J. (2006). Microbial diversity in the deep sea and the underexplored "rare biosphere". *Proceedings of the National Academy of Sciences USA*, 103(32), 12115–12120. doi:10.1073/pnas.0605127103

Sørheim, R., Torsvik, V. L., & Goksøyr, J. (1989). Phenotypical divergences between populations of soil bacteria isolated on different media. *Microbial Ecology* 17(2), 181–192.

Sorokin, Y. I., & Kadota, H. (1972). Techniques for the assessment of microbial production and decomposition in fresh water. I.B.P. Handbook No. 23. Oxford: Blackwell Publishing.

Stackebrandt, E., & Groebel, B. M. (1994). Taxonomic note: A place for DNA-DNA reassociation and 16S rRNA sequence analysis in the present species definition in bacteriology. *International Journal of Systematic Bacteriology*, 44, 846–849. doi:10.1099/00207713-44-4-846

Strickland, M. S., & Rousk, J. (2010). Considering fungal:bacterial dominance in soils–methods, controls, and ecosystem implications. *Soil Biology & Biochemistry*, 42(9), 1385–1395.

Sylvia, D. M. (2018). Vesicular-arbuscular mycorrhizal fungi. In R. W. Weaver, S. Angle, & P. Bottomley (Eds.), *Methods of soil analysis, Part 2. Microbiological and biochemical properties* (pp. 351–378). Madison, WI: Soil Science Society of America. doi:10.2136/sssabookser5.2.c18

Tabacchioni, S., Chiarini, L., Bevivino, A., Cantale, C., & Dalmastri, C. (2000). Bias caused by using different isolation media for assessing the genetic diversity of a natural microbial population. *Microbial Ecology*, 40(3), 169–176. doi:10.1007/s002480000015

Thiet, R. K., Frey, S. D., & Six, J. (2006). Do growth yield efficiencies differ between soil microbial communities differing in fungal:bacterial ratios? Reality check and methodological issues. *Soil Biology & Biochemistry*, 38(4), 837–844. doi:10.1016/j.soilbio.2005.07.010

Tilak, K. V., Ranganayaki, N., Pal, K. K., De, R., Saxena, A. K., Nautiyal, C. S., Mittal, S., Tripathi, A. K., & Johri, B. N. (2005). Diversity of plant growth and soil health supporting bacteria. *Current Science*, 10, 136–150.

Torsvik, V. J., Gokoyr, J., & Daae, F. L. (1990). High diversity in DNA of soil bacteria. *Applied Environmental Microbiology*, 56, 782–787. doi:10.1128/AEM.56.3.782-787.1990

Tunlid, A., & White, D.C. (1992). Biochemical analysis of biomass, community structure, nutritional status and metabolic activity of microbial communities in soil. In G. Stozky & J. M. Bollag (Eds.), *Soil biochemistry* (pp. 229–262). New York: Marcel Dekker, Inc.

Bruggen, A. H., & Semenov, A. M. (2000). In search of biological indicators for soil health and disease suppression. *Applied Soil Ecology*, 15, 13–24. doi:10.1016/S0929-1393(00)00068-8

Verma, P., Yadav, A. N., Kumar, V., Singh, D. P. & Saxena, A. K. (2017). Beneficial plant-microbes interactions: biodiversity of microbes from diverse extreme environments and its impact for crop improvement. In D. P. Singh, H. B. Singh, & R. Prabha (Eds.), *Plant-microbe interactions in agro-ecological perspectives. Volume 2, Microbial interactions and agro-ecological impacts* (pp. 543–580). Singapore: Springer. doi:10.1007/978-981-10-6593-4_22

Vieira, F. C. S., & Nahas, E. (2005). Comparison of microbial numbers in soils by using various culture media and temperatures. *Microbiology Research*, 160(2), 197–202. doi:10.1016/j.micres.2005.01.004

Wagg, C., Bender, S. F., Widmer, F., & van der Heijden, M. G. A. (2014). Soil biodiversity and soil community composition determine ecosystem multifunctionality. *Proceedings of the National Academy of Sciences USA* 111(14), 5266–5270. doi:10.1073/pnas.1320054111

Winding, A., & Hendriksen, N. B. (1997). Biolog substrate utilisation assay for metabolic fingerprints of soil bacteria: Incubation effects. In H. Insam & A. Rangger (Eds.), *Microbial communities: Function versus structural approaches* (pp. 195–205). Berlin: Springer. doi:10.1007/978-3-642-60694-6_18

Wakelin, S. A., Colloff, M. J., Harvey, P. R., Marschner, P., Gregg, A. L., & Rogers, S. L. (2007). The effects of stubble retention and nitrogen application on soil microbial community structure and functional gene abundance under irrigated maize. *FEMS Microbiology Ecology*, 59(3), 661–670. doi:10.1111/j.1574-6941.2006.00235.x

Wallenstein, M. D., & Vilgalys, R. J. (2005). Quantitative analysis of nitrogen cycling genes in soil. *Pedobiologia*, 49, 665–672. doi:10.1016/j.pedobi.2005.05.005

Wallenstein, M. D., Myrold, D. D., Firestone, M., & Voytek, M. (2006). Environmental controls on denitrifying communities and denitrification rates: Insights from molecular methods. *Ecological Applications*, 16(6), 2143–2152. doi:10.1890/1051-0761(2006)016[2143:ECODCA]2.0.CO;2

Wessén, E., Hallin, S., & Philippot, L. (2010). Differential responses of bacterial and archaeal groups at high taxonomical ranks to soil management. *Soil Biology & Biochemistry*, 42(10), 1759–1765. doi:10.1016/j.soilbio.2010.06.013

Woese, C. R., & Fox, G. E. (1977). Phylogenetic structure of the prokaryotic domain: The primary kingdoms. *Proceedings of the National Academy of Sciences USA*, 74(11), 5088–5090. doi:10.1073/pnas.74.11.5088

Xue, K., Wu, L., Deng, Y., He, Z., Van Nostrand, J., Robertson, P. G., Schmidt, T. M., & Zhou, J. (2013). Functional gene differences in soil microbial communities from conventional, low-input, and organic farmlands. *Applied Environmental Microbiology*, 79(4), 1284–1292.

Yadav, A. N., Verma, P., Sachan, S. G., Kaushik, R., & Saxena, A. K. (2018). Psychrotrophic microbiomes: molecular diversity and beneficial role in plant growth promotion and soil health. In D. G. Panpatte, Y. K. Jhala, H. N. Shelat, & R. V. Vyas (Eds.), *Microorganisms for green revolution. Volume 2: Microbes for sustainable agro-ecosystem* (pp. 197–240). Singapore: Springer. doi:10.1007/978-981-10-7146-1_11

Zelles, L., Bai, Q. Y., Beck, T., & Beese, F. (1992). Signature fatty acids in phospholipids and lipopolysaccharides as indicators of microbial biomass and community structure in agricultural soils. *Soil Biology & Biochemistry*, 24(4), 317–323. doi:10.1016/0038-0717(92)90191-Y

Zhu, S., Vivanco, J. M., & Manter, D. K. (2016). Nitrogen fertilizer rate affects root exudation, the rhizosphere microbiome and nitrogen-use-efficiency of maize. *Applied Soil Ecology*, 107, 324–333. doi:10.1016/j.apsoil.2016.07.009

# Epilogue

*Douglas L. Karlen*

Soil is a ubiquitous, fragile resource that most people never think about unless there is dirt on their cell phones, mud on their trucks, or muck in their favorite rivers and lakes. Soil may also garner some attention when grass and shrubs fail to grow in local green space because of that "hard clay" or when they hear the world is running out of sand (Torres et al. 2018). But in reality, since the time of Plato (~5000 BCE), public recognition and appreciation of soil has been sparce. Such generalizations are not true for everyone, and as a career Soil Scientist it is comforting to know there have been U.S. presidents, scientists, and literary writers who have pleaded with humanity to care for soil (Table 1).

I also find it interesting that several global soil proverbs (Yang et al., 2018) including many from our Native American brothers and sisters have embraced and stressed the intimate linkage between land, soil, water, air, and humanity for centuries (Table 2). Without question, it has been said many times by many inhabitants on the American continent and around the entire world that we must appreciate and take care of our fragile soil resources as if our very lives depend upon it – because it's true!

A multitude of techniques have been used to increase general awareness of our fragile, living soil resources. This includes the successful soil exhibit entitled "Dig It" developed by the Smithsonian's National Museum of Natural History with support from the Soil Science Society of America (SSSA) and its lead sponsor, the Nutrients for Life Foundation, as well as Bayer CropScience, LI-COR Biosciences, Syngenta, and the USDA. The exhibit was not only on display for two years (2008 to 2010) in the Smithsonian National Museum of Natural History in Washington D.C., but then traveled around the U.S. for seven years before being placed on permanent display at the at the Saint Louis Science Center in St. Louis, MO.

*Soil Health Series: Volume 2 Laboratory Methods for Soil Health Analysis*, First Edition.
Edited by Douglas L. Karlen, Diane E. Stott, and Maysoon M. Mikha.

**Table E.1** Selected quotes advocating for recognition and better management of soil resources.

| Quote | Source |
|---|---|
| A nation that destroys its soil, destroys itself | Franklin Delano Roosevelt |
| A thin layer of earth, a few inches of rain, and a blanket of air, make human life possible on our planet | John F. Kennedy |
| Civilization itself rests upon the soil | Thomas Jefferson |
| There can be no life without soil and no soil without life, they have evolved together | Charles E. Kellogg |
| A new day in agriculture will come if, and when, we get both our hands and our minds a bit deeper into our soils | William A. Albrecht |
| Take care of the land and the land will take care of you | Hugh Hammond Bennett |
| The real wealth of the Nation lies in the resources of the earth, — soil, water, forests, minerals, and wildlife | Rachel Carson |
| Of celestial body movement, far more do we know, then about soil underfoot | Da Vinci |
| [Soil is] the thin layer covering the planet that stands between us and starvation | W. E. Larson |
| What we do to the land, we do to ourselves | Wendell Berry |
| We do not inherit the earth from our ancestors, we borrow it from our children | Chief Seattle |

Another activity to build public awareness of soil resources was establishment of official "State Soils." In 20 states, these have been legislatively established and share the same level of distinction as official State Flowers and Birds. In Wisconsin, grassroot efforts resulting in the designation of the Antigo silt loam as the official soil in 1983 were led by Dr. Francis D. Hole[1] (Wikipedia, 2020). As an expression of his love for soils, Dr. Hole wrote lyrics,

1 Francis Doan Hole (1913–2002) was a soil science professor at the University of Wisconsin when I was an undergraduate. As an American pedologist, educator, and musician, his love for our fragile soil resources was documented by mapping the extent of soils and their properties throughout Wisconsin and using inventive lectures and musical performances to communicate and popularize the field of soil science. He is also recognized as the leader for establishing the Antigo silt loam as the official state soil of Wisconsin by 1983 Wisconsin Act 33. Hole and fellow advocates argued for many years that soils are a natural resource that took thousands of years to form after glaciers retreated from Wisconsin between 12,500 and 15,000 YBP. Dr. Hole relentlessly advocated that soil was essential to Wisconsin's economy and the very foundation of terrestrial life for residents of the State and elsewhere. Within the Soil Science Society of America and around the world, Francis is remembered as an "Ambassador of Soils" and "Poet Laureate of Soil Science". His life is indeed a tribute to the goals and aspirations of Soil Health.

**Table E.2** Selected Native American proverbs reflecting upon land and soil resources.

| Proverb | Source |
|---|---|
| The earth is the mother of all people,<br>All people should have equal rights upon it. | Attributed to Chief Joseph |
| Take only what you need,<br>Leave the land as you found it. | Attributed to the Arapaho |
| Mother Earth gave us an abundance of blessings,<br>To gather along life's path. | Unattributed Native American Proverb |
| Touching the Earth,<br>Equates to having harmony with nature. | Unattributed Native American Proverb |
| The ground on which we stand,<br>Is sacred ground. | Chief Plenty Coups, Crow Tribe |

played soft tunes on his fiddle, and sang songs about soil with students and others throughout his University of Wisconsin – Madison career (https://soils.wisc.edu/people/history/fdhole/). In his honor I have inserted words for three of his songs, reprinted from https://soils.wisc.edu/wp-content/uploads/2013/11/soil_songs.pdf.

***Where Have All the Bedrocks Gone?***

*Francis D. Hole, 1985 (Tune: Where Have All the Flowers Gone?)*

*Where have all the bedrocks gone?*
*Long time weathering.*
*Where have all the bedrocks gone?*
*That formed so long ago.*
*Where have all the bedrocks gone?*
*Gone to residuum. . .*
*and to sediments*
*and to the vital soils!*

***You Are My Soil, My Only Soil***

Francis D. Hole, 1985 *(Tune: You Are My Sunshine)*

You are my soil, my only soil;
You keep me vital night and day.
This much I know, friend,

You do support me;
Please don't erode my life's soil away!

***Some Think That Soil Is Dirt***

Francis D. Hole, 1985 *(Tune: Funiculi)*

Some think that. Soil is dirt and quite disgusting
This is not true.
This is not true.
Some think that it makes thee air all brown and dusty
Good dust's in me!
Good dust's in you!
Praise Mother Earth, she is our earthly Mother
She gives us bread.
She gives us bread.
Praise ground, the holy ground that softly under
Our feet that tread.
Our feet that tread.
Vigor, Vigor from the soil does flow;
Roots and life are teeming down below.
No wonder that the land's so green,
The ferns and flowers so fresh and clean!
Soil is everywhere;
From it sweet blessings gently flow.

The "Dig It" exhibit, State Soil designations, and soil science leaders such as Dr. Hole all helped build public awareness and interest in soil health, which has been defined as "the continued capacity of the soil to function as a vital living ecosystem that supports plants, animals, and humans" (USDA-NRCS, 2019). Throughout the second decade of the 21st century, soil health efforts were also enhanced by corporate investments, perhaps driven by economics of sustainability, but oh so important. Through these combined efforts, soil health has evolved from basic principles of soil conservation and stewardship, as well as from prior research and technology transfer efforts including those associated with soil pedology, soil tilth, soil condition, soil management, soil care, soil quality, soil productivity, soil resilience, soil security, and soil degradation. For me, there are two subtle differences between soil health and those efforts. They are: (i) increased emphasis on soil biology because of new tools, techniques, and significant advancements in our understanding of genetics, genomics, and organismal interactions; and (ii) an integration of soil biological, chemical, and physical property and process data and information through holistic assessments.

Recognizing the soil health advancements during the past two decades, these two books, *Approaches to Soil Health Analysis* (Volume 1) and *Laboratory Methods for Soil Health Analysis* (Volume 2) were proposed in 2017 to update two SSSA books [Defining Soil Quality for a Sustainable Environment (SSSA Special Publication No. 35) and *Methods for Assessing Soil Quality* (SSSA Special Publication No. 49)] published during the mid-1990s. Volume 1 is designed to provide an update on how soil health emerged from soil quality during the past two decades. Some of those changes included formation of the Natural Resources Conservation Service (NRCS) Soil Health Division by the USDA, creation of the Soil Health Institute (SHI) by the Noble and Farm Foundations, and numerous private-sector investments beginning with the National Corn Growers Soil Health Partnership (SHP). The second volume is intended to provide current (2020), science-based perspectives and methods for measuring selected soil properties and processes that are emerging as indicators for soil health assessments.

Without question, the soil health concept has been evolving steadily since it was first introduced in the 1970s. Therefore, these volumes are in no way viewed as the final word regarding soil health but rather "works in progress" fully recognizing that improvements and refinements in our knowledge-base and assessment methods are inevitable and continue to be discovered. Editing these Volumes has provided gratification following a professional career that attracted my attention nearly fifty years ago. I recognize any personal contributions to soil health efforts are minuscule considering all who have contributed to these efforts since the time of Plato. None-the-less, it is my hope that as my mentor and friend W. E. Larson (1921-2013) stated many times, each of us needs to do what we can to help humankind understand and appreciate the thin mantle that stands between us and starvation.

Building upon the inspiration of soil science leaders such as Hugh Hammond Bennett, William E. Larson, and John W. Doran, I decided to end these volumes with an Ode to Soil Health. My intent is to honor the living soil with its multitude of functions that influence life on earth in a multitude of ways. My hope is that by pausing, looking around, and seeing the unique beauty of the "Earth-only" living material known as soil, the importance of this fragile resource will be recognized. Perhaps by enticing those who have absolutely no idea of what lies beneath their feet (clay, sand, or dirt) as well as those who already know and recognize the connection between soil and life with a bit of humor, we will all pause, look more closely at what we walk on each day and truly strive to enhance soil health by mitigating soil erosion, nutrient depletion, loss of soil organic matter, compaction, contamination, or any other peril threatening these living, life-sustaining, fragile resources.

## References

Torres, A., J. Liu, J. Brandt, and K. Lear. 2017. The World is Running Out of Sand. Smithsonian Magazine (The Conversation). https://www.smithsonianmag.com/science-nature/world-facing-global-sand-crisis-180964815/

USDA-Natural Resources Conservation Service (NRCS). 2019. The Basics of Addressing Resource Concerns with Conservation Practices within Integrated Soil Health Management Systems on Cropland, by D. Chessman, B.N. Moebius-Clune, B.R. Smith, and B. Fisher. Soil Health Technical Note No. 450-04. Available on NRCS Electronic Directive System. Washington, DC. https://directives.sc.egov.usda.gov.

Wikipedia. 2020. Francis D. Hole. https://en.wikipedia.org/wiki/Francis_D._Hole

Yang, J. E., M. B. Kirkham, R. Lal, and S. Huber (eds.). 2018. Global Soil Proverbs: Cultural Language of the Soil. Catena-Schweizerbart, Stuttgart, Germany.

**An Ode to Soil Health**

Some may ask, why dirt is important to us,
Isn't soil something about which only a scientist would fuss?

Global wealth is derived from gold, silver, and oil.
Other than farmers, who cares about soil?

On fingers, windows, shirts, or in piles blocking our way,
Is soil anything more than sand and clay?

The answer is yes, for when an ancient scroll we do unroll,
From "dust we came and to dust we'll go," many extoll.

*Adama* and *Hava* refer to man and wife,
Reflect soil health, when translated to "Soil and Life."

The thin mantle in which humans for food do toil,
Is, what Earth Scientists call soil.

Sand, silt and clay particles, classified by size,
Bound by chemistry, physics, and biology before our eyes.

People think sand is found only on the beach,
For .05 to 2 mm particles, a morphologist does reach.

Silt at 2 to 50 microns in size, often moves by air,
Creating loess hills or getting caught in your hair.

Clay at less than 2 microns in size,
Its 1:1 or 2:1 mineralogy controls more than you realize.

Soil particles derived from rocks,
Link together as aggregate building blocks.

Bound by biological glues and chemical forces,
Aggregate soil structure has the strength of horses.

From the soil surface to underlying bedrock,
Pedons reflect all horizons in stock.

Granular, prismatic, and massive soil structure,
Soil Scientists view this, as intricate sculpture.

Water infiltration, air exchange, and runoff,
Enhancing soil health cannot be put off.

Acidity, salinity, and soil solution transport,
Productivity, this chemistry does support.

Soil chemistry and physics are fun to explore,
But dynamic soil biology adds much more.

In every gram of soil, many organisms are living,
Some cause problems, but others are life-giving.

In 1928 Fleming introduced penicillin to the scene,
Soil metagenomics revealed antibacterial violacein.

A symbiotic soil *Pseudomonad,* has for you,
Produced Pederi, an anticancer drug too.

Multitudes of bacteria and innumerable fungi that we can't see,
A myriad of algae and numberless nematodes also live free.

An army of earthworms and countless protozoa too,
Mychorrhizae and actinomycetes working just for you.

Millions of micro- and macro-fauna too,
Ants, termites, millipedes, earthworms, slugs, and snails to name a few.

Responding to temperature and drought,
Changes in soil carbon, these organisms bring about.

They protect food, feed and fiber from abiotic stress,
Without pathogen resistance, earth would be a mess.

Soil microbes decompose our waste,
Creating chitin, glomalin and other compounds in its place.

Glucosidase, PLFA and EL-FAME,
Soil Health indicators have several names.

Through tillage, irrigation, and fertilizing,
Microbes are also affected, by livestock grazing.

Carbon sequestration and humification,
Without those processes, GHG would put earth on permanent vacation.

Buffering a continuously changing climate,
Dynamic, living soil keeps earth in alignment.

A glimpse of living soil, I have given you,
But wait there is even more, these fragile resources do.

Soil supports plant roots and exchange of water and air,
Cycles nutrients, filters, buffers and sustains biological diversity with flare.

It produces food, feed, fiber and fuel when given good care,
But erodes and releases GHGs, if sustainable practices are not there.

To make living soil less stealth,
These Volumes were written to promote Soil Health.

To the end of this ode, we have now come,
Hopefully your quest for Soil Health has just begun.

Printed in the USA/Agawam, MA
September 16, 2026

910661.222